# Model-Based Reasoning

## Science, Technology, Values

# Model-Based Reasoning

## Science, Technology, Values

Edited by

Lorenzo Magnani
*University of Pavia*
*Pavia, Italy*

and

Nancy J. Nersessian
*Georgia Institute of Technology*
*Atlanta, Georgia*

Kluwer Academic / Plenum Publishers
New York, Boston, Dordrecht, London, Moscow

Proceedings of the International Conference on Model-Based Reasoning: Scientific Discovery, Technological Innovation, Values, held May 17–19, 2001, in Pavia, Italy

ISBN: 0-306-47244-9


233 Spring Street, New York, New York 10013

http://www.wkap.ml/

10 9 8 7 6 5 4 3 2 1

A C.I.P. record for this book is available from the Library of Congress.



Printed in the United States of America

Suppose a philosopher be given the concept of a triangle and he be left to find out, in his own way, what relation the sum of its angles bears to a right angle. He has nothing but the concept of a figure enclosed by three straight lines, and possessing three angles. However long he meditates on this concept, he will never produce anything new. He can analyse and clarify the concept of a straight line or of an angle or of the number three, but he can never arrive at any properties not already contained in these concepts. Now let the geometrician take up these questions. He at once begins by constructing a triangle. Since he knows that the sum of two right angles is exactly equal to the sum of all the adjacent angles which can be constructed from a single point on a straight line, he prolongs one side of his triangle and obtains two adjacent angles, which together are equal to two right angles. He then divides the external angle by drawing a line parallel to the opposite side of the triangle, and observes that he has thus obtained an external adjacent angle which is equal to an internal angle - and so on. In this fashion, through a chain of inferences guided throughout by intuition, he arrives at a fully evident and universally valid solution of the problem.

Immanuel Kant, *The Critique of Pure Reason*

# Preface

The volume is based on the papers that were presented at the International Conference *Model-Based Reasoning: Scientific Discovery, Technological Innovation, Values* (MBR'01), held at the Collegio Ghislieri, University of Pavia, Pavia, Italy, in May 2001. The previous volume *Model-Based Reasoning in Scientific Discovery*, edited by L. Magnani, N.J. Nersessian, and P. Thagard (Kluwer Academic/Plenum Publishers, New York, 1999; Chinese edition, China Science and Technology Press, Beijing, 2000), was based on the papers presented at the first "model-based reasoning" international conference, held at the same place in December 1998.

The papers explore how scientific thinking uses models and explanatory reasoning to produce creative changes in theories and concepts. Some address the problem of model-based reasoning in ethics, especially pertaining to science and technology, and stress some aspects of model-based reasoning in technological innovation.

The study of diagnostic, visual, spatial, analogical, and temporal reasoning has demonstrated that there are many ways of performing intelligent and creative reasoning that cannot be described with the help only of traditional notions of reasoning such as classical logic. Understanding the contribution of modeling practices to discovery and conceptual change in science requires expanding scientific reasoning to include complex forms of creative reasoning that are not always successful and can lead to incorrect solutions. The study of these heuristic ways of reasoning is situated at the crossroads of philosophy, artificial intelligence, cognitive psychology, and logic; that is, at the heart of cognitive science.

There are several key ingredients common to the various forms of model-based reasoning considered in this book. The term "model" comprises both internal and external representations. The models are intended as interpretations of target physical systems, processes, phenomena, or situations. The models are retrieved or constructed on the basis of potentially satisfying salient constraints of the target domain. Moreover, in the modeling process, various forms of abstraction are used. Evaluation and adaptation take place in light of structural, causal, and/or functional constraints. Model simulation can be used to produce new states and enable evaluation of behaviors and other factors.

The various contributions of the book are written by interdisciplinary researchers who are active in the area of creative reasoning in science and technology: the most recent results and achievements about the topics above are illustrated in detail in the papers.

The conference, and thus indirectly this book, was made possible through the generous financial support of the MURST (Italian Ministry of the University), University of Pavia, CARIPLO (Cassa di Risparmio delle Provincie Lombarde) and of Georgia Institute of Technology. Their support is gratefully acknowledged.

The editors express their appreciation to the other co-chair of the conference K. Knoespel (Georgia Institute of Technology, Atlanta, GA, USA), and to the members of the Scientific Committee for their suggestions and assistance: A. Bostrom, Georgia Institute of Technology, Atlanta, GA, USA; E. Gagliasso, University of Rome "La Sapienza", Rome, Italy; D. Gentner, Northwestern University, Evanston, USA; R. Giere, University of Minnesota, Minneapolis, MN, USA; M.L. Johnson, University of Oregon, Eugene, OR, USA; P. Langley, Stanford University, Stanford, CA, USA; B. Norton, Georgia Institute of Technology, Atlanta, GA, USA; Claudio Pizzi, University of Siena, Siena, Italy; Mario Stefanelli, University of Pavia, Pavia, Italy; P. Thagard, University of Waterloo, Waterloo, Ontario, Canada; Ryan D. Tweney, Bowling Green State University, Bowling Green, OH, USA; S. Vosniadou, National and Capodistrian University of Athens, Athens, Greece.

Special thanks to the members of the Local Organizing Committee R. Dossena, E. Gandini, M. Piazza, and S. Pernice, for their contribution in organizing the conference, to R. Dossena for his contribution in the preparation of this volume, and to the copy-editor L. d'Arrigo. The preparation of the volume would not have been possible without the contribution of resources and facilities of the Computational Philosophy Laboratory and of the Department of Philosophy, University of Pavia.

The more logically and computationally oriented papers deriving from the presentations given at the Conference will be published in the book *Logical and Computational Aspects of Model-Based Reasoning*, edited by L. Magnani, N.J. Nersessian, and C. Pizzi, which will appear in the Applied Logic Series, directed by. D.M. Gabbay and Jon Barwise, of Kluwer Academic, Dordrecht. The remaining selected papers will be published in five Special Issues of Journals: in *Foundations of Science*, Abductive Reasoning in Science; in *Foundations of Science*, Model-Based Reasoning: Visual, Analogical, Simulative; in *Mind and Society*, Scientific Discovery: Model-Based Reasoning, in *Mind and Society*: Commonsense and Scientific Reasoning, all edited by L. Magnani and N.J. Nersessian; *Philosophica*, Diagrams and the Anthropology of Space, edited by K. Knoespel.

Lorenzo Magnani, Pavia, Italy
Nancy J. Nersessian, Atlanta, GA, USA
December 2001

# Contents

# Model-Based Reasoning

Science, Technology, Values

# Metaphor-Based Values in Scientific Models

Mark Johnson
*Department of Philosophy, University of Oregon, Eugene, OR, USA,*
*markj@oregon.uoregon.edu*

Abstract: Theories of attention in cognitive psychology are based on various metaphoric models that determine how the phenomena of attention are identified, how they are studied experimentally, and how they are explained theoretically. Three of the central models in contemporary research programs are the Attention-Spotlight metaphor, the Attention As Limited Resource metaphor, and the Biased-Competition metaphor. Each of these metaphoric models involves a specific ontology of entities and processes, and each one entails a specific set of values that guide research on attention. These values are largely incompatible across the three different models. I argue that metaphor-based value systems are characteristic of all science and that this is not a lamentable problem, but rather the source of the insight and experimental fruitfulness of scientific models.

## 1. PAY ATTENTION: METAPHOR-BASED VALUES IN SCIENTIFIC MODELS

Most scientific research is model-based. There is a rapidly growing body of evidence that scientific models are largely metaphoric. Moreover, these model-defining metaphors give rise to crucial values that determine the nature and course of scientific research. I propose to give substance and credibility to these claims by focusing on one interesting field of contemporary metaphor-based science, the cognitive psychology of attention. An examination of the three most important contemporary theories of attention shows how and why the metaphors are indispensable. The first half of the present

*Model-Based Reasoning: Science, Technology, Values*, edited by
L. Magnani and N.J. Nersessian, Kluwer Academic/Plenum Publishers, New York, 2002

essay draws extensively from published and submitted work I have done with psychologist Diego Fernandez-Duque on the three major attention metaphors that shape research in cognitive psychology and cognitive neuroscience.[1] In the second half of the essay, I examine how the various metaphors specify some of the values embedded within different research programs for studying attention. I conclude with a speculation about the possible grounding of scientific values in our organic, embodied experience.

## 2. KEY METAPHORS GUIDING ATTENTION RESEARCH

We all think we know what attention is? It's whatever is involved in our "paying attention" to something, right? But precisely what's going on when we "pay attention"? As we'll see, even attention researchers cannot agree on which mental phenomena are properly cases of "attending to something". Even less do they agree about how to explain the phenomena. In fact, depending on which particular metaphoric model of attention researchers are assuming, the alleged phenomena of "attention" get individuated in quite different ways. It is not the case that we start with agreed upon phenomena of attention and then develop alternative explanations of them. Instead, which specific metaphoric model we adopt determines what we'll count as attention, how we'll go about studying it, and what we'll recognize as relevant evidence.

### 2.1 The Spotlight metaphor

In order to flesh out these claims, let's begin with what is probably the oldest and most used metaphor in the scientific study of attention - the conception of an internal mental "spotlight" that shines on particular perceptual contents or mental representations. What we are attending to is whatever is "illuminated" at the present moment of consciousness. Scientists have argued that "attention may be compared to a beam of light in which the central brilliant part represents the focus surrounded by a less intense fringe. Only the items located in the focus of attention are distinctly perceived whereas we are less aware of the objects located in the fringe of attention" (Hernández-Peón, 1964, cited in Watchel, 1967, p. 418). The Spotlight metaphor consists of the following conceptual mapping.

[1] The analysis of the Spotlight metaphor is taken from Fernandez-Duque and Johnson (1999), and the examination of the Limited Resource and Competition metaphors is taken from Fernandez-Duque and Johnson (forthcoming).

## 2.2 The attention spotlight metaphor

| *Source Domain (Spotlight)* | *Target Domain (Attention)* |
|---|---|
| Spotlight | Orienting System (mechanism of attention) |
| Agent who controls the spotlight | Executive System |
| Agent who sees | Awareness System |
| Seeing | Attending |
| Potential field of vision | Representational Space |
| Area illuminated by the spotlight | Attended representations |

The conceptual metaphor here is the entire mapping from the entities in the source domain (visual perception and devices of illumination) to constructed entities in the target-domain (attention). The source-to-target mapping allows researchers to use their knowledge of the source domain to construct a parallel knowledge of the target domain. For example, on the basis of how a spotlight works, researchers may reason that attentional "*scanning* [...] is a measure of how much the *beam* moves around the *field*, while the *focusing* [...] refers in some way to the *width* of the *beam*" (Watchel, 1967, p. 418) [italics added]. Such inferences come from the way we use our knowledge of the source-domain structure (a-f below) to construct a parallel knowledge of the target domain of attention (a'-f' below), as follows:

**Source-domain knowledge**

(a) There is a perceptual field with (b) objects in it. (c) The spotlight sheds (d) light over parts of the field, (e) making it easier to see specific objects. When the spotlight illuminates a target object (f) the target object is seen by the observer.

**Target-domain knowledge**

(a') There is a mental field with (b') unconscious ideas in it. (c') The attentional system directs (d') attention over the brain areas (or mental field), (e') making representations accessible by our awareness system. When the attentional system focuses on some target idea, (f') that target idea is acquired by the awareness system, and it becomes conscious. (taken, with modifications, from Fernandez-Duque and Johnson, 1999, pp. 93-94).

This metaphor-based knowledge about attention is not merely a folk-model. Rather, this metaphoric model has become the basis of a substantial body of empirical research within cognitive psychology (Cave and Bichot, 1999; Fernandez-Duque and Johnson, 1999). In recent years, the internal structure and logic of the Spotlight metaphor has also influenced research in cognitive neuroscience. Consider the following examples, in which scientists explored the "physiological correlates of the 'spotlight' of visual attention" (Brefczynski and DeYoe, 1999), by measuring hemodynamic and/or electrophysiological response in areas of the visual cortex:

1. Several areas of the visual cortex can create retinotopic maps of perceptual scenes. This means that objects close or adjacent to each other in the world activate brain areas close or adjacent to each other in the visual cortex. If attention "sheds light" over sensory areas, then cueing attention to more central areas of the visual field should acti-vate brain regions that map central locations, whereas cueing atten-tion to a peripheral part of the visual field should increase blood flow in peripheral areas that map that part of the visual field. There is some evidence for this prediction coming from research designed to test these metaphorical entailments (Brefczynski and DeYoe, 1999).
2. Since a spotlight moves in analog fashion, the target domain inference is that attention moves in an analog fashion. Therefore, the electrophysiological enhancement associated with the processing of attended stimuli should also move in an analog fashion, a prediction that has been tested and confirmed in the laboratory (Woodman and Luck, 1999).
3. Since it takes time for a spotlight to move from one location to another, the delay between the onset of a cue and the enhancement of the electrophysiological response at the cued location has been taken to be a measure of how long it takes the attentional spotlight to move to the cued location (Müller, Teder-Sälejärvi, and Hillyard, 1998).
4. In the source-domain, the spotlight is a different entity from whatever it shines on and from the agent who controls it. In the target domain, therefore, one expects to find something like an independent executive system that controls attention and is separate from both the orienting system and from the sensory areas that are attended to. This concept of the executive system as defined by the Spotlight metaphor led researchers to discover a network of cortical areas that participate in attentional control, moving attention from one location to another. Whether a stimulus is displayed at the attended location has no impact on the activation of these controlling areas. In other words, the perceptual systems that benefit from the attentional modulation appear to be separate from the neural system that controls the attentional spotlight and from the spotlight itself (Hopfinger, Buonocuore, and Mangun, 2000, Corbetta, Kincade, Ollinger, McAvoy, and Shulman, 2000; Martinez, Anllo-Vento, Sereno, Frank et al., 1999).

What these four examples of attention research show is precisely how the highly articulated internal structure of the source domain of the Spotlight metaphor generates entailments that shape our understanding of the target domain and gives rise to the ways we reason about attention. The entities in the source domain (such as a spotlight and an independent operator who directs the spotlight) have specific characteristics (e.g., the beam has a particular width at a given moment and moves in an analog fashion from one illuminated location to another) that structure scientists' conceptualization of the phenomena of attention. A different metaphor, with different source-domain entities and knowledge, would obviously give rise to a substantially different conception of attention and a different research program for studying it. As an example of this, let us consider a second major metaphor system in which attention is conceptualized as a valuable limited resource that is allocated for various cognitive tasks.

## 2.3 The attention as Limited Resource metaphor

One of the chief competitors to the Spotlight metaphor is the metaphor of attention as a Limited Resource that can be allocated by a general-purpose central processor in a graded fashion for the performance of different tasks. Pashler (1998) succinctly characterizes the folk model built upon this metaphor:

> Folk psychology postulates a kind of substance or process (attention) that can be devoted (paid) to stimuli or withheld from them. Whether or not attention is allocated to a stimulus is usually thought to depend on a voluntary act of will; in the metaphysics of folk psychology, this ultimately depends on the choice made by the self [...]. The available quantity of attention is assumed to be finite, and this finiteness is supposed to account for capacity limitations; this means having less attention available to pay to other things. Attention, according to folk psychology, can be devoted not only to stimuli, but also to activities, tasks, and behaviors (as in "pay attention to your driving"). Allocating more attention to a given task enhances performance (pp. 2-3).

Scientific versions of Resource models are most eloquently described in Kahneman's influential book *Attention and Effort* (1973) and have been further developed by several researchers, such as Norman and Bobrow (1975), Navon and Gopher (1979), and Hasher and Zacks (1977). Limited resource models are used to explain many psychological phenomena, such as dual task interference (Christie and Klein, 1996), automaticity (Norman and Bobrow, 1975; Schneider and Shiffrin, 1977), priming (Posner and Tudela, 1997), and mental rotation (Carpenter, Just, Keller, Eddy, and Thulborn, 1999).

The internal logic of the Limited Resource metaphor is made evident by the following mapping, in which cognitive operations are conceptualized as distributions of physical resources that are used up in various tasks.

### 2.3.1 The attention as a limited resource metaphor

| *Source Domain (Resources)* | *Target Domain (Attention)* |
|---|---|
| Valuable resource | Attention (Capacity/Arousal) |
| Quantity of resource | Amount of available attention |
| Allocation of resource | Allocation of attention to tasks |
| Seeing | Attending |
| Budget | Task strategy for allocating attention |
| Controller of the resource | Executive system |

The Limited Resource metaphor drives research strategies by generating questions that come from our experience of using *physical* resources to carry out various tasks. The logic of the source-domain thus gives rise to a corresponding set of questions about the nature and use of attentional resources, such as: What is the nature of such limited resources? How should they be quantified? Do those resources play a causal role, and if so, who or what controls their allocation? Answers to such questions can be developed only within a larger theoretical framework that includes models of the mind, which are irreducibly metaphoric (Fernandez-Duque and Johnson, 1999). Our beliefs about what attention *is,* and our understanding of how it works, make sense only in the context of those metaphoric models. Attention is not literally a limited substance (like water, soil, or gasoline) that can be quantified and parceled out. Neither is there a homunculus whose job it is to monitor this putative "scarce" and "valuable" resource. Yet the Limited Resource metaphor postulates an executive system that allocates attentional resources to various tasks on the basis of its monitoring of inputs from other systems.

In other words, it is the inferences generated by the individual submappings of the Limited Resource metaphor that permit researchers to design and carry out attention experiments. This can be seen quite clearly in the way our basic knowledge of the source domain carries over into our understanding of attentional phenomena. For instance, in the source domain there is

(a) a *limited commodity* (e.g., some substance or object, or its surrogate, money), that is
(b) *controlled* by someone who allocates it where he deems it most necessary, and
(c) does so in a flexible way, *applying resources* to different goals in *graded* fashion.
(d) The *amount of resource needed varies* for each task or product, and
(e) *affects its quality.*

When basic source domain knowledge like this is then applied to our conceptualization of the target domain, we get a series of corresponding knowledge claims (or entailments) about attention. Therefore, each piece of shared knowledge about the source domain (a through e above) correlates with a piece of knowledge about aspects of the target domain (a'-e'), and this shows up in scientific research, as can be seen in the following quotations from the research literature:

(a') "attention has been used to refer to [...] all aspects of cognition having to do with *limited resources* or capacity" [italics added] (Shiffrin, 1988, p. 739).
(b') Such attentional resources are *controlled by an executive system*, or general-purpose central processor, at the very top of the decision-making tree, that distributes the attentional resources: "[Attention] means a laborious process, whereby processing resources are *voluntarily allocated* to a particular task or activity at the expense of other tasks and activities" [italics added] (Umiltà, 1988, p. 175).
(c') "The *allocation* of attention is a matter of *degree.* [...] Attention is [...] *controllable.* It can be *allocated* to facilitate the processing of selected perceptual units or the execution of selected units of performance" [italics added] (Kahneman 1973, p. 201).
(d') "The *amount* of attention or effort exerted at any time *depends* primarily *on the demands* of current activities" [italics added] (Kahneman 1973, p. 201).
(e') Finally, the amount of attention given to some percept *affects its information processing*, that is, attention "causes" modulation in the product of information processing

The concept of limited mental resources has also been extended to neural systems: "Just as *physical energy* systems require resources, so do neural energy systems. It is the consumption of different types of resources that may be indexed by various metabolically based neuroimaging methods, including fMRI. Specifically, greater task demand translates into greater resource demand" (Carpenter et al., 1999, p. 18).

The Limited Resource metaphor has influenced theory and guided many research agendas. For example, the models predict that, as the system becomes overloaded, performance is expected to degrade in a smooth fashion (Norman and Bobrow, 1975). When a task requires more resources than are available, decreased performance should be expected: "Once the capacity limit is exceeded, selection of the information to be processed will be required" (Lavie, 1995, p. 452). Task interference would be expected to be non-specific, because all cognitive processes tap into the same unitary resource. Even if the tasks do not share any mechanisms of perception or response, interference will occur whenever the system's capacity is clearly overloaded.

The previous analysis shows how the understanding provided by the Limited Resource metaphor permits researchers to identify the phenomena, set research strategies, and test specific hypotheses about attention. It is the entailments of the metaphor that give rise to the knowledge we construct. This can be seen, for example, if we contrast the entailments of the Spotlight metaphor with those of the Limited Resource metaphor, which reveals why each is apt in certain contexts and for certain purposes.

The Limited Resource metaphor works well for modeling the graded allocation of attention to different tasks. In contrast, the Spotlight metaphor has difficulty explaining this dimension of attention, since a spotlight does not allow the person who controls it to split the beam or to reduce its intensity so that the light that is saved can be used to illuminate another location. On the other hand, as we have seen, the Spotlight metaphor works well for explaining spatial orienting and attention, whereas the basic Limited Resource metaphor doesn't account for the phenomena of spatial attention.

The entailments of the Limited Resource metaphor are specific enough to generate predictions that can be tested experimentally. For example, "effortful" tasks sometimes do not interfere with each other (perfect-time sharing) (Navon, 1984a), and there are cases where increasing the difficulty of one task is not followed by a cost in the performance of the secondary task (difficulty insensitivity). Moreover, sometimes changes in the modality of one of the tasks affect performance of the other task, even in the absence of difficulty changes (structural alteration effect) (Wickens, 1984). These experimental results challenge the initial Limited Resource metaphor and led to a replacement of the original single capacity model by a multiple resource model (Navon and Gopher, 1979; Wickens, 1984; Pashler 1998), which proposes that interference between two tasks occurs only when the same reservoir of limited resources is accessed, but not otherwise.

The versions of the Spotlight and Limited Resource metaphors analyzed above are what theorists call "causal" models, because they treat attention as somehow being efficacious in affecting our cognitive performances, that is, causing changes in perception and other cognitive activities. Cognitive psychologists using these "causal" models thus regard "attention" to be a term that refers to a selective process by which attended information is processed more efficiently that non-attended information. The chief problem for both theories is to explain what system allocates the precious attentional resource. What does the selecting of one stimulus over another? Both the Spotlight and Limited Resource metaphors (in the versions described) require some kind of executive system.

Causal theories can be contrasted with "effect" theories, which deny the existence of any central causal mechanism of attention. Whereas causal theories claim that attention "modulates" perception, effect theories conceptualize attention as an effect - a by-product - of the normal operation of various sensory and cognitive systems (James, 1890; Johnston and Dark, 1986). For example, representations that rise to a certain level of activation are given temporary priority in the functioning of the organism, but there is no central system that monitors this competition.

The principal metaphor underlying "effect" theories is a Resource Competition metaphor in which there is *no* central mechanism to determine who or what "wins" the competition, that is, there is no central system determining which stimuli will be attended to.

## 2.4 The Resource Competition metaphor

To avoid seemingly insoluble homunculus problems that confront "cause" theories, many researchers, spurred by recent developments in cognitive neuroscience, have adopted "effect" theories that treat attention as a by-product of information processing among multiple systems: "Attention is not a high-speed mental spotlight that scans each item in the visual field. Rather, attention is an emergent property of slow, competitive interactions that work in parallel across the visual field" (Desimone and Duncan, 1995, p. 217). In such models there is *no executive system* required. Instead, perceptual objects are seen as vying for limited attentional resources, that is, they compete for activation in areas of perceptual processing: "Objects in the visual field *compete* for processing in several cortical areas" (Rees, Frackowiak, and Frith, 1997, p. 835).

Competition models are typically based on the following metaphorical mapping.

### 2.4.1 The attention as Resource Competition metaphor

| *Source Domain (Resources)* | *Target Domain (Attention)* |
|---|---|
| Competing Individuals | Stimuli |
| Valuable Resource (e.g., food) | Mental Resources / Neuronal Receptive Fields |
| Goal: securing limited resources | Goal: securing resources for mental necessary for survival processing and aware perception |
| Competition for resources | Competition for neuronal activation |
| Survival of individual | Activation above a specified threshold |

In this mapping, perceptual inputs are conceived as competing for limited attentional resources. Stimuli that are not strong enough are eliminated, while strong stimuli succeed in gathering enough processing resources so that we become "conscious" of them. The competitive process gives rise to awareness, which is stable but also evolves (as in James's "stream of consciousness"). In this model awareness is not a natural kind. It is only a collection of inputs that succeed in rising above a certain 'awareness' threshold. Attention and awareness are not supramodal systems, but rather emergent processes of brain activity.

Dennett and Kinsbourne's (1992) Multiple Draft Model of consciousness provides a good illustration of the relation between awareness and attention in the context of Limited Resource metaphors. In that model, a conscious representation is not a mental entity viewed within some internal Cartesian theater, but rather a representation that becomes activated above a criterion level, by virtue of which it acquires a certain property (i.e., the property of being something of which we are consciously aware). Hence, "neuronal activation" is a *valuable resource*, since consciousness can only be achieved by gathering a certain amount of activation. Inputs *compete for* this *valuable* and *limited* "activation" *resource*.

Just as we saw earlier with the Spotlight metaphor, the entailments generated by the metaphor mapping give rise to specific predictions, experimental programs, and theoretical explanations. In each case, it is the logic of the source domain that determines our conceptualization of workings of attention and defines the relevant research programs. One good example of this is the literature on binocular rivalry. When different images are presented to the two eyes, perception alternates such that each image is visible for a few seconds at a time. This "binocular rivalry" phenomenon is usually understood as a competition between neuronal responses, so that at a given time one representation is dominant and the other is suppressed. Current views of binocular rivalry argue that "the neural representation of the two stimuli *compete* for visual awareness" (italics added) (Logothetis, Leopold, and Sheinberg, 1996, p. 621). Binocular rivalry is a process in which "each region of the retinal image in one eye seems to fight with the corresponding region in the other eye" (Wolfe, 1996, p. 588).

A second area where Competition models also provide an alternative interpretation of many findings is the neuropsychology of attention. When attention is cued to a location and the target occurs somewhere else, patients with right parietal lesions and hemispatial neglect respond very slowly. Spotlight models explain this pattern of results as a deficit in disengaging attention and "moving" it to the new location. In contrast, competition models argue for an attentional bias, which is brought about by competitive weights that favor stimuli represented in the undamaged areas over stimuli represented in the lesioned areas. Brain areas that code visual stimuli play a double role: they process perceptual, visuospatial and oculomotor information, but they also weight attention toward the stimulus location. Lesions in areas that code for visuospatial information, such as the parietal cortex, would lead both to information processing deficit and to attention imbalance. However, those areas do not constitute a specialized attentional system.

Other experimental support comes from the directional theory of attention, which argues that *rival* hemispheres have opposing orienting vectors which compete with each other (Kinsbourne, 1977). Lesions to one hemisphere lead to an imbalance in the opposite hemisphere. Thus, presenting a "competing stimulus" to the unaffected side exaggerates the bias, leading to an "extinction" of the stimulus represented by the impaired side. Due to the lack of competition from the lesioned hemisphere, perception in the unaffected side is better than normal (Làdavas, Pretronio, and Umiltà, 1990).

Once again, the message is clear - our very conception of attentional phenomena depends directly on which metaphor we are using. The submappings of a particular metaphor provide the details of the explanation of the relevant phenomena. The metaphors are the instruments of scientific reasoning and experiment design.

## 3. METAPHOR-BASED VALUES IN SCIENCE

The analysis given so far supports my claim that scientific research on attention is inescapably shaped by metaphor-based models. I've argued that each of the three principal models of attention is metaphoric and that the nature and structure of the source domain for each metaphor determines how researchers working with that particular model will conceptualize and explain the phenomena. There is a rapidly growing body of evidence suggesting that metaphor plays a central role in *all* science, not just in specific fields like the cognitive psychology of attention. Of course, one could never *prove* that every scientific model one could imagine must necessarily be metaphorical. The best strategy is to stand ready to investigate the workings of the key metaphors in any case of scientific research that might be brought before us. The more we do this, the more inclined we will be to give up the illusion of metaphor-free science.

An analysis of the mappings that constitute the metaphors in a given theory provides considerable insight into another controversial issue, namely, the role of values in science. Implicit throughout the previous metaphor analyses is the idea that the particular mappings give specific content to, and make it possible for us to apply, the values that guide scientific research.

Talk about the normative dimension of science routinely calls to mind a classic list of values recognized by most scientists as crucial to the growth and success of their science:

1. *Generalization:* Scientific explanations are expected to be framed as generalizations over data or phenomena. That generalization is a value becomes evident when it is contrasted with the demand in some disci-

plines today for highly particular "thick" descriptions of phenomena, in place of laws and other generalizations.

2. *Comprehensiveness:* Empirical science sets a high value on generalizations that cover and unify the widest possible set of phenomena. All other things being equal, the generalization with a broader scope of applicability is preferred, as giving a more powerful and productive explanation.
3. *Empirical testability:* Even though philosophers of science have argued for decades about such notoriously problematic notions as "confirmation," these arguments only exist against a background of shared agreement that scientific hypotheses must be subject to experimental test and empirical evaluation.
4. *Relevance:* Every science makes value judgments about what counts as a phenomenon, and this determines what the theory must explain and what is taken as appropriate evidence.
5. *Importance:* Even within the range of relevant phenomena, some phenomena are regarded as more important than others. Failure to explain fully the prototypical members of a category will be more damaging to a theory that a failure to explain non-prototypical or marginal cases.
6. *Simplicity:* Other things being equal (which they seldom or never are), the simplest explanation is typically preferred.
7. *Elegance:* Elegance is always the most elusive and, therefore, last-mentioned value of good explanation. It is relegated to an "aesthetic" category and then promptly ignored, yet one often hears scientists admiring the elegance of a certain theory.

This list is meant only to be indicative of typical values involved in science. It is neither exhaustive nor absolute, since there are other values in science, some of which may even conflict with, and perhaps even override, one or more of the values on this list. My interest with this list is to point out that these values, as stated, are too abstract to be useful in the actual practice of science. They get their meaning, power to constrain explanations, and applicability via the metaphors that define specific scientific models. I want to consider a few simple examples of this phenomenon, namely, the value-specifying role of the metaphors that shape science.

The adequacy of a given explanation is relative to the theory's determination of which phenomena are *relevant* (i.e., require explanation) and what form an explanation must take. So, for example, a model of attention that incorporates the Limited Resource metaphor will place a high explanatory value on *quantification.* The Limited Resource metaphor defines attention as a valuable quantifiable substance that can be allocated differentially to multiple cognitive tasks. It favors explanations that make it possible to measure the amount of attentional resource used for a given task. It also gives high priority (that is, *importance*) to explaining how these resources are allocated.

By contrast, Spotlight theories do quite poorly on this issue, since the ontology of the Spotlight metaphor lacks an appropriate concept of the "amount" of attentional light that can be directed to a particular cognitive task. In classic Spotlight models, the spotlight has homogeneous intensity and fixed size. Moreover, attentional light is not a quantifiable substance that can be divided up and differentially allocated to multiple tasks. So, experimental evidence suggesting that attentional fields are heterogeneous (Downing and Pinker, 1985) led to a revised model which posited a peak of attention at the center of the focal area, shading off in a graded continuous fashion as the field extends outward from the focus (LaBerge, 1995). This Attention Gradient metaphor (Fernandez-Duque and Johnson, 1999) is an attempt to satisfy the value of quantificational description of phenomena. Consequently, in the original Spotlight models quantification was taken to be relatively unimportant, whereas phenomena connected with our ability to move and reorient the attentional spotlight were given primary importance.

Another crucial issue - one that is tied closely to the value of *empirical testability* - is which entities are required for a satisfactory account. Causal theories (including the Spotlight metaphor and some versions of the Limited Resource metaphor) require some system or entity that directs and controls attention. Spotlight theories handle this by postulating an executive system (the controller of the attentional spotlight), while many Limited Resource theories require some type of homunculus that decides how the resource will be used.

Competition theories, in contrast, place a high value on the fact that no control system is required to account for attentional phenomena, since everything is supposed to work "from the bottom up." The Competition metaphor thus emphasizes the ontological *simplicity* of its account, which posits neither an attentional substance nor an executive mechanism for its allocation and control. In this way, the values of *relevance* and *importance* are also given significance, since the metaphors tell us what range of phenomena we must account for and which phenomena within that range are most characteristic of attention (and so most *important).*

*Empirical testability* requires a determination of what adequate experimental confirmation amounts to. We saw earlier that the internal logic of the Spotlight metaphor has no place for features of the object in determining what gets attended to - the "spotlight" shines on a location and illuminates everything within that location, so it does not differentially affect features of what is illuminated. Experiments revealing that features of objects play a role in attentional affects are thus regarded as challenging Spotlight theories.

The fact that the source domain of a particular metaphor determines the "ontology" of the target domain, and thereby determines what counts as relevant phenomena, *is a value judgment.* The metaphor tells you what to take as important. It tells you what the "data" are. It tells you what to look for in an explanation. We too easily forget, if we ever knew, that deciding what constitutes "the phenomena to be explained" is never a value-neutral endeavor. For example, Limited Resource models treat attention as a substance-like commodity. Attention is a "real" resource that can be "allocated" to this or that particular task. Having substantial existence is a positive value in "cause" theories, whereas attention does not even "exist" according to effect theories based on Competition metaphors. Causal theories thus regard themselves as superior, insofar as they purport to explain attention as a causally efficacious force or process, while effect theories claim superiority for precisely the opposite reason, namely, that they do not require the postulation of attention as a force or substantial reality. How we evaluate the importance of something alleged to have substantial reality is thus dependent on which metaphor of attention we have appropriated and which values it incorporates.

Similarly, in Spotlight theories the spotlight has a reality different from the light it sheds, different from the field where the light is shed (presumably sensory areas of the brain), and different from the controlling agent (presumably some executive system). In fact, much of the current debate in the brain imaging of attention revolves around the evaluative judgment of whether attention should be described as having a "source" (spotlight), a "substance" (light), and a "target" (field) that is illuminated. Once again, this stands in radical opposition to Competition theories that are taken by their proponents to be superior precisely for the opposite reason - that they assign attention an epiphenomenal status, rather than the status of a real entity, substance, or force.

The principal point I'm trying to make with these admittedly sketchy examples is that each metaphor will carry its own distinctive matrix of values that guide the research that flows from the metaphor. The metaphors do not necessarily generate the generic values that have been alleged to underlie the sciences. That is, we might think that generalization, comprehensiveness, empirical testability, relevance, importance, and simplicity are general theoretical values shared by all creatures with cognitive systems like ours. But *what the metaphors do is to give meaning, specificity, and force to those values. In other words, these generic values don't actually mean anything concrete and don't have specific implications for scientific research until they are given substance and application by the metaphors that organize our scientific models.* For example, consider what "comprehensiveness of explanation" in the abstract could mean? Nothing more than an empty ideal. But this ideal becomes a workable concept once we have metaphoric models that circumscribe the range of relevant phenomena that need to be explained. This was the point I was trying to make earlier when I said that we don't know what "attention" is, independent of some particular metaphorically structured model of it. Once a metaphoric model tells us what phenomena count as instances of "attention", only then is it possible to know what a more versus a less comprehensive account would be. Only then would we know what data are relevant for testing a particular hypothesis or model.

## 4. WHERE DO SCIENTIFIC VALUES COME FROM?

I have tried to provide some evidence that the metaphors tell us what the values are and give content and constraining force to the basic values that underlie particular sciences. I want to close with a brief speculation about the source of these values. My research (Johnson, 1987; Lakoff and Johnson, 1999) on the bodily grounding of meaning and on the way structures of our sensory-motor activities shape our conceptualization and reasoning suggests the hypothesis that our deepest scientific values come out of our organic interactions with our environment. In other words, our values result from the nature of our bodies, our brains, the environments we inhabit, and our ways of interacting with those environments. Such values are not, therefore, a priori givens that are intrinsic to some absolute, transcendent structure of scientific reasoning. Rather, they emerge out of our embodiment. Although I have no argument to make for this view, I want at least to suggest its reasonableness, as a basis for any naturalistic theory of scientific values.

John Dewey (1938) argued that both the values that most people believe to be intrinsic to science, as well as the alleged logic of scientific thinking, are the result of patterns of inquiry that have proved useful over extended periods of time for creatures with bodies like ours, who inhabit environments like the ones we encounter. We tend to distil from our methods of inquiry certain structures and assumptions that have made it possible for us to deal, more or less successfully, with certain types of problems that we have historically encountered, given our embodiment and our environments. There is nothing absolute about these methods and their embedded values, nothing that guarantees that they should stand as timeless standards of thought and action. Rather, they are what have so far proved reliable for our purposes in certain contexts. Most assuredly, situations might well change in ways that would make these methods less than satisfactory and would require us to rethink our science in light of changing conditions. The failure to retain such a critical flexibility could conceivably undermine the success of a given theoretical perspective.

Dewey would thus observe that our most cherished scientific methods, our most taken-for-granted epistemological values, and our most productive metaphors for doing science are historically conditioned products of human problem solving. They emerge, in other words, from our embodied practices and situations. They grow out of the very patterns of activity by which embodied organisms monitor and negotiate their environment so as to deal with problems that arise. Dewey puts this boldly when he connects physical problem solving with abstract inquiry: "there is no breach of continuity between operations of inquiry and biological and physical operations. "Continuity", on the other side, means that rational operations *grow out of* organic activities, without being identical with that from which they emerge" (Dewey, 1938, p. 26). These "organic activities" of which Dewey speaks are an organism's attempts to survive and function successfully within the environments it inhabits. Dewey saw inquiry as an activity (a rational practice) by which an organism seeks to resolve a problematic situation (that is, an indeterminate state of affairs in which the organism's fluid functioning is blocked or disrupted) into a determinate situation in which the organism can pursue its ends and grow. Inquiry at this bodily level thus involves actual changes in the organism-environment interaction. Inquiry alters the situation that gives rise to it and results in a different state of affairs than existed before the inquiry was initiated. Successful inquiries are those that lead to growth and fluid functioning.

Once you have Dewey's embodied, situated notion of inquiry, you see scientific inquiry as continuous with our embodied problem solving. Science may differ from common sense problem solving in the ways it employs language, symbol systems, and tools to aid in its search for solutions, but it shows the same pattern of inquiry that characterizes our organic, bodily functioning. Values of inquiry thus emerge out of our bodily engagement with our world.

Consider, for example, the value of comprehensiveness of scope in science. I suggest that such a value comes out of the fact that an organism has the best chance of growing and flourishing the more conditions of its environment that it can interact fluidly with, incorporate as part of its activities, and unify relative to its needs and purposes. Thus, the more comprehensive an hypothesis is, the more possibilities for action and growth it presents to us. That is why we value breadth of scope in an explanatory hypothesis. Now this certainly does not mean that comprehensiveness will always be an absolute value. There are other possible bases in organic activity that might recommend, say, narrow focus, or attention to a single, exclusive function, as a means for an organism's survival. That is why it is possible to have cases of competing values within a particular science, or even within a particular theory. The bodily grounding of values is no guarantee that those values will be mutually consistent or even compatible with each other.

Or consider the value of simplicity. Other things being equal, an organism will tend to seek the least expenditure of energy necessary to accomplish its purposes. Simplicity is a positive value just insofar as it leads to efficient functioning and ease of action. It is no surprise that such a value would be carried up into even our most abstract theoretical inquiries, since they are no less geared to the solving of problems and to successful interaction with our world. However, simplicity is not an absolute value, by any means. One can imagine situations in which complexity would be preferred, because of the way it might enhance growth of meaning and possibilities for action. Once again, the fact that our values emerge from our embodied experience does not guarantee that they are either absolute or fixed for all situations.

The value of empirical adequacy comes out of this account almost as a given, since, at the level of sensory-motor experience, what we strive for is fluid functioning *within our actual empirical situation.* In other words, we have to be in touch with our world. We have to take empirical conditions into account in planning our actions, or we will suffer the consequences.

## 4.1 The importance of metaphors for science

These abbreviated speculations about the possible bodily grounding of scientific values are offered only to suggest a direction for future inquiry into a naturalistic theory of scientific models and their values. The bodily basis of such values goes hand-in-hand with the bodily grounding of the metaphors that shape our thinking and our inquiry. An adequate account would obviously also have to examine the importance of historical and cultural factors that determine which values have held sway at different times, and why (Longino, 1996).

My principal focus has been on the way metaphors allow scientists to circumscribe the relevant phenomena, to define research strategies, to decide what will test their hypotheses, and to evaluate their explanatory progress. The metaphorical mappings are what allow us to carry over knowledge and inferences within the source domain to knowledge and inferences in the target domain. The corporeal, spatial logic of the source domain is the basis for our reasoning about the target domain. It also establishes most of the structure of the target domain.

This central role of metaphor in science should strike us as unremarkable, once we see that *all* of our abstract concepts are defined by one or more (often mutually inconsistent) metaphors (Lakoff and Johnson, 1999). Science, unsurprisingly and necessarily, draws on the same conceptual and cognitive resources of our ordinary commonsense reasoning, but science operates with highly refined symbolic representations and subjects its performance to rigorous standards that are not as strictly maintained in common sense evaluation.

Only a literalist about meaning and an objectivist about knowledge will find this claim about the centrality of metaphor disturbing. And well they should, since it undermines their pretension to absolute knowledge and disembodied meaning and thought. On the contrary, the inescapable working of metaphor in all science is not a lamentable shortcoming, as though it were a sign of an immature science. Rather, the metaphors are what make possible the great achievements of science, because they set research programs, structure scientific inferences, and give concreteness and explanatory force to our deepest scientific values.

## REFERENCES


Brefczynski, J.A and DeYoe, E.A., 1999, A physiological correlate of the "spotlight" of visual attention, *Nature Neuroscience* 2(4):370-374.

Carpenter, P.A., Just, M.A., Keller, T.A., Eddy, W., and Thulborn, K., 1999, Graded functional activation in the visuospatial system with the amount of task demand, *Journal of Cognitive Neuroscience* 11(1):9-24.

Cave, K.R. and Bichot, N.P, 1999, Visuospatial attention: Beyond a spotlight model, *Psychological Bulletin and Review* 6(2):204-223.

Christie, J. and Klein, R.M., 1996, Assessing the evidence for novel pop-out, *Journal of Experimental Psychology: Human Perception and Performance* 125(2):201-207.

Corbetta, J.D., Kincade, J.M., Ollinger, J.M., McAvoy, M.P., and Shulman, G.L., 2000, Voluntary attention is dissociated from target detection in the human posterior parietal cortex, *Nature Neuroscience* 3(3):292-297.

Dennett, D.C. and Kinsbourne, M., 1992, Time and the observer: The where and when of consciousness in the brain, *Behavioral and Brain Sciences* 15(2):183-247.

Desimone, R. and Duncan, J., 1995, Neural mechanisms of selective visual attention, *Annual Review of Neuroscience* 18:193-222.

Dewey, J., 1938/1991, *Logic: The Theory of Inquiry*, vol. 12 of *John Dewey, The Later Works, 1925-1953,* J.A. Boydston, ed., Southern Illinois University Press, Carbondale, IL.

Downing, C.J. and Pinker, S., 1985, The spatial structure of visual attention, in: *Attention and Performance*, M.I. Posner and O.S.M. Marin, eds., Lawrence Erlbaum Associates, Hillsdale, NJ, pp. 171-188.

Hernández-Peón, R., 1964, Psychiatric implications of neurophysiological research, *Bulletin of the Menniger Clinic* 59:151155.

Hopfinger, J.B., Buonocuore, M.H., and Mangun, G.R., 2000, The neural mechanisms of top-down attentional control, *Nature Neuroscience* 3(3):284-291.

Fernandez-Duque, D. and Johnson, M., 1999, Attention metaphors: How metaphors guide the cognitive psychology of attention, *Cognitive Science* 23(1):83-116.

Fernandez-Duque, D. and Johnson, M., forthcoming, Can't get enough attention: The metaphor of attention as a limited resource, *General Review of Psychology.*

Hasher, L. and Zacks, R.T., 1979, Automatic and effortful processes in memory, *Journal of Experimental Psychology: General* 108(3):356-388.

James, W., 1890/1950, *The Principles of Psychology*, Dover, New York.

Johnson, M., 1987, *The Body in the Mind: The Bodily Basis of Meaning, Imagination, and Reason,* University of Chicago, Chicago.

Lakoff, G. and Johnson, M., 1999, *Philosophy in the Flesh: The Embodied Mind and its Challenge to Western Thought*, Basic Books, New York.

Lavie, N., 1995, Perceptual load as a necessary condition for selective attention, *Journal of Experimental Psychology: Human Perception and Performance,* 21(3):451-468.

Logothetis, N.K., Leopold, D.A., and Sheinberg, D.L., 1996, What is revalling during binocular rivalry, *Nature* 380:621-624.

Longino, H., 1996, Cognitive and non-cognitive values in science: Rethinking the dichotomy, in: *Feminism, Science, and the Philosophy of Science*, L.H. Nelson and J. Nelson, eds., Kluwer Academic Publishers, Dordrecht, pp. 39-58.

Martinez, A., Anllo-Vento, L., Sereno, M.I., Frank, L.R. Buxton, R.B, Dubowitz, D.J., Wong, E.C., Hinrichs, H., Heinze, H.J., and Hillyard, S.A., 1999, Involvement of striate and extrastriate visual cortical areas in spatial attention, *Nature Neuroscience* 2(4):364-369.

Muller, M.M., Teder-Salejarvi, W., and Hillyard, S.A., 1998, The time course of cortical facilitation during cued shifts of spatial attention, *Nature Neuroscience* 1(7):631-634.

Navon, D., 1984a, Attention division or attention sharing, in: *Attention and Performance*, M.I. Posner and O. Marin, eds., Lawrence Erlbaum Associates, Hillsdale, N.J.

Navon, D. and Gopher, D., 1979, On the economy of the human-processing system, *Psychological Review* 86(3):214-255.

Norman, D.A. and Bobrow, D.G., 1975, On data-limited and resource-limited processes, *Cognitive Psychology* 7:44-64.

Pashler, H., 1998, *The Psychology of Attention*, MIT Press, Cambridge, MA.

Posner, M.I. and Tudela, P., 1997, Imaging resources, *Biological Psychology* 45:95-107.

Rees, G., Frackowiak, R., and Frith, C., 1997, Two modulatory effects of attention that mediate object categorization in human cortex, *Science* 275:835-838.

Schneider, W. and Schiffrin, R.M., 1977, Controlled and automatic human information processing: 1. Detection, search, and attention, *Psychological Review* 84:1-66.

Shiffrin, R.M., 1988, Orienting of attention, in: *Steven's Handbook of Experimental Psychology: Learning and Cognition*, R.C. Atkinson, R.J. Herrnstein, G. Lindzey, and R.D. Luce, eds., John Wiley and Sons, New York, pp. 739-811.

Umiltà, C., 1988, Orienting of attention, in: *Handbook of Neuropsychology*, F. Boller and J. Grafman, eds., Elsevier, pp. 175-192.

Wachtel, P.L., 1967, Conceptions of broad and narrow attention, *Psychological Bulletin,* 68(6):417-429.

Wickens, R.C., 1984, Processing resources in attention, in: *Varieties of Attention*, R.D. Parasuraman, ed., Academic, Orlando, FL.

Wolfe, J.M., 1996, Resolving perceptual ambiguity, *Nature* 400:867-869.

Woodman, G.F. and Luck, S.J., 1999, Electrophysiological measurement of rapid shifts of attention during visual search, *Nature* 400:867-869.

# Analogy in Scientific Discovery: The Case of Johannes Kepler

Dedre Gentner
*Department of Psychology, Northwestern University, Evanston, IL, USA, gentner@nwu.edu*

Abstract: Analogy is often linked with scientific discovery. In this paper I lay out a theory of analogical processing and apply it to the work of Johannes Kepler. Kepler is striking not only for the magnitude of his discoveries but because of the articulateness with which he laid out his reasoning processes, including his frequent and extended use of analogies. I discuss four analogical subprocesses - highlighting common structure, projecting inferences, re-representing relations, and noticing alignable differences - by which analogy brings about new ideas and show how they apply in Kepler's work. I focus particularly on a central extended analogy in which he used the phenomenon of the *light* from the sun, which travels to the planets and illuminates them, as a base domain for a new ontological entity (a precursor of gravity) - the *vis motrix,* that causes the planets to move.

Analogical reasoning can lead to change of knowledge - not only to enrichment of existing representations but also to true conceptual change. The work of Johannes Kepler (1571-1630) offers a clear example of the use of analogical processes in discovery. Kepler seems to have been a natural analogist. He used analogies constantly, both to make local points and to frame global theories. I focus here on one central analogy that played a formative note in this thinking: an analogy between light from the sun that illuminates the planets to something else - initially a spirit and later a force or power - emanating from the sun that causes the planets to move. He conceived this idea in its initial form early in his career, and over the next 20 years, used the analogy with light to develop the idea further of a *vis motrix* (a motive power between the sun and planets). As Toulmin and Goodfield (1961, p. 198) put it, "The lifelong, self-appointed mission of Johann Kepler

*Model-Based Reasoning: Science, Technology, Values*, edited by
L. Magnani and N.J. Nersessian, Kluwer Academic/Plenum Publishers, New York, 2002

[...] was to reveal the new, inner coherence of the Sun-centered planetary system. His central aim was to produce a 'celestial physics,' a system of astronomy of a new kind, in which the forces responsible for the phenomena were brought to light".

In this paper I will try to show how the cognitive processes inherent in analogy can promote conceptual change. I lay out a set of analogical mechanisms by which analogy can act to create changes in knowledge and consider how these analogical mechanisms could have brought about the kinds of theory change that Kepler experienced. I begin with a tour of Kepler's discoveries. Then I give a brief review of the slate of knowledge existing when Kepler began his work. Then I describe the evolution of Kepler's great analogy. Next comes a description of analogical processing followed by its application to Kepler. Finally, I consider the psychological status of Kepler's analogies.

Kepler's best-known discoveries are the three laws of planetary motion:

1. (1609) First Law: The planetary orbits are ellipses with the sun at our forces.
2. (1596, 1609) Second Law: A line between the sun and any planet sweeps out equal times.
3. (1634) Third Law: The square of the period of a planet's orbit is proportional to the cube of its radius: $T^2 = kR^3$

He made a number of other discoveries, including an incremental calculus for computing the volume of a wine barrel; the modern theory that comet's tails consist of ejected matter; the proposal that the moon causes tides; a detailed theory of optics; writings on the comic sections and, in 1627, the Rudolphine Tables for predicting planetary motions, which were roughly 30 times more accurate than prior tables (Gingerich, 1993, p. 50).

But by far Kepler's most important discovery was his causal theory of planetary motion. As Gingerich (1993) puts it, Kepler's most consequential achievement was the mechanizing and perfecting of the world system. By the *mechanization* of the solar system, I mean his insistence on a "new astronomy based on the causes, or the celestial physics", as he tells us in the title of his great book [the *Astronomia Nova*]. By the *perfection* of the planetary system, I mean the fantastic improvement of nearly two orders of magnitude in the prediction of planetary positions (p. 333).

To see the magnitude of Kepler's discovery, we begin by reviewing the system of cosmology that Kepler inherited when be began his work in 1590. The ontology and explanatory system of medieval astronomy followed a line laid down by Plato and Aristotle and culminating in Ptolemy's system of the II century AD:

1. The earth is at the center of the universe and is itself unmoving.
2. The earth is surrounded by physically real crystalline spheres, containing the heavenly bodies, which revolve around the Earth.
3. The heavenly bodies move in perfect circles at uniform velocity. (However, epicycles and eccentrically positioned circles were used to account for the observed motions).
4. Heavenly bodies and their spheres are composed not of the four terrestrial elements - earth, air, fire and water, but of a fifth element, the *quintessence* - a crystalline aether that is pure, unalterable, transparent, and weightless. The farther from Earth, the purer is the sphere. By the same token, celestial phenomena must be explained in different terms from earthly phenomena.
5. All motion requires a mover. In the Aristotelian universe, the outermost sphere, containing the fixed stars, is moved by an "unmoved mover," the *primum mobile.* Each sphere imparts motion to the next one in.
6. Celestial bodies have souls. In particular, each planet is controlled by its own spirit, which mediates its motion. (This last principle, descended from the Stoics, enjoyed a resurgence in the XVI century for reasons explained below).

When this Aristotelian-Ptolemaic system was integrated with Catholic theology in the early XIII century by Albertus Magnus and Thomas Aquinas, angelic spirits were assigned to the celestrial spheres in order of rank, from Seraphim in the outermost and purest sphere, that of the *primum mobile,* inward to Cherubim (controlling the sphere of fixed stars) Thrones, Dominations, Virtues, Powers, Principalities, Archangels, and finally Angels, (controlling the sphere of the moon). The resulting conceptual scheme, dominant until the XVI century, was one of extreme intricacy and cohesion.

The Stoic view overlapped with the Aristotelian-Ptolemaic view in being geocentric, with a sphere of fixed stars. However, it differed in that it postulated that the heavens were filled not with pure aether but with a kind of intelligent pneuma (a combination of fire and air), which became more pure with distance from the earth. The heavenly bodies, also made of pneuma, were intelligent and capable of self-direction (Barker, 1991). There were thus two explanations for celestial motion: (1) the transmission of motion through interlocking crystalline spheres, and (2) the action of planetary intelligences or souls. Kepler dismissed the first because of two recent astronomical events. The first was a new star - a nova (or supernova) in 1572 - evidence against the Aristotelian doctrine of the unchanging and incorruptible firmament. The second was a comet in 1577 (and others not long after), whose path ran through the planetary spheres - which should therefore have been cracked. This was evidence against the view that the planets moved by attachment to crystalline spheres. This left the idea of guiding spirits moving the planets. Throughout his career Kepler wrestled with the idea that the planets move themselves intelligently, gradually paring away aspects of intelligent thought in search of a more purely physical description.

The other major event that set the stage was Nicolaus Copernicus's publication of *De revolutionibus orbium celestium* in 1543, the year of his death - thirteen centuries after Ptolemy's model and thirty years before Kepler's birth - proposing the idea that the earth and other planets moved around the sun. Copernicus argued for his system on grounds of mathematical elegance and sufficiency, not on the basis of physical causation. Indeed, for mathematical reasons, he placed the center of the solar system at the center of the Earth's orbit, rather than the center of the sun itself.

Kepler embraced Copernicus's heliocentric characterization of planetary orbits. But from the start, he changed this conception radically, infusing it with causal significance. This causal interpretation of Copernicus's theory led to a reaxiomitization of astronomy. As Gingerich (p. 333) notes, "Copernicus gave the world a revolutionary helio*static* system, but Kepler made it into a *heliocentric system.* In Kepler's universe, the Sun has a fundamental physically motivated centrality that is essentially lacking in *De revolutionibus.* We have grown so accustomed to calling this the Copernican system that we usually forget than many of its attributes could better be called the Keplerian system." In part this is Kepler's own doing: he saw himself as simply making clear the real significance of Copernicus's work and titled one of his great works *The Epitome of Copernican Astronomy.*

With this as background, in 1596 the 25-year-old Kepler posed a simple but profound question: Why do the outer planets move slower than the inner planets? He noticed that the periods of the outer planets were longer, relative to those of the inner planets, than could be predicted simply from the greater distances they had to travel - that is, they traveled slower. From within the Stoic cosmology of planetary spirits, he asked whether the "moving souls" were simply weaker the further the planet. Instead, he conjectured the planets might be moved not by their own individual spirits but one spirit residing in the sun - the *anima motrix.* In this conjecture, he drew on an analogy with the sun's light (Kepler, *Mysterium cosmographicum*, 1596, p. 199):

> [...] one of two conclusions must be reached: either the moving souls [*motricis animae*] are weaker the further they are from the Sun; or, there is a single moving soul [*motricem animam*] in the center of all the spheres, that is, in the Sun, and it impels each body more strongly in proportion to how near it is.

He reasons that just as light grows fainter with distance, so might this motivating spirit or power (Kepler, *Mysterium cosmographicum,* 1596, p. 201).

> Let us suppose, then, as is highly probable, that motion is dispensed by the Sun in the same proportion as light. Now the ratio in which light spreading out from a center is weakened is stated by the opticians. For the amount of light in a small circle is the same as the amount of light or of the solar rays in the great one. Hence, as it is more concentrated in the small circle, and more thinly spread in the great one, the measure of this thinning out must be sought in the actual ratio of the circles, both for light and for the moving power [*motrice virtute*] [...]

Kepler was well aware of the implausibilities in his proposal. In the *Astronomia Nova* (1609), Kepler challenged his theory with the thorny question of action at a distance:

> For it was said above that this motive power is extended throughout the space of the world, in some places more concentrated and in others more spread out...This implies that it is poured out throughout the whole world, and yet does not exist anywhere but where there is something movable (Kepler 1609/1992, p. 382).

He answers this challenge by invoking the light analogy:

> But lest I appear to philosophize with excessive insolence, I shall propose to the reader the clearly authentic example of light, since it also makes its nest in the sun, thence to break forth into the whole world as a companion to this motive power. Who, I ask, will say that light is something material? Nevertheless, it carries out its operations with respect to place, suffers alteration, is reflected and refracted, and assumes quantities so as to be dense or rare, and to be capable of being taken as a surface wherever it falls upon something illuminable. Now just as it is said in optics, that light does not exist in the intermediate space between the source and the illuminable, this is equally true of the motive power (Kepler, 1609/1992, p.383).

It can be seen here that already by 1609 the idea of a motive *spirit* in the sun was giving way to a motive *power* or *force*. By the time of his 1621 revision of the *Mysterium cosmographicum*, Kepler had fully re-represented the "soul" of the sun to be a physical force or power (Kepler, *Mysterium cosmographicum*, 1621, p. 201):

> If for the word "soul" [*Anima*] you substitute the word "force" [*Vim*], you have the very same principle on which the Celestial Physics is established [...]. For once I believed that the cause which moves the planets

> was precisely a soul, as I was of course imbued with the doctrines ... on moving intelligences. But when I pondered that this moving cause grows weaker with distance, and that the Sun's light also grows thinner with distance from the Sun, from that I concluded, that this force is something corporeal, that is, an emanation which a body emits, but an immaterial one.

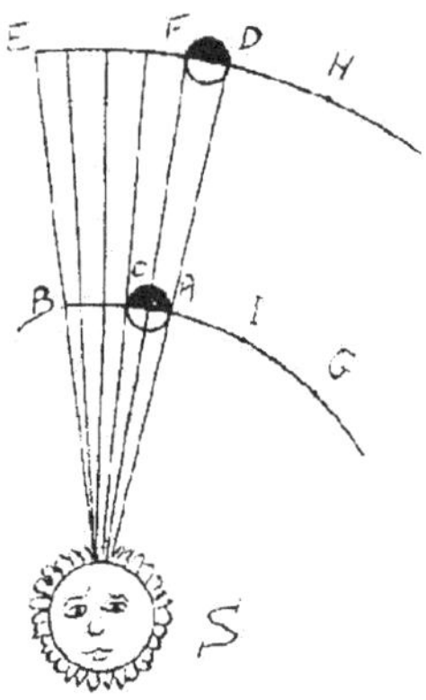

*Figure 1.* Kepler's depiction of the sun's light spreading outward.

Kepler returned repeatedly to the light-*vis motrix* analogy, analyzing and extending it across several works. From it he derived that (like light) the *vis motrix* travels invisibly from the sun to the planet through space and is detectable only when it contacts a planet. As with light, the concentration of the *vis motrix* should decrease with distance, this explains why the speed of the planet decreases with distance from the sun (analogous to the diminishing of brightness with distance from a light source). He delved into the base domain of this analogy - the behavior of light. He published a treatise on astronomical optics in 1604 (*Astronomiae pars optica*) and another in 1610 (the *Dioptrice*). With this considerable knowledge of the behavior of light, Kepler had a base domain that was systematic and well-understood - ideally suited to provide inferential resources for the target.

The *vis motrix* or *virtus motrix* (roughly, *force* or *power*) was a true precursor of gravity - a power that acts at a distance, in a constrained and lawful fashion. It was a fundamental change in the ontology of the solar system. Kepler did not arrive at the full notion of gravity; that was left for Newton some 80 years later; Kepler still believed that the sun's power pushed the planets around instead of attracting them inward. But he had something very close to an inverse square law for the strength of this central power, and he applied this attractive power (like gravity) not only between the sun and planets, but between the earth and moon and tides.

The magnitude of the conceptual change that occurred over the course of Kepler's work is hard to grasp, in part because we now live in a cosmology defined by the changes brought by Kepler and his contemporaries (see Table 1). One major change Kepler's abandonment of the idea that the planets move in perfect circles at uniform speed. Uniform speed was the first to go, replaced early in his career by the idea that the planets move faster the closer to the sun (the "equal areas in equal times" law). But it took years of struggle to give up the circle and move to the oval and finally the ellipse. As Hanson (1958, p.4) notes, "Before Kepler, circular motion was to the concept of a planet as 'tangibility' is to our concept of 'physical object'".

Kepler also abandoned the idea of an externally perfect universe set in motion once and for all and now running externally in unchanging paths. He abandoned both the idea of the planets' orbits as crystalline spheres containing the planets and that of the orbits as eternal paths traveled by planetary intelligences. Instead he came to see them as paths continually negotiated between the sun and the planets. As Toulmin and Goodfield (1961) noted, "One cannot find before Kepler any clear recognition that the heavenly motions called for an explanation in terms of a *continuously* acting physical force" (p. 201, emphasis in original). Another major conceptual change is Kepler's change from the *anima motrix* - a spirit in the sun that could move the planets - to the *vis motrix* or *virtus motrix* - a physical force or power. This is clearly an ontological change, an instance of what Thagard (1992) calls "branch jumping."

| BEFORE | AFTER |
|---|---|
| Planetary system is governed by mathematical laws | Planetary system is governed by physical causality |
| Planets' orbits are crystalline spheres containing planets or eternal circles traveled by planetary intelligences | Planets' orbits are paths continually negotiated between the Sun and the planets |
| Celestial phenomena are separate from earthly physics | Terrestrial knowledge extends to astronomical phenomena |
| Planetary paths are perfect circles of uniform speed | Planetary paths are ellipses, faster when closer to the Sun and slower when further from the Sun |
| *Anima motrix* as "spirit" in Sun that moves planets | *Vis motrix* as "force" from Sun that moves planets |

*Table 1.* Conceptual Change.

Analogical cognition. It seems clear that analogical thinking played a big part in Kepler's discoveries. How do these processes work? The defining characteristic of analogy is that it involves an alignment of relational structure. There are three psychological constraints on this alignment. First, the alignment must be structurally consistent: i.e., it must observe parallel connectivity and one-to-one correspondence. Parallel connectivity requires that matching relations must have matching arguments; and one-to-one correspondence limits any element in one representation to at most one matching element in the other representation (Gentner and Gentner, 1983). A second characteristic of analogy is relational focus: analogies must involve common relations but need not involve common object descriptions The final characteristic of analogy is systematicity: analogies tend to match connected systems of relations (Gentner, 1983). The systematicity principle captures a tacit preference for coherence and causal predictive power in analogical processing. This implicit preference for aligning connected systems permits the generation of spontaneous *candidate inferences*. By aligning a causal system in the base domain with a (typically less complete) system in the target domain, further statements that are connected to the system in the base can be projected into the target.

There are at least six ways in which the process of analogical comparison can lead to knowledge change (Gentner and Wolff, 2000; Gentner, Brem, Ferguson, Markman, Levidow, Wolff, and Forbus, 1997): (1) highlighting and schema abstraction - extracting common systems from representations, thereby promoting the disembedding of subtle and possibly important commonalities (including common relational systems); (2) projection of candidate inferences from one domain to the other; (3) noticing alignable differences - becoming aware of contrasts on dimensions or predicates that are present in both analogs and/or that are connected to the common structure; (4) re-representation - altering one or both representations so as to improve the match (and thereby, as an important side effect, promoting representational uniformity); (5) incremental analogizing: extending the mapping by returning to the base domain for more material to add to the analogy; and last, the rarest of these, (6) re-structuring - altering the domain structure of one domain in terms of the other 9 though this process probably requires processes besides analogy, as discussed by Nersessian (1992). I focus here on the first four processes, which enable a learner to notice abstract commonalities, to make new inferences, and even to reconceive the concepts in the domain. Let us take these in turn, using the operation of the Structure-mapping engine (SME) as a specific model of the kinds of mechanisms that may be operating (Falkenhainer, Forbus, and Gentner, 1989).

*Highlighting commonalities* may seem like a rather mundane learning process, but this is not true in the case of common relations. SME's alignment process, taken as a model of human processing, suggests that the act of carrying out a comparison promotes structural alignment and renders the common structure more salient. There is considerable evidence for this claim

(Gentner and Wolff, 2000; Gick and Holyoak, 1983). When a learner is induced to compare two things - whether because of a direct instruction to compare, or because of common labels, or chance juxtaposition - the alignment process renders the common relational structure more salient and prompts a re-representation at a more abstract level (Gentner and Medina, 1998; Loewenstein, Thompson, and Gentner, 1999).

Achieving a structural alignment not only invites abstracting the commonalities but also invites *inferences* from the more structurally complete system to the less-understood system. This kind of inference projection can add new knowledge. Inspired by the analogy with light, Kepler mapped a precursor of the inverse square law from light to the motive power.

*Alignable differences* are differences connected to commonalities - typically, the same dimension or predicate with two different values and emerge from the alignment process. Interestingly, it is easier to find differences for high-similarity pairs than for low-similarity pairs (Markman and Gentner, 1993). For example, participants given a high-similarity pair like *hotel* and *motel* readily list (alignable) differences: "hotels are in cities, motels are on the highway; you stay longer in hotels than in motels; hotels have many floors, motels only one or two;" and so on. In contrast, when given a low-similarity pair like *magazine/kitten,* subjects tend to list nonalignable differences, such as "you pet a kitten; you don't pet a magazine;" or "kittens have fur and magazines don't;" This is because they are useful in forcing the clarification and delineation of the analogy and in revealing significant differences. For example, Kepler asked whether the motive power could undergo an eclipse. He decided not, and used this difference to conclude that the motive power cannot be the same thing as the sun's light.

*Re-representation* occurs most typically when two analogs that share considerable structure nonetheless contain a stubbornly different pair of predicates which, if they matched, would allow a much larger march. In this case the learner is motivated to try to recast one or both predicates so as to find an identity. For example, given "the sun propels Jupiter forward" and "the sun pushes Venus around" one might re-represent these relations as "the sun moves (Jupiter/Venus) around". In SME, this is expressed as the principle of *tiered identity.* An analogy must begin with some identical predicates in order to be viable. (Otherwise we could entertain vacuous analogies such as "Fred loves New York" and "The planets revolve around the sun"). given a set of initial identities, people tend to seek additional identities, so that the end result is a larger match. One effect of this implicit drive to increase the match is that it promotes representational uniformity. Two phenomena are likely to be represented more similarly after an analogy than they were before. Finally, analogical change sometimes involves *re-structuring*, in which significant changes of structure occur. This is of course quite rare - if it were not there would be no stability to mental life. It is doubtful that restructuring occurs in one pass of an analogy. Rather, I suspect it requires repeated working through either the same analogy or, as in Kepler's case, a set of interrelated analogies.

*Induction and abduction.* As noted above, the analogical processes of highlighting and schema abstraction provide a mechanism by which inductive abstraction can take place. Structure-mapping permits the induction of structured abstractions, not merely sets of common features. This is important for modeling scientific reasoning. Structure-mapping can also capture analogical abduction - that is, it can be used to infer the explanation for a set of observations in a domain (Peirce, 1935; see also Magnani, 2001).

Kepler's analogy between the light and the *anima motrix* led to a massive abduction - the conjecture that the causal relation between the *anima motrix* and the planets's motion was analogous to the causal relation between the sun's light and the illumination of the planets by the sun. Table 2 shows SME's mapping of this analogy, with the abduction in bold (see Gentner et al., 1997, for details).

| LIGHT | ANIMA MOTRIX |
|---|---|
| Planet | Planet |
| Sun | Sun |
| Space | Space |
| *Distance* (planet, sun) | *Distance* (planet, sun) |
| *Brightness* (planet) | *Speed* (planet) |
| | |
| PROMOTE (*Brightness* (planct)) | **PROMOTE (*Speed* (planet))** |
| PRODUCE (sun, light) | **PRODUCE (sun, anima)** |
| Concentration | **Concentration** |
| (light, planet) | **(anima, planet)** |
| Distance *(planet, sun)* | ***Distance* (planet, sun)** |
| TRAVEL | **TRAVEL** |
| (light, sun, planet, space) | **(anima, sun, planet, space)** |
| REACH (light, planet) | **REACH (anima, planet)** |
| | |
| CAUSE | **CAUSE** |
| (TRAVEL | **(TRAVEL** |
| (light, sun, planet, space), | **(anima, sun, planet, space)),** |
| REACH (light, planet), | **REACH (anima, planet))** |
| CAUSE | **CAUSE** |
| (REACH (light, planet), | **(REACH (anima, planet),** |
| PROMOTE (*Brightness* (planet)) | **PROMOTE (*Speed* (planet))** |
| QROP- | QROP- |
| (*Brightness* (planet), | (*Speed* (planet), |
| *Distance* (planet, sun)) | *Distance* (planet, sun)) |

*Table 2.* Simulating Kepler's analogy: the mapping produced by SME. N.B: Bold text marks the abduction. *QPROP* means "qualitatively proportional to" (Forbus, 1984).

Kepler used a large number of analogies for the sun's power, returning again and again to three main analogies. The first viewed the sun's power as analogous to light, as in Table 2 - a power that propelled the planets around in their orbits. A second analogy viewed the sun's power as magnetism - a new phenomenon just being explored by Gilbert in *De magnete* (1600). In this analogy, Kepler likened the sun and planet to two magnets that approach or repel each other depending on which poles are proximate. The third analogy viewed the sun's power as a kind of current in which the planets navigated as boatmen. This last was used to explain why the planets move closer and further on their orbits instead of maintaining a constant distance from the sun. He postulated that as the sun rotates around its axis, it creates a whirling circular river of motive power that pushes the planets around. (In Kepler's pre-Newtonian physics, the sun was required to push the planets around in their orbits, not merely to attract them.) Then, as a boatman can steer his boat sidewise to the river's current, so the planets could move in and out across this current of motive power. But although Kepler worked with this analogy for decades, he was never satisfied with it; it seemed to require an excessive degree of sentience on the part of the planets to sense how to steer. He articulated this challenge in the *Astronomia nova* (1609; quoted in Koestler, 1978, p. 342):

> Kepler, does't thou wish then to equip each planet with two eyes? Not at all. For it is not necessary, either, to attribute them feet or wings to enable them to move [...]. Our speculations have not yet exhausted all Nature's treasures, to enable us to know, how man senses exist [...]. We are discussing natural matters of much lower rank: forces which do not exercise free will when they change their activities, intelligences which are by no means separate from, but detached to, the stellar bodies to be moved, and are one with them.

*Was Kepler's use of analogy apparent or real?* I have suggested that analogical processes of highlighting, projection, re-representation and drawing alignable differences operate to bring about conceptual change in Kepler's work. But were the analogies in Kepler's works and journals integral to his thought? Might they not have been designed after the fact, for purely communicative or persuasive purposes? There are reasons to think Kepler's analogies were just what they seem - tools central to his thought. First, Kepler was an open, inclusive writer, whose writings are unusually rich in descriptions of his thought processes. Many of Kepler's commentators have noted the exceptional - at times even excessive - candor and detail of his scientific writing. Holton, 1973, p.69-70 notes that

> [Modern scientists are] [...] taught to hide behind a rigorous structure the actual steps of discovery - those guesses, errors, and occasional strokes of

> good luck without which creative scientific work does not usually occur. But Kepler's embarrassing candor and intense emotional involvement force him to give us a detailed account of his tortuous process [...]. He gives us lengthy accounts of his failures, though sometimes they are tinged with ill-concealed pride in the difficulty of his task.

Kepler was explicit in his intention to share the difficulties of discovery: "I therefore display these occasions [errors and meanderings] scrupulously, with, no doubt, some attendant difficulty for the reader. Nevertheless, that victory is sweeter that was born in danger, and the sun emerges from the clouds with redoubled splendour" (*Astronomia nova*, 1609, p. 95). He frequently included long, tedious sections of calculations that turned out to lead nowhere, informing the reader only afterward that the reasoning had been wrong from the start. In the midst of one such section he wrote, "If this wearisome method has filled you with loathing, it should more properly fill you with compassion for me, as I have gone through it as least seventy times" (p. 256). This is not to say that Kepler's writings are pure diaries; some filtering and organizing surely took place. But his fascination with the process of discovery led him to preserve much of the trail. Indeed, Kepler had the rather touchingly optimistic belief that readers would wish to follow "the roads by which men arrive at their insights into celestial matters." In the introduction to the *Astronomia nova* (Kepler, 1609/1992) he states this agenda - characteristically, in an extended analogy:

Here it is a question not only of leading the reader to an understanding of the subject matter in the easiest way, but also, chiefly, of the arguments, meanderings, or even chance occurrences by which I the author first came upon that understanding. Thus, in telling of Christopher Columbus, Magellan, and of the Portuguese, we do not simply ignore the errors by which the first opened up America, the second, the China Sea, and the last, the coast of Africa; rather, we would not wish them omitted, which would indeed be to deprive ourselves of an enormous pleasure in reading (p. 78).

Kepler's openness extended beyond sharing the process of discovery. He also provided a running account of his feelings about the work, including the kind of emotional remarks that no modern scientist would consider publishing. The open spontaneity of Kepler's style suggests that his writings were in good measure directly reflective of his thinking. For example,

If I had embarked upon this path a little more thoughtfully, I might have immediately arrived at the truth of the matter. But since I was blind from desire [to explain the deviation from a circular orbit] I did not pay attention to each and every part [...] and thus entered into new labyrinths, from which we will have to extract ourselves. (Kepler, 1609/1992, pp. 455-456) or from the same work,

> Consider, thoughtful reader, and you will transfixed by the force of the argument [...]. And we, good reader, will not indulge in this splendid triumph for more than one small day [...] restrained as we are by the rumours of a new rebellion, lest the fabric of our achievement perish with excessive rejoicing (p. 290).

Beyond the general openness, there are several specific signs that Kepler used analogics to think with and not simply to explain. First, Kepler pursued his major analogies with intensity within and across his major works - spelling out commonalities and differences, inferences, and incremental extensions. Not only did he push the analogical mappings, he also sought further information about both sides of the analogy. For both his major analogies, he delved energetically into the base domain. For the magnetism analogy, he studied Gilbert's *De magnete*. For the light analogy, he investigated optics and wrote two treatises (the *Astronomiae pars optica* in 1604 and the *Dioptrice* in 1610). He used this information both to challenge and to extend the analogy. For example, he devoted many pages to exploring whether the magnetism-*vis motrix* comparison was purely an analogy or might instead be the causal means by which the Sun influenced the planets.

Further, Kepler made several direct statements that he used analogy. For example, he justified his analogy between light through a lens and the geometry of conic sections and as follows:

> But for us the terms in Geometry should serve the analogy (for I especially love analogies, my most faithful masters, acquainted with all the secrets of nature) and one should make great use of them in geometry, where - despite the incongruous terminology - they bring the solution of an infinity of cases lying between the extreme and the mean, and where they clearly present to our eyes the whole essence of the question (Kepler, *The Optics*, 1904; quoted in Vickers, 1984, pp. 149-150).

Another indication that Kepler actually used analogy in reasoning is that he wrote explicitly on how best to use analogy in scientific thinking. In fact, the improper use of analogy was one of the few areas in which Kepler - normally generous towards other scientists - showed a sharply critical side. For example, he criticized the use of symbols as informative analogies by hermeticists and alchemists (for instance, the claim that iron must be corrosive since its (al)chemical symbol has a sharp point). "I have shown that Ptolemy luxuriates in using comparisons in a poetical or rhetorical way, since the things that he compares are not real things in the heavens". He also wrote: "I too play with symbols, and have planned a little work, *Geometric Cabala*, which is about the Ideas of natural things in geometry; but I play in such a way that I do not forget that I am playing. For nothing is proved by symbols [...] unless by sure reasons it can be demonstrated that they are not merely symbolic but are descriptions of the ways in which the two things are connected and of the causes of this connexion." This is essentially the modern conception of analogical reasoning.

A further indication that analogy was a natural mode of thought for Kepler is that he used analogy prolifically. Kepler drew on a variety of base domains in his work - a boat in a current, magnets, a balance beam, an orator, and so forth. He also used them in informal writing, in personal letters and journals. Complaining of having to make astrological forecasts to please his patrons he wrote (Kepler, 1606, *De Stella nova in pede Serpentarii)*:

> A mind accustomed to mathematical deduction, when confronted with the faulty foundations [of astrology] resists a long, long time, like an obstinate mule, until compelled by beating and curses to put its foot into that dirty puddle.

He often engaged in analogical play, as for example in the preface to the *Astronomia nova* (1620; 1992 pp. 30-33). Describing his solution to the long-unsolved problem of the orbit of Mars, he analogized Mars to a captive of war:

> I am now at last exhibiting for the view of the public a most Noble Captive, who has been taken for a long time now through a difficult and strenuous war waged by me under the auspices of Your Majesty [...]. It is he who is the most potent conqueror of human inventions, who, ridiculing all the sallies of the Astronomers, escaping their devices, and striking down the hostile throngs, kept safe the secret of his empire, well guarded throughout all ages past, and performed his rounds in perfect freedom with no restraints [...]. In this place chief praise is to be given to the diligence of Tycho Brahe, the commander-in-chief in this war, who [...] explored the habits of this enemy of ours nearly every night for twenty years, observed every aspect of the campaign, detected every stratagem, and left them fully described in books [...] I [...] directed the Brahean machines thither, equipped with precise sights, as if aiming at a particular target, and besieged each position with my enquiry [...]. This campaign did not, however, succeed without sweat [...] [because of] the enemy's enterprise in making sallies, and his vigilance for ambuscades [...] whenever he was driven or fled from one castle, he repaired to another, all of which required different means to be conquered [...]. At last, when he saw that I held fast to my goal, while there was no place in the circuit of his kingdom where he was safe or secure, the enemy turned his attention to plans for peace [...] and, having bargained for liberty within limits subject to negotiation, he shortly thereafter moved over most agreeably into my camp with Arithmetic and Geometry pressing closely by his sides.

*Kepler contrasted with modern scientists.* As noted above, Kepler's way of using analogy contrasted sharply with that of the alchemists who preceded him. How does it compare with the approach of modern scientists? Empirical studies of learning and problem - solving have shown that analogies to prior knowledge can foster insight into new material (Bassok and Holyoak, 1989; Gentner and Gentner, 1983; Gick and Holyoak, 1983). These laboratory results are supported by observations of the scientific process (Dunbar, 1995; Nersessian, 1984, 1986; Thagard, 1992).

Dunbar's (1994, 2001) detailed observations of the online process of discovery in four different microbiology laboratories have led him to a characterization of what makes for a creative laboratory. He suggests that the highly productive microbiology laboratories are those that use analogies in quantity and take them seriously. In the successful lab groups, analogies are extended and "pushed" in group discussions. This is the most direct evidence to date that the process of working through an analogy contributes to scientists' on-line creative thinking, and it lends force to Kepler's introspection that analogy furthered - and perhaps even engendered - his theories.

Attention to inconsistencies is another common factor. Dunbar noted that creative laboratories seize on inconsistencies between results and predictions and between observations that should have agreed. and work on explaining them. Kepler shows the same aesthetic. A well-known instance is his rejection of his own early model of Mars's orbit after several years of work due to a discrepancy of only 8 minutes of arc with Tycho Brahe's observations – which led him with great trepidation to abandon the idea of circular orbits. But there were many others. His initial reformulation of Copernicus's model was driven in part by its inconsistent treatment of Mercury's orbit and its continued use of epicycles. In both cases, I suggest that attention to inconsistencies acts as a *motivator* for conceptual change, whereas analogy acts as a *process* by which conceptual change occurs.

Dunbar's third claim is that creative labs have interaction patterns that bring together heterogeneous knowledge, partly because it increases the range of different analogies that are used. Here too historical period is a factor. There was no clear division of disciplines in 1600. A Kepler or a Galileo could command enough broad knowledge to create his heterogeneous pool of analogies. Kepler's published works, besides his great works on celestial physics, included papers on optics, the nature of comets, the birthdate of Jesus, and a new method of measuring wine casks (in which he developed a method of infinitesimals that took a step towards the calculus). Such diversity within an individual might be analogous to the heterogeneity of background Dunbar noted in his successful lab groups.

There are also some interesting differences between Kepler and Dunbar's microbiologists. By far the vast majority of the analogies Dunbar observed were close literal similarities (what he calls *local analogies*), typically involving the same kind of organism or species, similar diseases or genetic materials, etc. Kepler also used close analogs on many occasions. When he first noticed the regularity that speed diminishes as distance from the sun increases, he immediately asked whether this between-planet pattern would apply within planets. This led to the inference that each planet moves fastest when it is closest to the sun, and this in turn led to the first statement of the equal-areas in-equal-times law. By this analogy, early in his career he was able to give up the assumption of uniform speed. Another close analogy is a kind of reverse analogy he used in the calculation of Mars's orbit - he imagined how the earth's orbit would appear from Mars. A third instance is a mapping from the sun's power affecting the planets to infer that the earth's pwer would affect the moon. This led him to suggest that the moon causes the earth's tides.

But despite these similarities, Kepler's analogizing is strikingly different from that of the microbiologists in one important way. Unlike Dunbar's microbiologists, who use distant analogies only for local illustrative points, Kepler used many distant analogies as integral to his work. We could attribute this difference to the contrast between a rare genius and a group of merely excellent scientists. But it could also stem from a difference in the evolutionary stage of science. Kepler's self-appointed task was to invent the science of astrophysics - or as he put it: "Ye physicists, prick your ears, for we are about to invade your territory". The task was all the more challenging in that there was no well-established physics. Given this underdeveloped state of affairs, distant analogies were in many cases his only option. There was no literal similarity to be had. In contrast, in the microbiology laboratories that Dunbar studies, the historical stage is one of a fairly well developed (but not yet fully explored) framework. At this stage, there are many close analogs that are likely to be fruitful. In sum, I suggest that close analogies and far analogies may be characteristic of within-paradigm and between-paradigm stages in the history of a field, respectively (Kuhn, 1962). Local analogies fill in an established framework; distant analogies create new frameworks.

*Creativity, fluidity and structure.* Creativity is often characterized as a kind of conceptual fluidity in which concepts flow together and boundaries are constantly shifting. But if we take the notion of "fluidity" seriously, it seems exactly the wrong idea. I suggest that a better model of the creative process begins with a representational structure and alters that structure, sometimes locally and sometimes radically. To bear down on the fluid analogy for a moment, moving a bit of liquid from one part of a large body to another yields no discernible difference. In contrast, moving or replacing part of a structure leads to a noticeably different structure. In other words, structured belief representations before and after the change are what makes the change matter.

Kepler's conceptual changes proceeded from alignment and mapping between highly structured representations. His discoveries are best modeled by a structure-sensitive process. For example, SME, using structured representations, behaves in what could be considered a creative manner when it derives cross-dimensional structural matches, projects candidate inferences, infers new entities and incrementally extends its mapping. Further, this model predicts that minor alternations - like noting the abstraction or mapping an inference between two close analogs - should be frequent and easy, whereas radical restructuring is rare but possible. I suggest that structured representations and structure-sensitive alignment and mapping processes are instrumental in conceptual change and discovery. Analogy is an engine of creativity in part because it preserves a fair degree of structure while inviting alterations.

One more aspect of analogical creativity needs comment. I have focused here on the use of analogies in online thought - that is, the processes of analogical reasoning once one has both analogs in mind. But it is obviously crucial to ask how potential analogs come to mind. The processes that govern access and reminding to prior potential analogs are far less under the control of the reasoner than those that govern online mapping. To use an analogy I've used before, gaining access to long-term memory remindings is like baiting a hook and throwing it into a vast ocean. There is no telling which fish, if any, will bite. Indeed, remindings can occur any time, even when the person is otherwise occupied. However, the example of Kepler suggests that there are practices that foster analogical remindings. The first point is rather obvious: thinking intensely about a domain increases the probability of remindings to that domain. (I experienced a striking (but non-productive) instance of this phenomenon the day of the World Trade enter attack when, seeking distraction, I weeded the garden and experienced roughly 15 remindings from pulling weeds to terrorism). The second point is that pushing hard on the analogies that do occur increases both the likelihood and the quality of further remindings. The more one articulates and abstracts the relational structure of a domain, the more it can call forth remindings to other relationally similar knowledge. I speculate that creativity involves a kind of iterative process of pondering, retrieval, intensive mapping, and re-retrieval. From this point of view, Kepler's genius owes as much to passionate labor as to creative leaps.

# REFERENCES

Barker, P., 1991, Stoic contributions to early modern science, in: *Atoms, Pneuma, and Tranquility: Epicurean and Stoic Themes in European Thought*, M.J. Osler, ed., Cambridge University Press, Cambridge, MA, pp. 135-154.

Bassok, M. and Holyoak, K.J., 1989, Interdomain transfer between isomorphic topics in algebra and physics, *Journal of Experimental Psychology: Learning, Memory, and Cognition* 15:153-166.

Dunbar, K., 1995, How scientists really reason: Scientific reasoning in real-world laboratories, in: *The Nature of Insight*, R.J. Sternberg and J.E. Davidson, eds., MIT Press, Cambridge, MA, pp. 365-395.

Falkenhainer, B., Forbus, K.D., and Gentner, D., 1989, The structure-mapping engine: Algorithm and examples, *Artificial Intelligence* 41:1-63.

Forbus, K.D., 1984, Qualitative process theory, *Journal of Artificial Intelligence* 24: 85-168.

Gentner, D., 1982, Are scientific analogies metaphors, in: *Metaphor: Problems and Perspectives*, D.S. Miall, ed., Harvester Press, Brighton, England, pp. 106-132.

Gentner, D., 1983, Structure-mapping: A theoretical framework for analogy, *Cognitive Science* 7:155-170.

Gentner, D., Brem, S., Ferguson, R., Wolff, P., Levidow, B.B., Markman, A.B., and Forbus, K., 1997, Reasoning and conceptual change: A case study of Johannes Kepler, *Journal of the Learning Sciences* 6(1): 3-40.

Gentner, D., Brem, S., Ferguson, R.W., Wolff, P., Markman, A.B., and Forbus, K.D., 1997, Analogy and creativity in the works of Johannes Kepler, in: *Creative Thought: An Investigation of Conceptual Structures and Processes*, T. B. Ward, S. M. Smith, and J. Vaid, eds., American Psychological Association, Washington, DC, pp. 403-460.

Gentner, D., and Gentner, D.R., 1983, Flowing waters or teeming crowds: Mental models of Electricity, in *Mental Models*, D. Gentner and A. L. Stevens, eds., Erlbaum, Hillsdale, NJ, pp. 99-129.

Gentner, D., and Jeziorski, M., 1993, The shift from metaphor to analogy in western science, in *Metaphor and Thought*, A. Ortony, ed., Cambridge, England, pp. 447-480.

Gentner, D., and Medina, J., 1998, Similarity and the development of rules, *Cognition* 65: 263-297.

Gentner, D., and Wolff, P., 2000. Metaphor and knowledge change, in: *Cognitive Dynamics: Conceptual Change in Humans and Machines*, E. Dietrich and A. Markman, eds., Lawrence Erlbaum Associates, Mahwah, NJ, pp. 295-342.

Gick, M.L., and Holyoak, K.J., 1983, Schema induction and analogical transfer, *Cognitive Psychology* 15: 1-38.

Gingerich, O., 1993, *The Eye of Heaven*, The American Institute of Physics, New York.

Hanson, N.R., 1958, *Patterns of Discovery: An Inquiry into the Conceptual Foundations of Science*, Cambridge University Press, Cambridge.

Holton, G., 1973, *Thematic Origins of Scientific Thought*, Harvard University Press, Cambridge. MA.

Holyoak, K.J., and Thagard, P.R., 1995, *Mental Leaps: Analogy in Creative Thought*, MIT Press, Cambridge, MA.

Kepler, J., 1969, *Epitome of Copernican Astronomy*, (tranlated by C.G. Wallis), Kraus Reprint Co., New York.

Kepler, J., 1981, *Mysterium Cosmographicum*, (translated by A.M. Duncan, 2nd ed.), Abaris Books, New York, (original work published 1596, annotated 1621).

Kepler, J., 1992, *New Astronomy* (translated by W. Donahue), Cambridge University Press, Cambridge, (original work published 1609).

Koestler, J.L., 1968, *The Sleepwalkers*, Grosset and Dunlap, The Universal Library, New York.

Kuhn, T.S., 1962, *The Structure of Scientific Revolutions*, University of Chicago Press, Chicago.

Loewenstein, J., Thompson, L., and Gentner, D., 1999, Analogical encoding facilitates knowledge transfer in negotiation, *Psychonomic Bulletin and Review* 6(4): 586-597.

Magnani, L., 2001, *Philosophy and Geometry: Theoretical and Historical Issues*, Kluwer Academic Publishers, Dordrecht.

Markman, A.B., 1997, Constraints on analogical inference, *Cognitive Science* 21(4): 373-418.

Markman, A.B., and Gentner, D., 1993, Splitting the differences: A structural alignment view of similarity, *Journal of Memory and Language* 32: 517-535.

Nersessian, N.J., 1984, *Faraday to Einstein: Constructing Meaning in Scientific Theories*, Nijhoff, Dordrecht.

Nersessian, N.J., 1986, A cognitive-historical approach to meaning in scientific theories, in: *The Process of Science: Contemporary Philosophical Approaches to Understanding Scientific Practice*, N. J. Nersessian, ed., Nijhoff, Dordrecht.

Nersessian, N.J., 1992, How do scientists think? Capturing the dynamics of conceptual change in science, in: *Cognitive Models of Science*, R.N. Giere and H. Feigl, eds., University of Minnesota Press, Minneapolis, pp. 3-44.

Peirce, C.S., 1934-35, *Collected Papers*, Harvard University Press, Cambridge, MA.

Thagard, P., 1992, *Conceptual Revolutions*, Princeton University Press, Princeton, NJ.

Toulmin, S. and Goodfield, J., 1961, *The Fabric of the Heavens*, Harper, New York.

Vickers, B., 1984, Analogy versus identity: The rejection of occult symbolism, 1580-1680, in: *Occult and Scientific Mentalities in the Renaissance*, B. Vickers, ed., Cambridge University Press, Cambridge, pp. 95-163.

# Model Experiments and Models in Experiments

Mary S. Morgan
*London School of Economics, Department of Economic History, London, UK, and University of Amsterdam, Faculty of Economics and Econometrics, Amsterdam, The Netherlands*
*m.morgan@lse.ac.uk*

Abstract: This paper explores how model-based reasoning might be characterized in terms of making experiments. I outline the characteristic steps of such model experiments or "experiments on models", and apply the notions to a diagrammatic model in everyday use in economics. I compare these model experiments with two other kinds of experiments (using examples from the same economic domain): laboratory experiments and a hybrid form. This hybrid form involves simulations or "experiments with models". The paper concludes with a comparison of the role that models play in each case.

In *Models as Mediators* (1999), Margaret Morrison and I suggested a framework for thinking about the way models were used in economics and physics. We suggested that they operated as instruments of investigation which scientists used to find out about both theories and the world. We associated this role of models with their characteristic functional autonomy in the realm of scientific argument (that is, models function partly independently of the theories or data that they involve), and linked this in turn with certain features of their construction. However, instruments come in many sizes and guises, and function in many different ways, making any more specific definition of the nature of models difficult. Magnani's (2002) all-embracing term "epistemic mediators" nicely captures the active and various roles that models play in modern sciences, but the question of how we reason with models remains open to argument.

How can we characterize the way models work as instruments to help us to learn about our theories or the world? Should we treat the way we work

*Model-Based Reasoning: Science, Technology, Values*, edited by
L. Magnani and N.J. Nersessian, Kluwer Academic/Plenum Publishers, New York, 2002

with models as a matter of abductive logic? Or should we treat model reasoning as fundamentally analogical? And are these two, and other characterizations, antagonistic or complementary? The answer I will explore in this paper is to treat model-based reasoning as akin to extended thought experiments. This thesis is motivated by two observations about the role of models in science. First, using models necessarily involves the scientist as an active participant: not just as thinker and as cognitive instrument, but as experimenter. Second, modern science has seen a recent explosion of different epistemic mediators, hybrids wherein the two notions of experiment and model go hand in hand under the general name of simulations (see Sismondo, 1999).

The material used in this paper comes from economics. On the one hand, economists build mathematical models and experiment *on* them. On the other hand, economists undertake laboratory experiments. In between, they carry out various kinds of hybrid experiments in which they experiment *with* models. My agenda is to present the characteristic use of mathematical models as a method of experiment and to compare this with the method of laboratory experiment. The next step is to contrast these two end cases with an example of an in-between hybrid. I end the discussion by considering how economists use these model experiments to cross the inference gap and to say something about the economic world.

## 1. MODEL EXPERIMENTS = EXPERIMENTS ON MODELS

Economics is now a modeling discipline and has been for the last few decades. That is to say that economists' discourse is concerned with "models" rather than with "theories", in theoretical work, in applied work, and in giving policy advice. Economists' reasoning is centred on models which are often mathematical in form, sometimes statistical or diagrammatic, but only very rarely material or physical. But the terms "discourse" and "reasoning" do not fully capture the nature of economists' research activity in using models. The characteristic steps of modeling can be taken as follows:

1. Build a model to represent the world.
2. Ask questions about the model.
3. Manipulate the model to answer questions.
4. Relate answers to the phenomena of the world.

First, building a model in economics involves the economist in representing or denoting his/her theories and ideas about a particular aspect of the economic world into a mathematical model. Typically, these models are small in scale and the problems of accurately describing the immensely complicated real economic world are assumed away. The typical economic model consists of a few equations or lines on a two-dimensional diagram, and the symbols used represent "agents" (people or firms) or goods or other economic elements such as prices, quantities and so forth. There are various accounts about how such models are built, and how either theories or observations about the world or both are embedded in the model, but that is not my topic here (see rather Hamminga and De Marchi, 1994, and the papers on economics in Morgan and Morrison, 1999).

My object here is to discuss how such mathematical models are *used*, and this is given in the next three steps. These might appear to be self-evident, but there are various ways in which they can be characterized. In an earlier paper (Morgan, 1999), I gave the following account: The starting point (the second step in the list above) is provided by the scientist who asks questions about the world portrayed in the model. These questions take the form : "What happens if …?" or "Let us assume that …? and each question requires the model to be used to provide the answer. The questions are neither random nor general, rather they are prompted by the theoretical issues or economic events or policy agenda the model was built to address. These questions constitute the "external dynamic" or "intervention" in the model.

The third step is that the model is manipulated to provide answers to the question. The model does not manipulate itself - it requires the scientist as question-maker to determine the starting point and direction of solution, and the scientist as manipulator to deduce or derive the answer by manipulating the model. The model itself provides two resources. One is its form which in turn dictates the kind and rules of mathematical reasoning (algebraic, geometric, diagrammatic or whatever) appropriate to the case. The other is its structure, which must have sufficient flexibility and sufficient elements (variables and relations) to enable questions of interest to be answered. R.I.G. Hughes (1997) has called this internal quality the "internal dynamic", referring to the quality of manipulability that the model must have.

The final or fourth step is to relate the answers given by the model back to the original prompt for the questions, back, that is, to the scientists' theories and queries about the way the economic world works. In my earlier (Morgan, 1999) account, I discussed the ubiquitous use of narratives (or story-telling) in the provision of answers and in making the link between the model and the world.

This paper provides a complementary characterization of these three latter steps of model usage as a kind of glorified thought experiment. I call them "glorified" only in that such thought experiments are too complicated to be done in the head - they are rather pen-and-paper experiments. The term experiment refers here to the combination of the external and internal dynamics, the activity of asking questions about a circumscribed and limited model world and deriving answers about that world. This is a process in which scientist and model are jointly active participants, neither is passive. The final step of relating the answers back to the real world, I treat as an inference problem.

I take as my exemplar for the model experiment one of the most common and well-used models in economics, namely the supply and demand model, which appears either as a diagram with supply and demand curves or as a set of three equations (the two functions and an equilibrium condition). The effective use of this model is associated with Alfred Marshall, an English economist of the late XIX and early XX centuries, who was famous for his writings about the nature and workings of industry. In the chapter that I analyze here, Marshall (1930, book 5, chapter 13) used the standard supply and demand diagram shown in Figure 1, in which the DD' curve represents the potential demand by consumers for a good at various prices and the SS' curve the potential supply by producers over the same price range (by convention, prices are given on the vertical axis, and quantities on the horizontal). In an analysis which was typical for him (the argument went on in the text, and the model manipulations were shown in footnotes), Marshall asked four questions, conducted nine model experiments and three thought experiments (all in less than ten pages) and then used the answers to provide a commentary on both policy and theoretical implications of the answers.

The first question Marshall asked is: What happens in an industry if there is some "great or lasting" change in normal demand? To answer this, the model is manipulated as follows: an increase in demand at all prices means that the demand curve shifts upwards to the right to the position dd' (see the left-hand diagram (a) in Figure 1). The experiment shows that at the curves' new intersection point (the point at which all exchanges are believed to take place, or the "equilibrium point"): $e^*$, compared to the old intersection point: $e$, quantity has risen ($q$ to $q^*$) and price has risen too ($p$ to $p^*$). However, according to Marshall's text, there are two alternative shapes which the supply schedule can take: either horizontal or downward sloping like the demand curve (neither are shown here - the reader will have to imagine them or draw the model). In the former case, the experiment shows that equilibrium quantity rises but price does not change and in the latter case that quantity rises but price falls. Thus, one question and three similar experiments with diagrammatic models reveals that equilibrium quantity always rises, but that price changes depend on the shape of the supply function. The first two of these three experiments could have been done mentally as thought experiments, but only if the model diagram was already known to the economist and its rules for manipulation understood. But, as Marshall remarked, the third case is difficult to treat without the actual diagram and its manipulation.

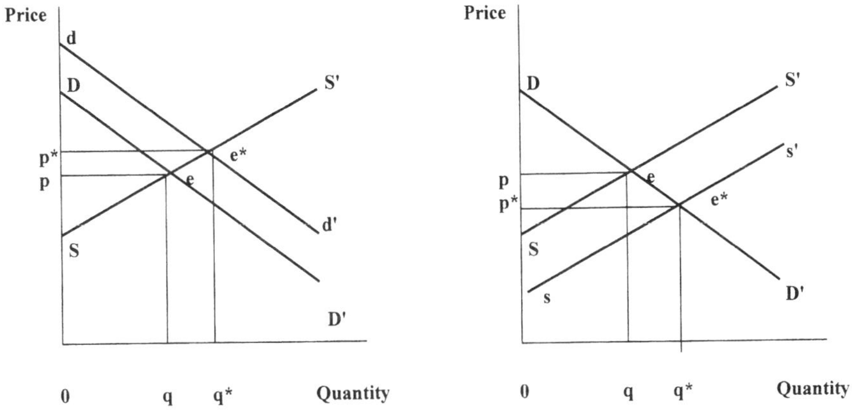


*Figure 1.* Marshall's model experiments to explore effects of (a) change in normal demand and (b) change in facilities of supply.

Marshall then asks: What happens if there is a decrease in normal demand? He does no model experiments here; knowing the answers to the first question provides immediate answers to thought experiments for each case.

The third question Marshall asked was: What happens if there is an increase in the facilities of supply? This question prompts a further three model experiments in which there is a shift to the right or downwards of the supply curve from *SS'* to *ss'* (as shown in the right-hand diagram (b) of Figure 1). These model experiments allow him to answer that regardless of the shape of the supply curve, equilibrium price falls and quantity rises, though there is a range of price changes in the three cases.

His fourth question is: What happens if a tax is placed on the price of the good? Here the reasoning necessary to follow through the answers to the question requires quite a complicated model experiment, but using exactly the same set of diagrams. The answer hinges on what happens to something called the "consumers' surplus" which is the triangle defined by the points *Dpe* on Figure 1. If a tax is placed on a good, the price paid by consumers will rise, and their share of this "surplus" (the difference consumers would have been willing to pay and the amount they actually pay at the market equilibrium price, *p*) will consequently fall. The answers given by the model experiments lead to two further sets of observations in which Marshall relates the findings to wider issues. On the one hand, the experiments prompt a discussion of the principles of taxation in relation to the model outcomes and in relation to the indirect changes in elements in the model revealed by the experimental manipulations. On the other hand, the model work leads to theoretical discussions on the validity of general claims about the nature of the equilibria involved and to certain general issues of ethics and distributive justice.

It is evident that these experiments on the model take much for granted. For economists using Marshall's book, much could be assumed by virtue of the familiarity of the model and from Marshall's tradition of arguing (known as the method of comparative statics), as well as from his own earlier chapters. Nevertheless, this brief report shows how model reasoning - the role of questions and the manipulation of the model to ascertain answers - involves a kind of thought experiment and begins to indicate how the experimental results are used in arguments about the world. There are two particular points which bear further analysis, namely: how experimental controls are instantiated, and what type of demonstration is involved in such model experiments. That analysis requires we have the comparison case of laboratory experiments ready to hand.

## 2. LABORATORY EXPERIMENTS

Economists have typically assumed that market outcomes will have certain characteristics, in particular that a group of buyers and sellers in a market will arrive at an equilibrium price, that is, at the intersection of the demand and supply curves in Marshall's diagram. This assumption has driven almost a century of research into the conditions under which this assumption will hold and increasing reliance on mathematics to investigate the features of this theory in the kind of model experiments I discussed above. But many questions remain about how markets work and how the independent individual buyers and sellers arrive at a price in the real world as opposed to the idealized markets portrayed in economic theory and in mathematical models. These questions were the first ones to be investigated in a laboratory (or rather, classroom!) experiment in economics and there is now a record of fifty years of such experiments on this topic.

The, now classic, first market experiment in economics was reported by Edward Chamberlin (1948) who described a set of forty-six classroom ("laboratory") experiments. In these, class participants were divided into groups of "buyers" and "sellers", each of whom was given a card showing either the maximum price they, as a buyer, would be willing to pay (their "reservation" price) for a unit of a good or the minimum price, as a seller, they would be willing to accept for a unit. Each participant could trade one unit during a short period when the "market" was in operation by circulating through the marketplace (the classroom) and trying to strike a bargain to buy or sell privately with another participant. Once a contract was concluded, the contract price was written on the class board, but not the reservation prices.

Experimentalists in economics follow the same ambitions as in other laboratory sciences, seeking to remove or control interfering factors in the environment which might invalidate the experimental results and to control the inputs and intervention so that certain behavior can be isolated and its experimental variation become known. These controls are enforced as much through experimental design choices as through direct physical means. Chamberlin used the experimental design to set separate limits on the contract price for each person; he controlled the distribution of those limits (see below), the amount which could be contracted in each period by each person and the amount of information made publicly available; but he left open how bargaining negotiations were conducted. We can think of many of these controls as in the form of "rules" (some of which are known to economists as "institutions") which participants had to follow.

In these experiments, the prices on the cards, when ranked in order, indicated potential demand and supply schedules which were conventional in shape (i.e. downward sloping demand and upward sloping supply) and neither particularly steep or flat. If drawn out however, as Chamberlin did in his report, we see immediately that unlike the smooth continuous curves drawn by Marshall and assumed in most mathematical model work, these schedules are steps, for these reservation prices are set at particular whole units, so a schedule will not provide a smooth line. Since the experiment often went on with a less than full set of price cards handed out, due to a limited number of class participants, the schedules sometimes also had considerable gaps (or larger steps) in them. In most of the experiments, the average price of contracts was lower than the "equilibrium" price predicted by Marshallian theory (ie at the intersection of the demand and supply curves constructed from the reservation prices on the cards) and sales were higher than the amount predicted. Most of Chamberlin's paper was given over to explorations about why these findings might have arisen, particularly to explain the difference between average prices found in the experiments and the "equilibrium price" expected (from theory). Chamberlin came to doubt that there was even a tendency towards this equilibrium.

If we compare Chamberlin's laboratory experiment with Marshall's model experiments, we can see that the latter's outcomes depended on the assumption that before the experiment (as it were) trading was at the intersection point of a supply and demand curve and that there was an automatic adjustment to the new intersection (or equilibrium) point whenever one of the curves was shifted in the experimental manipulation. The smoothness of the curves and the automatic adjustment to equilibrium were two of the hidden but controlling assumptions of the model experiments. In these classroom experiments, the stepped curves were controlled by design, but while those curves limited each participant's range of behavior, there was nothing in the experimental design which enforced buyers and sellers to trade at the intersection point of those curves.

Later laboratory experiments by Vernon Smith (1962) followed a very similar experimental design, using class participants, each labeled as a buyer or seller and each with a reservation price. But each experiment had several periods during which the market operated, and in each time period, each person could make one new trade. In addition, contracting was conducted openly by students raising their bids and offers in public so that everyone knew what all the bids and offers were. These two features meant that students had more chance to learn about the demand and supply reservation prices held by others in the "market". Smith carried out ten sets of experiments, varying the shapes of the demand and supply schedules, sometimes changing their levels in mid-experiment, and sometimes letting participants trade two units. With these design features, Smith found a much greater evidence of convergence of exchange prices (than had Chamberlin) towards the "equilibrium" price indicated by the supply and demand schedules constructed from the set of reservation prices.

## 3. COMPARISON: MODEL EXPERIMENTS AND LABORATORY EXPERIMENTS

Our first two dimensions of contrast between model experiments and laboratory experiments are "controls" and "demonstration methods" (see Table 1). In the laboratory experiment, elements are controlled physically or materially. Control over the environment, inputs, and intervention all have to be enforced by the scientist either directly within the experiment or by careful choice of experimental design. Chamberlin and Smith exerted little control over the environment of the open classroom, but exerted control over the behaviour of the participating subjects by limiting what they could do and know. They imposed this control by setting rules of behavior and the distribution of the reservation prices throughout the group of traders and by limiting the flow of information.

In contrast, in the model experiment, controls are made by assumption. There were certain things in Marshall's model experiments which he did not bother to make any assumptions about. For example, he motivated his first experiment by listing five reasons why normal demand might have risen: change in fashions; new use for the good; new market for the good; decline in supply of a substitute good; and increase in wealth or income. But it made no difference to the model experiment which of these were relevant for he assumed that each cause had the same effect, namely, a rise in normal demand which was the starting point of the experiment. There were a lot of hidden assumptions that he did not bother to make evident, such as the automatic adjustment to equilibrium, that the curves were smooth, cut the axes and so forth, which were required to make the model experiments work neatly. He also assumed away all sorts of potentially disturbing factors (such as events in closely related markets):- in effect such factors are either assumed absent or negligible, or to be constant. Modeling economists are used to using the single term *ceteris paribus* conditions (i.e. all other things being equal) for this set of assumptions. Where the laboratory economist has to enforce these different types of conditions (usually in the design stage) to ensure an adequate experimental procedure, the model experimentalist can simply make the necessary assumptions so that the model will be tractable. (A fuller analysis of controls is given in Boumans and Morgan, 2001).

| | Ideal Laboratory Experiment | Hybrid Experiment: Experiment with Model | Model Experiment: Experiment on Model |
|---|---|---|---|
| Controls on: Inputs, Interventions, Environment | experimental | experimental on inputs; assumed on intervention and environment | assumed |
| Demonstration Method | experimental in laboratory | simulation: exp'l/math'l using model | deductive in model |
| Materiality of:<br>Inputs<br>Intervention<br>Outputs | <br>material<br>material<br>material | <br>quasi-material<br>non-material<br>non-material | <br>mathematical<br>mathematical<br>mathematical |
| Control; Representation; and Inference | same material; specific, narrow and limited; relies on accurate replication | | different material; casual, approximate and general; relies on accurate representation |

*Table 1.* Three types of experiment.

This difference in experimental control procedures may be aptly stated as follows: the laboratory scientist *creates an artificial environment within a real world* while the mathematical modeler *creates an artificial world.* (Indeed, Boumans, this volume, shows that this latter idea is now the self-conscious aim of some economists.) Posing the difference in this way suggests that, as epistemic mediators, the power of laboratory experiments is potentially greater than that of model experiments. Nevertheless, economists often find themselves surprised by the results of their mathematical model experiments. Even though economists know the elements that they put into the model (so that the outcomes of model experiments are already built into the model), the answers to questions are not fully known, or fully understood, in advance because we cannot think through how a change to one thing in the model will affect all the other elements and relationships. In contrast, in the laboratory, there is always the possibility of not only being surprised but of being confounded, for the world in the laboratory is one where not only are we ignorant of the outcomes, we don't even know in advance everything about the behavior of the material elements being used.

Another fundamental point of difference between model experiments and laboratory experiments lies in the nature of their demonstration. It is easier to recognize, and to label this difference, as that between experimental demonstration and logical or mathematical demonstration than it is to provide any further characterization of the difference. It is tempting to portray mathematical model demonstration as superior on the view that it is grounded in logic, whereas with modern science studies, we have grown wary of any simple description of the way experimental science works. But it is well recognized that mathematical argument has its own informal nature. In any case, it is difficult to present the kinds of diagrammatic model reasoning that Marshall used as mathematical logic when that logic depends critically on all sorts of shared disciplinary views (some labelled "theory") about the elements represented in the model and, more critically, on what constitutes a valid manipulation.

This issue of the nature of demonstration might well be related to the purpose of experiments. It is sometimes thought that experiments are primarily devices for theory testing, yet the aim of much of the experimental work in modern chemistry, for example, is to make new substances, indeed hundreds of thousands of new things are made each year in laboratories. Perhaps, just as the bulk of chemical experimentation is creative, exploratory work, not theory testing at all, we should regard mathematical modeling experiments in economics in the same way. Modeling work is creative and exploratory: model worlds are created by the economist and subsequent model experiments provide a way of exploring the attributes and limits of the model world and how it works.

Seen in this light, our question about the form of the demonstration becomes less one about the nature of experimental proof versus a truth preserving logic, and more a question about what both laboratory and model experimenters mean when they say of their experiment or model that "it worked". This shared colloquialism refers to the immediate success in answering questions (in a model experiment) or gaining results (from a laboratory experiment), that is, to step 3 in my list. Making inferences from that answer or result to things in the world ("theory testing" of a limited sort) is a separate step, step 4.

# 4. MODELS IN EXPERIMENTS = EXPERIMENTS WITH MODELS

I come now to the case of models in experiments, where experiments use models as an experimental instrument, rather than as the object of experiment. I take as my case here the work of two of my colleagues at the University of Amsterdam: Cars Hommes (from CeNDEF - Centre for Nonlinear Dynamics in Economics and Finance) and Joep Sonnemans (at CREED - the experimental unit in the economics faculty). In a series of three experiments (see Sonnemans et al., 1999 and Hommes et al., 1999), they have joined Hommes' mathematical experiments and numerical simulations on the "cobweb model" to Sonnemans' laboratory experiments on learning.

The individuals in their laboratory experiments are told that they are each advising a hypothetical supplier who operates in a market where (because of production lags) he must make output decisions based on the expected price while buyers made decisions on the actual price. Thus, the experimental subjects are asked to predict the price for the following year (and to design strategies for predicting future prices) for the supply of the good, knowing only the current and previously realized prices (that is, not knowing how anyone else in the market will act). These one-period-ahead predictions (or the alternative written down strategies) are then used, either individually, or in randomly selected groups, as inputs into a market model, for unknown to the experimental participants, the hypothetical market participants have demand and supply functions which are already specified by the researchers in mathematical form. The model is used as the calculation device, or instrument, in the experiment which takes the experimental subjects' expected prices as inputs and uses them to determine the "realized" market prices for each period. These are then taken as existing prices relevant for the prediction of the next period's price, as the experiment runs over several periods. The values of the parameters in the two functional relations in the model are chosen in such a way that, according to the participants' predictions, a sequence of such prices over time may behave nicely (converge towards a stable market equilibrium level), or get stuck in a cycle, or result in chaotic behaviour, but in theory at least, the participants should be able to learn from their experience to reach the stable level.

We can label these economic investigations as "experiments" or "simulations", but the labels are unhelpful in denoting the differences between their hybrid status and the two end cases of laboratory and model experiments. These experiments embody elements of both laboratory and model experiments at the same time: see the middle column of Table 1. The "market" consists of real (laboratory) inputs operating in conjunction with a mathematical model. Each part is subject to its own kinds of "control". Hommes and Sonnemans instituted laboratory control where material could be isolated and manipulated and substituted model-based controls for experimental control in those areas which could not be isolated by experimental means.

It is important to the status of these experiments that the input strategies were proposed by experimental subjects, under experimental conditions, so that even when they are joined up according to the model requirements, they maintain some kind of real world, or quasi-material, status. Like all experiments, these ones have an air of artificiality about them stemming from the combination of strict rules and structured responses together with the natural variation that comes from the participants. The experimental subjects were students given small monetary incentives to predict prices in the experimental environment rather than industry managers whose jobs might depend on their abilities to predict prices in the market. The controls in the experimental situation might also render their behavior even more artificial: for example, participants were not free to write down just any price prediction - it had to be within a fixed range determined by the mathematical model; and although they could write down any pricing strategy they liked - it had to satisfy certain conditions that enabled it to be programmed. Yet, it is surely more important that, just as in the real world, experimental subjects had to make decisions about pricing without knowing either the demand relation in the market or about their rival suppliers' pricing strategies. The presence of such quasi-material inputs has lead me elsewhere to refer to such experiments as "virtually experiments" (see Morgan, 2000, for a fuller classification which also includes "virtual experiments").

The mathematical model used is called a "cobweb model" and takes its form from empirical work in the 1930s on agricultural goods markets. Its form therefore has a certain empirical, if not material, quality. But, it is constructed under the *ceteris paribus* conditions that befit mathematical-model experiments, and the parameter values which govern its shape are chosen more because they allow for a range of outcomes according to the participants' responses than because they have been empirically validated.

The demonstration involved a mixture of experimental and mathematical (by calculation rather than by reasoning) methods. It is perhaps this combination, rather then the mix of real world and model inputs or the mixture of experimental and model controls, which marks out this kind of experiment as a simulation. We see again this same mixed demonstration method when the experimental participants wrote down their strategies for responding to realized prices, and these strategies were fed into the model to calculate long period dynamics of the equivalent realized market prices.

The economists were not just interested here in understanding how the individuals responded to situations but in the effect of their different predictions and strategies when combined together in the mathematical model and perhaps saw this as the element of the real world they were seeking to capture in conducting the experiment. It is the unexpectedness which results from these material inputs, known to the experimenters but not designed by them, and their interaction in combination, which creates the possibility of confounding the experimenters' expectations. Thus, the model device which brought these individuals together was also a critical part of the experiment. While economists can reproduce reasonably complex situations and induce certain kinds of economic behavior in the laboratory, they can not recreate the open environment of market forces in which those individuals act. The best they can do is either accurately describe the behavior of inputs, or gain such knowledge by experiment, and fit these inputs together in a model of a market. In these cases, the particular problems economists wanted to learn about have a complexity or scale at a level beyond that which can be investigated experimentally and the combination of models and quasi-material inputs provided alternative resources for experiments.

These three types of experiment, laboratory, model and hybrid experiments, have all used the same basic supply and demand model. In the hybrid experiments of Hommes and Sonnemans, the model itself is passive or fixed whereas in Marshall's earlier model experiments, the experiment consists of changing something in the model to see what happens. Marshall's experiments were *on* models: the model world was the object of experiment. The hybrid experiments were experiments *with* a model, it was used it as an instrument to calculate the outputs depending on the various material inputs. Although the model plays a more active role in the hybrid experiments than in Chamberlin's and Smith's classroom experiments, even there it played two roles, namely it placed rather loose limits on each participant's behavior and it acted as a benchmark to assess their experimental results against those obtained from Marshall's model experiments.

## 5. EXPERIMENTAL VALIDITY: THE INFERENCE GAP

The validity of experimental results is a question of considerable concern for economists conducting both laboratory experiments and mathematical model experiments. Their worries appear different, because couched in different terms, but both are directly related to issues of control and representation. (I touch on only these two issues, but of course recognize that there are other problems to be overcome.) For laboratory scientists, the question is seen as one of "parallelism" (see Guala 1999 and this volume). Do results obtained in the laboratory hold true in the uncontrolled world? Is the artificial, controlled world created for the experiment sufficiently like the real world to justify inference from one to the other? For economists using models for extended thought experiments, the issue is portrayed in terms of the realism of assumptions. Is the model sufficiently realistic in its assumptions for inferences to be made from the model to the world?

First, the inference gap can be readily understood in terms of different types of control. The mathematical modelers understand the problem of inference from model experiments as directly related to the problem of the kinds of assumptions involved in building their model. Mathematical model building involves a process of abstraction and simplification of economic events being modelled (such as in Marshall's cases), and experimental control in such model experiments is gained by the economist making assumptions about connecting, confounding and disturbing factors in the relationships. These processes create a simple model world of supply and demand relations which seems far from the complicated real world. The assumptions and simplifications make the mathematical model tractable so that experiments on it will produce results, but these same controls limit the validity of the results in direct application to specific events in the world. Paradoxically, these same qualities seem to make it easier for economists to "apply" model experiment results "approximately", or even as Gibbard and Varian (1978) suggest, "casually", to a wide number of objects and circumstances in the economic world which seem to share some common traits with the model situation. Thus, Marshall's experiments on models provide us with general inferences about the direction of change in quantity in response to a shift in the demand curve due to a whole range of causes. But, unless we know the shape of the supply curve, we would be hard pressed to make inference about any particular case. Strangely perhaps, the most evident element in the inference gap for model experiments, the validity of any inference between two such different media - from a mathematical to a material world - does not seem to bother economists.

In experimental economics, the validity of experimental results is defended by referring to the design of the experiment. Control is dependent on the choice of experimental set up, circumstances and procedures (institutional rules, rewards and so forth). These choices are guided by the experimenter's need to design the experiment in such a way that *economic* behavior is made manifest in the experiment. Although there are arguments as to whether experimental subjects (such as the students in the class experiments) really do behave "naturally" in the artificial situation like, for example, managers in industries, nevertheless, they share the quality of being humans, and so that part of the inference gap is surely less than for mathematical model experiments. These design features are then adduced by experimentalists as reasons why the results they find in their controlled situations carry over to the uncontrolled world. This may make it possible to infer to the same type of situations (in terms of the objects, structure and circumstances) in the world, but, that very same tightness of controls and high levels of specificity involved in the laboratory experimental set-up makes inferences to related or similar situations in the world more problematic.

Thus, Chamberlin could make some tentative inferences about the behavior of people acting under the same market rules and market model in the world, but could not say much about Marshall's type of market situation where the assumed rules are very different. In contrast Smith found it easier to make some more general inferences about behavior under his rules because he found stable results over a set of experiments with considerable variation in supply and demand curves. If we take the Chamberlin and Smith cases together, we might perhaps make limited comparative inferences about how people behave, and the effects of this, under two different kinds of market rules of interaction.

In contrast, the hybrid laboratory experiments with models by Hommes and Sonnemans were characterised by a pragmatic mixture of control techniques. Recall that the investigators used experimental control over the inputs to the experiment but substituted model-based controls for the manipulation regime. This mixture makes it difficult to make general comments on the experimental validity of these hybrid experiments. It seems reasonable to suggest that the more the hybrid experiment leans towards a model experiment, with abstract objects subject to controls by assumption, the more difficult is a well-founded inference from model results to specific detailed material cases, but general inferences remain open. The more material the inputs and the interventions remain, the more something like a laboratory experiment is involved, supporting inference from specific experimental conditions to a limited range of equivalent situations in the world (see Table 1). Such was the level of natural variation in the material elements (the predictions and strategies) in the Hommes and Sonnemans cases, that it proved difficult even to make specific inferences.

How is this inference gap in the case of model experiments to be overcome? In Morgan (1999), I suggested that narrative or story-telling is a favourite device of economists to relate the model world to the real world. But that account gives the form of the inference argument, rather than the basis for why or when inference might be reasonable. Let us recall how and why we make model experiments. The method of modeling involves postulating a representation - a model - to capture certain elements of the world and manipulating this representation - experimenting with the model - to see if we can demonstrate in the world of the model the same phenomena that we observe in the world. But remember, we don't know the mechanism or situation responsible for the phenomena we observe - otherwise we wouldn't need to build a model of the world to try to reproduce it. We are forced to hypothesize the structure because we cannot analyze it via laboratory experiment. We use this hypothesized structure to explore particular sets of circumstances and to demonstrate processes, events and outcomes in the world of the model.

Whether we can carry back something from our manipulations or demonstration in the world of the model to learn something about the domain of the real world (or whether we can interpret the demonstrated results in terms of the original domain) depends on how well we have represented the world and whether we have been able to capture to some degree the structure of the problem.

Thus, the second main issue to be addressed with respect to the inference gap lies in the importance of representation. A useful entrance to this issue comes in May Brodbeck's aphorism: "Model ships appear frequently in bottles; model boys in heaven only" (1968 [1959], p. 579). In my account, neither of her models can answer many questions about the world: we can't get much help in understanding how ships can transport goods or explain why boys are naughty from these models. Why not? One reason is that they have no internal dynamic - they are not very manipulable - so they don't contain the resources for model experiments. The other reason is because the nature of the representation provided by these particular models is not useful for the kinds of questions we want to answer about the real world.

These two examples help us to see the connection between our four steps of model experiments: representation, questions, demonstration and the possibilities of inference. If we want to understand exactly how a ship can float, simplifying a ship into a model which accurately represents the exact details of the portholes of the ship doesn't help us but simplifying to capture the relations between length, draft and displacement may help us. We have to represent the structure of the elements and relations relevant to the question asked if we are to provide an experimental demonstration involving that structure in answer to that particular question. And that representation has to be a descriptively accurate one if we are to make inferences from it to the world. An idealization process which provides us with a model of the behavior of a boy in heaven may not be very useful if we are trying to answer questions about real boys in the world; rather, we need to represent in a model some of the essential characteristics relevant for understanding boys in the real world if experiments with the model are to teach us something about such real boys.

This is not to argue that economic models have to be completely realistic: of course not. Nor does it meant that models shouldn't involve simplifications and idealizations: of course they can. Passive irrelevant factors can and should be simplified away, and we can go with idealized accounts of motivations, provided we can give a rather more accurate representation of the behavior, situations or interactions relevant for our questions. Control via assumptions is a sensible way to limit the world in the model to make it experimentally tractable. The realism of assumptions does matter, but not of all of them at the same time. The hybrid cases may throw some further light on the issue. We want the model world to accurately represent the real world, to be descriptively accurate if you like, at precisely that point where our questions are to be answered. If our questions are about how different pricing strategies affect a supply and demand market outcome, then we want our input strategies to be "realistic". If our questions are about what happens when the demand curve shifts, we want both demand and supply curves - but particularly the latter, since the experimental manipulation of the demand curve traces out the shape of the supply curve on which the particular results depend - to be accurately represented.

Like laboratory experiments, it is not always clear whether you can carry over the results of model experiments to the real world. Both laboratory experiments and model experiments depend on controls to make the experiment work, but this makes some aspects of the world artificial. The other aspects are the ones of interest for the experiment, and whose (controlled) variation we seek to use when we make the experimental manipulation; these are the ones which must remain as "real" as possible. In the case of laboratory science, successful experiments depend on *accurate replication in the laboratory* of the elements, changes and outcomes in that part of the world relevant to the question. Only then can inferences be made from the results of experiment to teach us something about the world. In the case of modeling science, successful experiments depend on *accurate representation in the model* of the parts of the world relevant for our questions. Otherwise, we can only learn about the world in the model, which may be rewarding for theory development but not for learning about the world.

## ACKNOWLEDGEMENTS

This paper, prepared for "Model Based Reasoning" (Pavia, May 2001), develops several ideas first presented at the workshop "Towards a More Developed Philosophy of Scientific Experimentation" (Amsterdam, June 2000) and at the "Language, Logic and Logistics: Modeling and Cross-Disciplinary Discourse" (State University of New Mexico, January 2001). I thank Cars Hommes for discussions about his experiments, and Marcel Boumans, Francesco Guala, Hans Radder, Rom Harre, and the participants at all three workshops for helpful comments relevant to this paper. I thank the British Academy for supporting my research during this period.

## REFERENCES

Boumans, M., 2002, Calibration of models in experiments, in this volume, pp. 75-93.

Brodbeck, M., 1968 [1959], Models, meaning and theories, in: *Readings in the Philosophy of the Social Sciences*, M. Brodbeck, ed., Macmillan, New York, pp. 579-601.

Chamberlin, E.H., 1948, An experimental imperfect market, *Journal of Political Economy* 56(2):95-108.

Gibbard, A. and Varian, H.R., 1978, Economic models, *The Journal of Philosophy* 75(11):664-677.

Guala, F., 1999, The problem of external validity (or "parallelism") in experimental economics, *Social Science Information* 38(4):555-73.

Guala, F., 2002, Models, simulations, and experiments, in this volume, pp. 59-74.

Hamminga, B. and De Marchi, N., 1994, *Idealization in Economics,* Rodopi, Amsterdam.

Hommes, C., Sonnemans, J., Tuinstra, J., and Van de Velden, H., 1999, Expectations driven price volatility in an experimental cobweb economy, University of Amsterdam, CeNDEF working paper, 99-07.

Hughes, R.I.G., 1997, Models and representation, PSA 1996, vol II, pp. 325-336.

Magnani, L., 2002, Epistemic mediators and model-based discovery in science, in this volume, pp. 305-329.

Marshall, A., 1930 [1890], *Principles of Economics* (VIII edition), Macmillan, London.

Morgan, M.S., 1999, Models, stories and the economic world, Research Memoranda in the History and Methodology of Economics, 99-1, Faculty of Economics and Econometrics, University of Amsterdam; forthcoming in *Journal of Economic Methodology*, 2001.

Morgan, M.S., 2000, Experiments without material intervention: Model experiments, virtual experiments and virtually experiments, Research Memoranda in the History and Methodology of Economics, 00-1, Faculty of Economics and Econometrics, University of Amsterdam, forthcoming in *Scientific Experimentation and Its Philosophical Significance*, H. Radder, ed., Pittsburgh University Press, Pittsburgh.

Morgan, M.S. and Morrison, M., eds., 1999, *Models as Mediators: Perspectives on Natural and Social Science,* Cambridge University Press, Cambridge.

Sismondo, S., ed., 1999, *Modeling and Simulation, Science in Context*, Special Issue 12:2.

Smith, V.L., 1962, An experimental study of competitive market behavior, *Journal of Political Economy* 60(2):111-137.

Sonnemans, J., Hommes, C., Tuinstra, J., and Van de Velden, H., 1999, The instability of a heterogeneous cobweb economy: A strategy experiment on expectation formation, University of Amsterdam, CeNDEF working paper, 99-06.

# Models, Simulations, and Experiments

Francesco Guala
*Centre for Philosophy of the Social Sciences, University of Exeter, UK, f.guala@ex.ac.uk*

Abstract: I discuss the difference between models, simulations, and experiments from an epistemological and an ontological perspective. I first distinguish between "static" models (like a map) and "dynamic" models endowed with the capacity to generate processes. Only the latter can be used to simulate. I then criticise the view according to which the difference between models/simulations and experiments is fundamentally epistemic in character. Following Herbert Simon, I argue that the difference is ontological. Simulations merely require the existence of an abstract correspondence between the simulating and the simulated system. In experiments, in contrast, the causal relations governing the experimental and the target systems are grounded in the same material. Simulations can produce new knowledge just as experiments do, but the prior knowledge needed to run a good simulation is not the same as that needed to run a good experiment. I conclude by discussing "hybrid" cases of "experimental simulations" or "simulating experiments".

## 1. INTRODUCTION

Empiricist philosophies of science draw a sharp distinction between descriptive or representational devices (scientific theories) and what is described or represented (the natural or social world). Models and simulations are customarily placed among the representational tools, whereas experiments are considered parts of the natural or social world that have been carefully designed in order to answer some specific question. There are, however, bits of science that do not fit neatly, and for which a different scheme

*Model-Based Reasoning: Science, Technology, Values*, edited by
L. Magnani and N.J. Nersessian, Kluwer Academic/Plenum Publishers, New York, 2002

of classification may be more appropriate. In this paper I shall try to show that it is sometimes useful to think of models, experiments and simulations as tokens of the same kind, somehow located between our statements about the world (call them scientific laws, principles, theories, axioms), and the world itself (see also Guala, 1998). Borrowing from Margaret Morrison and Mary Morgan (1999), we may say that such entities "mediate" between theory and reality.

First, let us notice that everyday scientific talk often does treat experiments, models and simulations as tokens of the same kind. In one of the earliest papers in the field of experimental economics, for example, the term "simulation" appears three times only in the first page (Smith, 1991, p.8), alongside other expressions such as "experiment" and "experimental game". Or take medicine. Experimental physiologists make extensive use of animals in their investigations, for well known (although controversial) ethical reasons. Most often, these activities fall under the label of "animal experimentation". But it is not uncommon to hear or read the expression "animal models", especially when experimenters fear that the findings will not be easily transferable from animal subjects to human beings.

Why do scientists slip from "experiment" talk, to "model" and to "simulation" talk? A plausible answer is that the difference is purely epistemic in character: "experiment" and "theory" being the pillars upon which all proper science should stand, scientists signal their epistemic doubts using a special terminology. An incomplete or less than certain theory becomes a "model"; a dubious experiment becomes a "simulation", and so on. However, perhaps there is something deeper to be said, and the rest of the paper is devoted to explore this possibility.

## 2. MODELS AND SIMULATIONS

Models have been at the forefront of research in the philosophy of science for at least two decades now. Indeed, the latest orthodox "theories of scientific theories", the so-called "Semantic View" of theories, *identifies* theories with sets of models. The Semantic View is more a family of doctrines than a single, unified philosophical theory, but all its versions share a distaste for the older "syntactic" approach, according to which theories are basically sets of statements or laws. In the semantic approach the fundamental component of a theory, the model, is in contrast a *structure* - a set of objects with properties and relations among them and/or their parts - that satisfies the linguistic components of the theory. The latter are secondary, in the sense that they can be formulated in various equivalent ways, as long as they are satisfied by the models. The axioms, laws, etc., may change depending on the language and system of axioms scientists choose, but the models won't. The models must be put at work by means of a "theoretical hypothesis", stating that they stand in a certain relation (of similarity, isomorphism, analogy, etc., depending on which version of the Semantic View one subscribes to) with real-world entities or systems. Since the Semantic View is presently the received explication of the concept and role of scientific models, I shall take it as my point of departure here. The next question is: what is a simulation?

Mario Bunge (1969) defines simulation as a relation between two entities, *x* and *y*, where *x* simulates *y* if (1) there exists a correspondence relation between the parts or the properties of *x* and *y*; and (2) the analogy is valuable to *x*, or to another entity (*z*) that controls *x* (p. 20). The first striking feature of this definition is its anthropocentrism. It makes no sense to say that a natural geyser "simulates" a volcano, as no one controls the simulating process and the process itself is not useful to any one in particular. I shall assume for the sake of this paper that the second part of Bunge's definition captures some important connotations of the term simulation. But the first part is unsatisfactory, because it leads to include things that we would not intuitively call "simulations" at all. Consider a map: if it has been drawn adequately, there must exist some correspondence relation between its parts and parts of the territory it is aimed at representing.[1] Since the map is also somehow "controlled" by someone (its user or its creator), and is certainly valuable to her, it does fulfil all of Bunge's criteria. Yet, it would be odd to say that a map "simulates" the territory.

Now consider a map together with a set of flags and miniaturized soldiers and tanks, of the sort you find in military head-quarters or in games such as "Risk". If the toy-flags, mini-soldiers and mini-tanks are moved on the map according to the appropriate rules, we can properly claim that a battle or a military campaign is being simulated. Why? Whereas the map alone is somehow "inert", the same map, plus the miniatures, plus the players or officials moving the miniatures according to the rules, make a "dynamic" system. I shall here follow Stephan Hartmann (1996) and distinguish *static* from *dynamic* models. A static model can only represent a system at rest. A dynamic model can also represent the time-evolution of the system (p. 82).[2] A dynamic model, then, can be in different states at different times, and usually each state will correspond to a specific combination of values taken by the variables in the model. Such a model will be able to be in as many different

[1] I am here referring standard maps on paper only. Giere (1999, pp. 44-47) provides a detailed discussion of the function of maps as models.

[2] I am paraphrasing Hartmann here, for he speaks of models as if they were linguistic entities (made of "assumptions", for example), whereas in this paper I follow the Semantic approach and take them to be objects.

states as all logically or physically possible permutations of the values its variables can take. Only "dynamic" systems of this sort can properly speaking simulate. "*A simulation imitates one process by another process*" (Hartmann, 1996, p. 83), where a "process" is a time-ordered sequence of states a system takes in a given time period.[3]

This characterization opens some interesting questions. Consider my previous example: in order for the map-plus-miniatures to be a simulating device, the system must be capable of taking different states (the miniatures must change their position on the map, for instance). This means that there must be an agent prompting the changes in the system itself. Such a role may be played for instance by the officials in the army's head-quarters. Thus, counter-intuitively perhaps, the officials must *belong to* the simulating device itself. If "simulation" is an anthropomorphic or more in general agent-dependent notion, as Bunge seems to suggest, we should not be troubled by this. It is just natural that what is to be included and what to be excluded in a simulating system is partly arbitrary and/or dependent on one's interest. Simulations are not in nature, it is us who "see" them and often build them according to our purposes. Similarly, a checkerboard and some pawns cannot by themselves simulate anything - although they can *represent* something: for example the state of a given battle at time *t*. A checkerboard, some pawns, *and two players* can simulate a battle or a war (albeit at a very high level of abstraction) by representing a sequence of states of that battle or war. Most often, a simulating device will have some mechanism built into it, which once triggered will make the system go through a series of states automatically. The agent's role, then, will be merely that of setting the initial state and starting the process, which will keep running until it is exogenously interrupted or runs out of steam.

## 3. SIMULATIONS VS. EXPERIMENTS: THE EPISTEMIC ACCOUNT

The distinction between simulations and experiments is more tricky than the one between models and simulations. In everyday scientific talk, such a distinction is certainly loaded with epistemic connotations: simulations are supposed to be somehow less fertile than genuine experiments for the production of scientific knowledge. Their results are often qualified as "mere" simulations not to be mistaken for the "real thing" (i.e. the real-world system whose behaviour is being simulated, or an experiment on the real-world

[3] I am here modifying slightly Hartmann's (1996) own definition of "process" as "some object or system whose state changes in time" (p. 83).

system). The interesting question, however, is whether the epistemic difference is fundamental, or whether it is just a by-product of some more basic difference between experiments and simulations.

I should make clear that I am not interested in conceptual distinctions *per se*. My primary aim is to make sense of some tools that are widely used in science. And this is no mere philosophical quibble: scientists worry about the same issues - probably even more than philosophers do. Take the sort of laboratory work done by psychologists and economists interested in behavioral decision making. The psychologist Baruch Fischhoff represents practitioners' worries by means of a graphic example. In the psychology lab, choices look like this:

> **Choice A.** In this task, you will be asked to choose between a certain loss and a gamble that exposes you to some chance of loss. Specifically, you must choose either: Situation A. One chance in 4 to lose $200 (and 3 chances in 4 to lose nothing). OR Situation B. A certain loss of $50. Of course, you'd probably prefer not to be in either of these situations, but, if forced to either play the gamble (A) or accept the certain loss (B), which would you prefer to do? (Fischhoff, 1996, p. 232).

But in the real world, choices look like *this*:

> **Choice B.** My cousins [...] ordinarily, I'm like really close with my cousins and everything. My cousin was having this big graduation party, but my friend – she used to live here and we went to [...] like started preschool together, you know. And then in 7th grade her stepdad got a job in Ohio, so she had to move there. So she was in Ohio and she invited me up for a weekend. And I've always had so much fun when I'd go up there for a weekend. But, it was like my cousin's graduation party was then, too – like on the same weekend. And I was just like I wanted to go to like both things so bad, you know. I think I wanted to go more to like up Ohio, you know, to have this great time and everything, but I knew my cousin – I mean, it would be kind of rude to say, "Well, my friend invited me up, you know for the weekend." And my cousins from out of town were coming in and everything. So I didn't know what to do. And I wanted mom to say, "Well, you have to stay home", so then I wouldn't have to make the decision. But she said "I'm not going to tell you, you have to stay home. You decide what to do". And I hate when she does that because it's just so much easier if she just tells you what you have to do. So I decided to stay home basically because I would feel really stupid and rude telling my cousin, well, I'm not going to be there. And I did have a really good time at her graduation party, but I was kind of thinking I could be in Ohio right now (Fischhoff, 1996, p. 232).

What do choices in environments like the first one tell us about behavior in environments like the latter? And what *are* environments like the former anyway? Are they *simulations* of real-life situations, or are they *experiments* on human decision-making? One possible answer is that experiments like "Choice A" test subjects' "pure" cognitive capacities. But this would be unsatisfactory: decision processes may be completely different in the two circumstances, and "purity" is a poor consolation if it is unlike anything we are ultimately interested in explaining and understanding. A more reasonable answer is that in situations like "Choice B" there is just too much going on, and simplified settings like "Choice A" are just intermediary steps on the way towards the understanding of complicated "real-world" decision making. Indeed, it is always useful to think of experimental work as involving (at least) three different systems and two distinct hypotheses (see Figure 1).

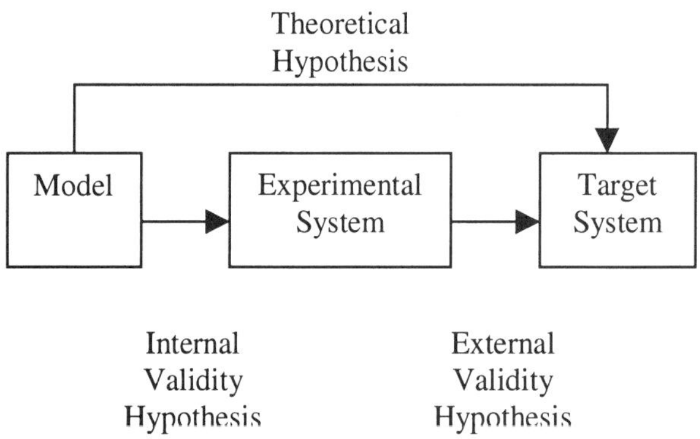


*Figure 1.* From models to target systems.

As semantic theorists point out, models must be put in correspondence with their "real-world" target by means of a theoretical hypothesis specifying what sort of relation (isomorphism, analogy, similarity, etc.) is supposed to hold between them. But the path between a model and the real world can be split into a number of sub-steps. Most often, the relation between a model and an appropriately designed experimental system is tested first (in the lab). Then it is further hypothesized that the experimental system stands in some specific relation with the target. This is typical of sciences, like medicine or economics, which are ultimately aimed at explaining phenomena in the unconstrained world outside the lab. Since the quality of field data is usually too poor to answer specific questions concerning causal mechanisms and processes, physiologists and social scientists find it useful to try and answer them first in the laboratory. Then, they show (or argue, or assume) that the laboratory was a correct replica of the real-world systems they were interested in the first place.

What kind of correspondence relation is specified by the external validity hypothesis? Given the amount of controversy concerning the nature of theoretical hypotheses in the semantic approach, one may be pessimistic about the prospects of finding a general answer to this question. But, minimally, an external validity hypothesis should map elements of the experimental system onto elements of the target system. These elements should be entities, properties, and relations between these entities and properties. (Causal relations should be prominent, one would think.) Experimental psychologists like Fischhoff are worried because the fact that X causes Y in system A (the lab), does not imply that the corresponding entity/property X* is a cause of Y* in system B (the target). An external validity hypothesis or assumption bears the weight of any inference from the social science laboratory to the real social world.

"Model" and "simulation" talk is more common in the experimental branches of sciences, like economics or medicine, in which external validity worries are widespread and taken seriously. Indeed, you do not find in physics the sort of *a priori* skepticism towards experiments that you find, for instance, among economists. But the question is: should the distinction between experiments and other mediating devices like simulations be based on such epistemic differences? At least two arguments can be leveled against this solution. Firstly, epistemic degrees of confidence in a particular scientific tool or device change in time. Thus, Galileo's experiments were not immediately greeted with enthusiasm, but it would seem odd to claim that for this reason they should have been labeled as "simulations" until they were accepted by the scientific community. What a scientific device is should not depend on whether we are confident in the reliability of its results.

Secondly, simulations require an external validity hypothesis too, which may or may not be true depending on the circumstances. If I simulate the battle of Waterloo using toy soldiers and horses, I work on the hypothetical assumption that, for example, the speed of horse miniatures stands approximately in the same relation with the speed of little soldiers on my map as the speed of infantry units stood with the speed of cavalry units in 1815. Only under this hypothesis can I use the simulating model to investigate, for instance, what would have happened if Napoleon had chosen a different strategy. The difference then must lie elsewhere. Perhaps in the fact that psychologists and economists use human beings (students, for example) as subjects in their experiments. The intuition is that, unlike pawns and armies, toys and troops, human beings are *the same* in and out of the lab. But how exactly?

## 4. SIMULATIONS VS. EXPERIMENTS: THE ONTOLOGICAL ACCOUNT

A material model of the propagation of light, according to the wave theory, can be built with the aid of water in a ripple tank. At a general level of analysis any kind of wave can be modeled as a perturbation in a medium determined by two forces: the external force producing the perturbation, and the reacting force working to restore the medium at rest. General relationships such as Hooke's law or D'Alembert's equation may hold for *all* kind of waves. More fundamental relationships, such as Maxwell's equations, describe the properties of the electric and the magnetic field only. The D'Alembert wave equation belongs to electromagnetic theory because electricity behaves *like* a wave, although the fundamental forces at work are different from those at work in case of, e.g., water waves. The terms appearing in the equation describing the target and the model-systems are to be interpreted differently in the two cases: the forces are different in nature, and so are the two media in which waves travel. The similarity between the theoretical model of light waves and the ripple-tank model holds at a very abstract level only. The two systems are made of different "stuff": water waves are not light waves. Because of the formal similarity, though, the behavior of light waves can be *simulated* in a ripple tank. Both light waves and water waves obey the same non-structural law, despite their being made of different "stuff". This is due to different reasons in each case: different underlying processes produce similar behavior at an abstract level of analysis.[4] Similarly, human behavior can to a certain extent be simulated by means of computerized models, but arises from "machines" made of flesh, blood, neurons, etc. rather than silicion chips.

Herbert Simon (1969, pp. 15-18) puts it as follows: simulations rely on a process of abstraction from the fundamental principles governing the behaviour of the simulating and the target systems. If similar "organizational properties' arise at a given non-fundamental level from different substrata, it is possible to abstract from the substrata and simulate the behavior of a system A by observing the behavior of another system B which happens to (or which is purposely built so as to) display those non-fundamental properties. Working on this idea, we can devise a criterion to demarcate genuine experiments from "mere" simulations. The difference lies in the kind of rela-

[4]Of course, if one believes in the reductionist story according to which everything physical is made of the same fundamental sub-atomic particles, then both light and water waves are "made of the same stuff". But the reductionist story is controversial (photons seem to have different properties from other particles), and at any rate the fact that everything is made of the same stuff does not play any relevant role in explaining why both systems display certain non-fundamental relations.

tionship existing between, on the one hand, an *experimental* and its *target* system, and, on the other, a *simulating* and its target system. In the former case, the correspondence holds at a "deep", "material" level, whereas in the latter the similarity is admittedly only "abstract" and "formal". It is tempting to claim that in a simulating device the simulated properties, relations, or processes are generated by *different* (kinds of) *causes* altogether. Such a claim is problematic if you endorse a formalistic view of (type-level) causation – for example a view that defines causation in terms of purely probabilistic relations. If, in contrast, one takes causation to be a substantive property of specific kinds of systems (à la Wesley Salmon, for instance),[5] the problem may disappear. In a genuine experiment the same "material" causes as those in the target system are at work; in a simulation they are not, and the correspondence relation (of similarity or analogy) is purely formal in character.[6]

## 5. THE METHODOLOGY OF "PURE" SIMULATIONS

Because of the different nature of the correspondence relation, simulations and experiments are appropriate research tools in different contexts. Typically, simulations are used in one of two different ways: either (1) to bootstrap from the fact that a given effect (which we have observed in system A) can be produced by means of simulation B, to the fact that the relations governing the behavior of B also govern the behavior of A. Or (2) to argue that a certain effect observed by simulating with B will also be observed in the case of A because the two are governed by similar relations.[7] Both procedures are knowledge-producing ones. The point to be stressed is that in both cases the relationships have to be fully specified for the simulations to be carried on. Systems of this kind are "transparent boxes", to which the old *dictum* applies: "a simulation is no better than the assumptions built into it" (Simon, 1969, p. 18).[8]

Geologists working on stratigraphy, for instance, study the structure of rock layers below the earth's surface. They also investigate the process of strata formation, but have to face very serious obstacles, such as the impos-

[5] The latest versions of Salmon's theory of causation, however, are of little help outside the realm of physics.

[6] See also Ernst Nagel's (1961, p. 110) distinction between "substantial" and "formal", or Mary Hesse's (1963, p. 63) "material" and "formal" analogies.

[7] One may be unable to experiment with A, or the equations describing A may be so complicated that they can be solved only by means of some "brute-force" solution in B.

[8] Which does not mean that they are *just* as good as that: to run a "good" simulation involves the use of approximations, computer implementation, etc. and thus requires more knowledge and skills than simply specifying the correct basic equations.

sibility of doing controlled experiments (processes of sedimentation last for millennia, and of course the target systems are too large to be manageable in a lab), the difficulty to gather data even about the present geography of the strata, the strong theory ladeness of the interpreted data, the complex interdependencies within geological systems, and so on. In order to solve at least some of these problems, geologists have devised simulation techniques like STRATAGEM, a computer-based modeling package used by large companies such as Shell Oil.[9] This simulation device works on the basis of a number of structural equations taken from the theory of "sequence stratigraphy". The equations model the system's outcome (the actual sedimentation) as a function of a number of variables including the hydrodynamics of sediment deposition, the subsidence patterns, the global sea level, the amount of sediment supplied to the basin, etc. The outcome of the simulation is dependent on the approximate validity of the theory of sequence stratigraphy, and also on the correct specification of the initial conditions and of the values assigned to the free parameters in the equations (incidentally, these are all problematic assumptions to be made in the specific case). Geologists try to simulate systems A (real-world geological structures) by means of a computer-model B, and all the fundamental relations in B must be known to be approximately correct and specified in advance.

In Figure 2, and in the diagrams that follow, I represent the presently unknown features of a target system by means of dotted lines.[10] The question mark denotes an aspect of a system whose nature or functioning the scientists is investigating in a given case.

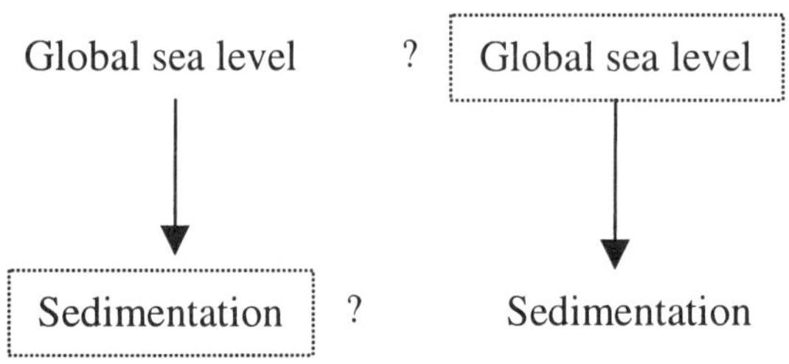


*Figure 2.* Pure simulations in geology.

[9] I have learned about simulation techniques in geology from Francis Longworth and his unpublished paper on the methodology of STRATAGEM.

[10] In fact, a more precise representation would involve at least *two* systems (say, A and B), each one made of entities/properties and relations (say, $X_A \rightarrow Y_A$, $X_B \rightarrow Y_B$), plus some "external validity' hypothesis stating that the two systems stand in a certain relation to each other. The dotted lines in my figures are shorthands for cases in which such a hypothesis is *not* known or well-confirmed.

Notice that some features of simulating systems like STRATAGEM is exactly opposite from that envisaged in the epistemic account. In order for these simulations to be successful, geologists must be very confident that the (formal) correspondence between relations in the simulating device and relations in the target holds to a certain degree of approximation. Experiments, as we shalls see in the next section, do not require as much.

## 6. THE METHODOLOGY OF "PURE" EXPERIMENTS

The crucial presumption behind experiments is that relevant components of the laboratory system are made of the same material as those of the target system. (We shall see in a moment what "relevant" means.) The experiment should feature the same causal processes that are at work in the real world, rather than just display some formal relation by means of a device made of different "stuff". Experiments are useful when one has an imperfect understanding of some basic causal mechanism of the system under study. They are useful in these contexts precisely because the laboratory "stuff" is the same as the non-laboratory "stuff".

What is unknown is often (but not always) what is under test. An experiment can give us more confidence in a theoretical model, if the theory makes some contestable assumption about some component of the target system, and if the experiment includes the real component (for example real human behaviour, as in experimental psychology and experimental economics). An experiment that merely reproduces all the assumptions of the model, for example by paying subjects to act according to the behavioral theory of the model, does not test anything at all (except perhaps the incentive system). But notice that not all that is imperfectly understood needs to be under test. For instance, one can test the efficacy of a drug without a detailed understanding of the mechanism of propagation of a disease. The efficacy of the drug rather than, say, the process of infection, is what is under test. Or you can do experiments on market behavior even without a proper understanding of the mechanisms of individual choice and belief formation. Market institutions, instead of individual behavior, are under test in these experiments. Subjects may trade at a certain equilibrium price because they are acting in a fully rational way, or perhaps because they are following some rule of thumb, or even by sheer imitation. Whatever the real causal process, we can use laboratory tests to study selected aspects of specific real-world economies as long as we are confident that the same (unknown) basic principles of behaviour apply in both cases.

Considerations of this kind have sometimes been proposed by experimentalists to defend their methodology. The economist Vernon Smith (1991), for example, argues that "the laboratory becomes a place where real people earn real money for making real decisions about abstract claims that are just as "real" as a share of General Motors". For this reason, "Laboratory experience suggests that all the characteristics of "real world" behavior that we consider to be of primitive importance [...] arise naturally, indeed inevitably, in experimental settings" (pp. 100-1). This reasoning supports experimenters" confidence in their results. To them, the "real" character of experimental markets helps to bridge the gap between a theory and its intended target of application. "Laboratory microeconomies are real live economic systems, which are certainly richer, behaviorally, than the systems parametrized in our theories" (pp. 254-5). Experimental economies are indeed supposed to work according to the same principles as the target systems in the intended domain of economic theory, because the *relevant* components of the laboratory system are made of the same "stuff".

Thus, both experiments and simulations are knowledge-producing devices. But the knowledge needed to run a good simulation *is not quite the same* as the one needed to run a good experiment. When reproducing a real-world system in the laboratory, the relationships describing the behavior of both systems may not be known in advance. But one does not have to specify the full set of structural equations governing the target system. The trick is to *make sure* that the target and the experimental system are similar in most relevant respects, so as to be able to generalize the observed results from the laboratory to the target. Experimenters make sure that this is the case by using materials that resemble as closely as possible those of which the parts of the target system are made. They also make sure that the components of the mediating device are put together just like those of the target, and that nothing else is interfering. Of course, quite a lot of knowledge is required in order to do so, but no fundamental theory of how the target system works is required. Parts of the laboratory system can be put between brackets and used as "black boxes". Experimental systems are reliable if they are made of the same "stuff" as real world economies. No process of abstraction from the material forces at work is needed in order to draw the correspondence from the laboratory to the outside world. One may abstract from "negligible" causal factors, but not from the basic processes at work. The similarity is not merely *formal*, but holds at the *material* level as well.

## 7. HYBRIDS: SIMULATING EXPERIMENTS, EXPERIMENTAL SIMULATIONS

The distinction between simulations and experiments taken from Simon seems to be of the "black-or-white" sort. Either the "stuff" is the same as that of the target system, or it is not. But as a matter of fact there are intermediate cases between the two extremes. Mary Morgan (in this volume) discusses "hybrid" entities that are neither entirely simulations nor entirely experiments, but a little bit of both. She focuses on cases in which some "materiality" is transferred from the target system to the mediating entity, and reproduced therein, only to a certain extent. These hybrids are "quasi-material" entities, or a mixture of "modeling and experiment", as she puts it. The second case, which I would like to explore in this paper, is that of hybrids which combine purely experimental and purely simulating components. In fact, if simulations and experiments produce novel scientific knowledge in different ways, they must be partly complementary, and we should be able to combine them in the same project to exploit the potential of both.

My example comes once again from the social sciences. Experimental psychologists and economists are often concerned with designing experiments that reproduce in all relevant respect real-world decision situations. Subjects, for example, are invited to trade goods in an environment governed by the rules of a real-world institution (say, a certain auction system). Even where realism is sought, however, experimenters may have to make use of artificial devices. Take, for example, experiments on so-called "common value" goods - items whose value is the same, but unknown, to all traders. In experiments of this kind uncertainty is customarily implemented by means of a random draw. The subjects trade lottery tickets, in other words, which will be played out at the end of the experiment. Here uncertainty is *simulated* by means of a random draw. Uncertainty arises from the interaction of experimental subjects with a lottery device, rather than with a "real" good of unknown value (say, a concession for an oil tract or a license for mobile phone frequencies). Before you use such a device, you need to be confident that such a way of modeling *that* particular aspect of the target system is legitimate. One has to be reasonably sure, in other words, that *that* part of the theory is right. Here more "paradigmatic" sciences like economics are seemingly (but misleadingly, perhaps) better off: economists are confident to make such a move because auction theory prescribes to model agents' uncertainty as a probability distribution of this sort. If challenged, they reply that the two phenomena (uncertainty faced by oil companies bidding for real tracts, and uncertainty faced by subjects in a laboratory auction) are particular instances of the same phenomenon - *uncertainty*, full stop. To ask whether uncertainty has been produced correctly in the laboratory would be like asking whether in an experiment to investigate the properties of steam the boiler has been heated by burning coil rather than by a Bunsen burner: heat is heat however it is produced, and obeys the laws of thermodynamics.[11] Psychologists, in contrast, tend to be suspicious of assumptions of this sort and argue that human behavior (people's reactions to the same stimuli) vary widely from context to context. Of course there may be mechanisms general enough to support inferences to a wide range of circumstances, but that is an empirical matter to be settled *after*, not before experimentation takes place.

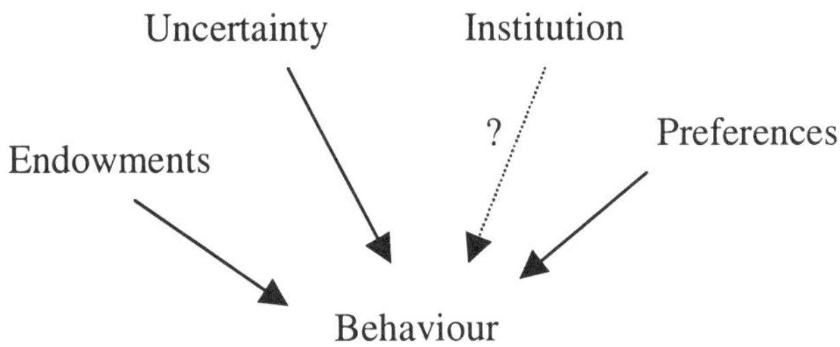


*Figure 3.* Experimental simulations.

The structure of such "experimental simulations" may be represented as in Figure 3. The hypothesis under test here is signaled by the question mark. Other causal factors are part of the target system, but do not necessarily have to be reproduced identically in the experimental design. For the experiment to work, the effect of the environment may be simulated by a device that "does the same thing" as the corresponding real-world factor. But for that to be the case, we must *know* that the relation between "simulating" and "simulated" factors is one of formal similarity. Alternatively, if you are not confident that the simulated features really closely match those of the target system, you must make sure that they falsify reality in the "right" way. If you want to test the cognitive capacities of human beings, for example, you may want to create an environment that is much less ambiguous than the real world. The reasoning being that if subjects cannot read the situation correctly there, surely they will not be able to do it in messier circumstances. Sometimes "hyper-realism" is better than no realism or even tentative realism.

## 8. CONCLUSION

Experiments, simulations, and models belong to the same category of scientific tools. They all are somehow "in between" what we say about the

[11] I owe this example to Bob Sugden.

world and the world itself. When we use them, we must be careful that they stand in the right relation both with what we say, and with the part of the world we are interested in. This is quite different from standard empiricist accounts, according to which models and simulations stand on the theory's side at one end of the spectrum, and experiments stand on the world's side at the other end. It is also different from the standard view of testing, according to which scientists ask very general questions about the world, design very tight and controlled experiments to answer such questions, and the data will hit the theory no matter what. For the data to hit the target, the experiment, simulation, or "simulating experiment" must mirror the target in the right way. Whether the mirroring should be purely formal or material in character, depends on the kind of question we are asking, and on the amount and quality of background knowledge we have accumulated about the target system itself.[12] Take Fischhoff's "Choice A": is it an experiment or a simulation? In the abstract, there is no answer to such a question. It depends what the scientists were aiming at in the first place, what sort of target systems they had in mind, which aspects of the target they were investigating, and crucially what they already knew about the relationship between their targets and "experiments" (or "simulations") like "Choice A".

## ACKNOWLEDGEMENTS

Several helpful criticisms and suggestions were provided by Dan Hausman, Mary Morgan, Bob Sugden, John Worrall, Sang Wook Yi, and members of the audience at MBR'01. All the remaining mistakes are mine.

## REFERENCES

Boumans, M., 2001, Calibration of models in experiments, in this volume, pp. 75-93.

Bunge, M., 1969, Analogy, simulation, representation, *Revue internationale de philosophie* 87:16-33.

Fischhoff, B., 1996, The real world: What good is it?, *Organizational Behavior and Human Decision Processes* 65:232-248.

Giere, R., 1999, Using models to represent reality, in: *Model-Based Reasoning in Scientific Discovery*, L. Magnani, N.J. Nersessian, and P. Thagard, eds., Kluwer Academic/Plenum Publishers, New York, pp.41-57.

Guala, F., 1998, Experiments as mediators in the non-laboratory sciences, *Philosophica* 62:901-918.

[12] Marcel Boumans's paper in this volume discusses another problematic case of economic experiment/simulation.

Hartmann, S., 1996, The world as a process: Simulations in the natural and social sciences, in: *Modelling and Simulation in the Social Sciences from the Philosophy of Science Point of View*, R. Hegselmann, U. Mueller, and K. Troitzsch, eds., Kluwer, Dordrecht, pp. 77-100.

Hesse, M.B., 1963, *Models and Analogies in Science*, Sheed and Ward, London.

Longworth, F., 1998, Computer modeling in the earth sciences, working paper, Imperial College of Science, Technology and Medicine.

Morgan, M.S., 2001, Model experiments and models in experiments, in this volume, pp 41-58.

Morrison, M. and Morgan, M.S., 1999, Models as mediating instruments, in: *Models as Mediators*, M.S. Morgan and M. Morrison, eds., Cambridge University Press, Cambridge, pp. 10-37.

Nagel, E., 1961, *The Structure of Science*, Hackett, Indianapolis.

Simon, H.A., 1969, *The Sciences of the Artificial*, MIT Press, Boston.

Smith, V.L., 1991, *Papers in Experimental Economics*, Cambridge University Press, Cambridge.

# Calibration of Models in Experiments

Marcel Boumans
*Department of Economics, University of Amsterdam, Amsterdam, Netherlands, boumans@fee.uva.nl*

Abstract: The assessment of models in an experiment depends on their material nature and their function in the experiment. Models that are used to make the phenomenon under investigation visible - sensors - are assessed by calibration. However, calibration strategies assume material intervention. The experiment discussed in this paper is an experiment in economics to measure the influence of technology shocks on business cycles. It uses immaterial, mathematical instruments. It appears that calibration did not work for these kinds of models, it did not provide reliable evidence for the facts of the business cycle.

## 1. INTRODUCTION

The way in which models used in experiments can be assessed depends on at least two characteristics: their material nature and their function in the experiment. The kind of materiality not only determines the nature of control and inference in the experiment, but also the confidence one can have in the experiment's outcomes (Boumans and Morgan, 2001; Morgan, 2000). Traditionally, models are defined in terms of their logical and semantic connections with theories. So, usually no methodological distinction is made between the assessment of models and theories. However, by answering the question "What role do models play?" Morrison and Morgan (1999) showed that models function as autonomous agents, that is they are partially independent of both theories and the world, and therefore can be used as instruments of investigation in both domains. Hence, models should be assessed as

*Model-Based Reasoning: Science, Technology, Values*, edited by
L. Magnani and N.J. Nersessian, Kluwer Academic/Plenum Publishers, New York, 2002

instruments and not as theories. Moreover, the kind of assessment depends on the specific function the particular model has in the experiment.

This paper discusses two different models used in Kydland and Prescott's (1996) "Computational Experiment" to measure the percentage of the business cycle fluctuations caused by technology shocks. Borrowing Morgan's (2000) taxonomy, their experiment can be considered as a "virtual experiment": a non-material experiment on non-material objects. Both models are used as sensors to make facts about the business cycle phenomenon visible. One sensor functions as a filter, the other as a measuring instrument. Although business cycles are detected through the use of data, they are in general not directly observable, which in general applies to all phenomena, that is all potential objects of explanation by a theory (Bogen and Woodward, 1988; see below). In Kydland and Prescott's experiment, the phenomenological facts are unobservable components of the observed data. To display them we need a filter. The filter's role is to extract a specific signal from the available data by removing noise and other irrelevant components. The role of the measuring instrument is more specific, namely to generate quantitative facts about the phenomenon. Considering models as "mediating" instruments between theory and the world (Morrison and Morgan 1999), both sensors are located on the theory-world axis as "mediators" between facts about the phenomenon and data.

Theory → Facts about Phenomenon $x$ ← SENSORS ← Data $y = x + noise$

To understand this position both instruments take on the theory-world axis, it is helpful to use the distinction Woodward (1989) makes between data and phenomena. Phenomena are relatively stable and general features of the world which are potential objects of explanation and prediction by general theory. Data, the observations, play the role of evidence for claims about phenomena. The contrast between data and phenomena can be characterized in three ways. In the first place, the difference between data and phenomena can be indicated in terms of the notions of error applicable to each. In the case of data the notion of error involves observational mistakes, while in the case of phenomena one worries whether one is detecting a real fact rather than an artifact produced by peculiarities of one's instruments or detection procedures. A second contrast between data and phenomena is that phenomena are more "widespread" and less idiosyncratic, less closely tied to the details of a particular instrument or detection procedure. A third way of thinking about the contrast between data and phenomena is that scientific investigation is typically carried on in a noisy environment, an environment in which the observations reflect the operation of many different causal factors.

> The problem of detecting a phenomenon is the problem of detecting a signal in this sea of noise, of identifying a relatively stable and invariant pattern of some simplicity and generality with recurrent features - a pattern which is not just an artifact of the particular detection techniques we employ or the local environment in which we operate (Woodward, 1989, pp. 396-7).

Underlying the contrast between data and phenomena is the idea that theories do not explain data, which typically will reflect the presence of a great deal of noise. Rather, an investigator first subjects the data to analysis and processing, or alters the experimental design or detection technique, all in an effort to separate out the phenomenon of interest from extraneous background factors. "It is this extracted signal rather than the data itself which is then regarded as a potential object of explanation by theory" (p. 397).

Instruments located between data and phenomena on the theory-world axis are not assessed in the same way as theories are by confronting the outcome of the model with facts about the phenomenon. Confronting sensors with phenomenological facts rather means comparing these instruments with other instruments that generate these facts and are chosen to act as standards. In other words, assessment of filters and measuring instruments is done by calibration, that is standardization in the sense of comparing the instrument's performance with a standard. But - as will be shown - the validation problems of filters differ from those of measuring instruments. Both problems were caused by the lack of materiality, but in one case with respect to the standard, and in the other case with respect to the instrument itself.

## 2. ROBERT LUCAS'S PROGRAM

Kydland and Prescott's experiment was an implementation of Lucas's program for business-cycle research. This program not only framed their account of models and theories, but also advanced the view that business cycles should be considered as phenomena in the above-described meaning of the word.

Lucas's contribution to macroeconometric evaluation of economic policy - the "Lucas critique" - implied that estimated parameters which were previously regarded as "structural" in econometric analysis of economic policy actually depend on the economic policy pursued during the estimation period. Hence, the parameters may change with shifts in the policy regime. The Lucas critique was an implicit call for a new research program. This program involves formulating and estimating macroeconometric models with parameters that are invariant under policy variations, so that they can be used for evaluating alternative policies. According to Lucas, the model invariance is located on the level of the parameters describing tastes and technology. Lucas was the first who emphasized stable parameters instead of stable relationships.

Lucas characterized the business cycle by enumerating seven "qualitative features" of economic time series:

> i) Output movements across broadly defined sectors move together. [...] ii) Production of producer and consumer durables exhibits much greater amplitude than does the production of nondurables. iii) Production and prices of agricultural goods and natural resources have lower than average conformity. iv) Business profits show high conformity and much greater amplitude than other series. v. Prices generally are procyclical. vi. Short-term interest rates are procyclical; long-term rates are slightly so. vii). Monetary aggregates and velocity measures are procyclical (Lucas, 1977, p. 9).

By defining the business cycle in this way, Lucas indicated that he considered the business cycle as a general phenomenon of capitalist economies:

> There is, as far as I know, no need to qualify these observations by restricting them to particular countries or time periods: they appear to be regularities common to all decentralized market economies. Though there is absolutely no theoretical reason to anticipate it, one is led by the facts to conclude that, with respect to the qualitative behavior of co-movements among series, business cycles are all alike. To theoretically inclined economists, this conclusion should be attractive and challenging, for its suggests the possibility of unified explanation of business cycles, grounded in the general laws governing market economies, rather than in political or institutional characteristics specific to particular countries or periods (Lucas, 1977, p. 10).

Borrowing this definition from Lucas, Prescott (1986) preferred to refer to business cycles as "business cycle phenomena", "which are nothing more or less than a certain set of statistical properties of a certain set of important aggregate time series" (p. 10). By explicitly treating the business cycle as a general phenomenon not restricted to particular countries or time periods, the business cycle was stabilized and its "qualitative features" considered as "stylized facts".

Lucas also introduced a "new standard for what it means to understand business cycles":

> One exhibits understanding of business cycles by constructing a model in the most literal sense: a fully articulated artificial economy which be-

> haves through time so as to imitate closely the time series behavior of actual economics (Lucas, 1977, p. 11).

In the "equilibrium business-cycle program" dominated by Lucas' instructions, it became standard practice to run an experiment with an artificial economy:

> One of the functions of theoretical economics is to provide fully articulated, artificial economic systems that can serve as laboratories in which policies that would be prohibitively expensive to experiment with in actual economies can be tested out at much lower costs (Lucas, 1980, p. 696).

But not all models will be equally useful, to test them as "useful imitations of reality" we should subject them to shocks "for which we are fairly certain how actual economies, or parts of economies, would react. The more dimensions on which the model mimics the answer actual economies give to simple questions, the more we trust its answer to harder questions" (Lucas, 1980, pp. 696-7).

## 3. CALIBRATION

If theories do not provide systematic explanations of facts about data, as Bogen and Woodward (1988) convincingly have argued, how do data constitute reliable evidence for some phenomenon? Because facts about phenomena are actually created by instruments for which data function as input, assessing reliability implies the investigation of these instruments. Beside this strategy, Woodward (1989) also mentions other possibilities to increase reliability, but these are not applicable in the case of business-cycle research: control of possible confounding effects and systematic error, replicability, and data reduction. Control and replication are impossible in a macroeconomic environment and data reduction is a luxury economists cannot afford. In the case of business cycle research, reliability can only be achieved by investigation of the equipment used. Because Woodward only discusses material instruments, he confines himself to the empirical investigation of the equipment, and calibration is seen as falling under this heading. But the experiment discussed here is "virtual" and thus empirical investigation is not possible. Fortunately, calibration is defined in such a broad sense that it does not exclude the investigation of nonmaterial instruments. This broader meaning will be used for the validation of the two sensors. But one should be warned, calibration does not guarantee a correct result; though its successful performance does argue for the validity of the result (Franklin, 1997, p. 76).

An important problem of instruments used to make unobservables visible is how to distinguish between the facts about the phenomenon and the artifacts created by the instrument. Allan Franklin (1986) discusses nine epistemological strategies to distinguish between a valid observation and an artifact. One of these nine strategies is calibration, "the use of a surrogate signal to standardize an instrument". "If an apparatus reproduces known phenomena, then we legitimately strengthen our belief that the apparatus is working properly and that the experimental results produced with that apparatus are reliable" (Franklin, 1997, p. 31).

When the model functions as a measuring instrument a more specific definition of calibration is needed, namely the one used in metrology. In metrology, calibration is the "set of operations that establish, under specified conditions, the relationship between values of quantities indicated by a measuring instrument or measuring system, or values represented by a material measure or a reference material, and the corresponding values realized by standards" (VIM, 1993, 6.11). A standard is the "material measure, measuring instrument, reference material or measuring system intended to define, realize, conserve or reproduce a unit or one or more values of a quantity to serve as a reference" (VIM, 1993, 6.1). The general philosophy adopted for the creation of standards is that they be based upon some principle that is known to be as invariant as can possibly be found. Standards are entirely of man's choice, nothing about the natural world defines them, they are a contrivance of man. However, they are often based upon naturally occurring phenomena when these possess the required degree of definition.

An example of this approach to base standards on a set of proper invariants is the way in which the international metric organization aims to link the base units of the International System of Units (SI):

| Quantity | SI Base Unit | Symbol |
|---|---|---|
| length | metre | m |
| mass | kilogram | kg |
| time | second | s |
| electric current | ampere | A |
| thermodynamic temperature | kelvin | K |
| amount of substance | mole | mol |
| luminous intensity | candela | cd |

to the real world not through prototypes, but through the fundamental constants of physics, which are supposed to be universal and unchanging.

| Fundamental quantity | Symbol |
|---|---|
| velocity of electromagnetic radiation in free space | $c$ |
| elementary charge | $e$ |
| mass of the electron at rest | $m_e$ |

| Fundamental quantity | Symbol |
|---|---|
| Avogadro constant | $N_A$ |
| Planck constant | $h$ |
| universal gravitational constant | $G$ |
| Boltzmann constant | $k$ |

Only the kilogram is still equal to the prototype kept at the International Bureau of Weights and Measures under conditions specified in 1889 which has consequent effects on the practical realization of the definitions of the ampere, mole and candela. The ampere's definition, for example, involves the kilogram, but an alternative link used in practice but not yet officially accepted is the Josephson-effect constant ($K_{J\text{-}90}$) and von Klitzing's quantum-Hall resistance ($R_{K\text{-}90}$), both of which were given fixed, conventional values in 1990.

## 4. COMPUTATIONAL EXPERIMENT

Kydland and Prescott (1982) introduced calibration to macroeconomics as a means of reducing the free parameters of their paradigm business-cycle model in their paper "Time to build and aggregate fluctuations". However, since Kydland and Prescott's application, calibration has been controversial in economics. One reason is its ambiguous meaning. The purpose of the parameterization is not sufficiently clear, allowing for three different interpretations: estimation, testing and standardization. In economics calibration is mostly seen as a simulation-based method of estimation, and sometimes as testing, but Kydland and Prescott explicitly consider models as measuring instruments and therefore interpreted calibration as standardization.

The ambiguity around calibration methods in economics was probably the reason the *Journal of Economic Perspectives* published a symposium on this subject in the winter 1996 issue. The journal gave Kydland and Prescott the opportunity to explicate the "econometric tool" they used in their (1982) "Time to build" paper, which they now called "The computational experiment". According to the authors, any economic computational experiment involves five major steps: "pose a question; use a well-tested theory; construct a model economy; calibrate the model economy; and run the experiment" (Kydland and Prescott, 1996, p. 70). The question they had answered in 1982 (but not explicitly posed before 1996) was "What is the quantitative nature of fluctuations induced by technology shocks?" (p. 71). In explaining what they mean by calibration the authors refer to the graduation of measuring instruments, like a thermometer (p. 74). Discussing business cycle research, Prescott (1998) explicitly specifies a model as "a measurement instrument used to deduce the implication of theory" (p. 2). Referring to Lucas, he defines a theory as "an implicit set of instructions for constructing a model economy for the purpose of answering a question" (p. 2), so that the "quantitative answer to the question is deduced from the model economy" (p. 3). Equating economic models with measuring instruments, Kydland and Prescott (1996) arrive at an interpretation of calibration that comes very close to the one given by Franklin above: "Generally, some economic questions have known answers, and the model should give an approximately correct answer to them if we are to have any confidence in the answer given to the question with unknown answer" (p. 74). The answer to this latter question, the one they treated in their 1982 paper, is that "the model economy displays business cycle fluctuations 70 percent as large as did the US economy" (p. 74). In other words, the answer is supposed to be the measurement result carried out with a calibrated instrument.

## 5. CALIBRATION OF THE MEASURING INSTRUMENT

The problem is what are the economic questions for which we have known answers? Or, what are the phenomena with which the model is calibrated? The answer is most explicitly given by Cooley and Prescott (1995). They describe calibration as choosing parameters for the model economy "so that it mimics the actual economy on dimensions associated with long term growth" by setting parameter values equal to certain "more or less constant" ratios. These ratios are the so-called "stylized facts" of economic growth, "striking empirical regularities both over time and across countries", the "benchmarks of the theory of economic growth" (Cooley and Prescott, 1995, p. 3). The naming refer to Kaldor's (1958) "stylised facts" of growth, but the ones that are used in the business-cycle literature are those as characterized by Solow (1970) and summarized by Cooley and Prescott as follows:

1. Real output grows at a more or less constant rate.
2. The stock of real capital grows at a more or less constant rate greater than the rate of growth of the labor input.
3. The growth rates of real output and the stock of capital tend to be about the same.
4. The rate of profit on capital has a horizontal trend.
5. The rate of growth of output per-capita varies greatly from one country to another.
6. Economies with a high share of profits in income tend to have a high ratio of investment to output. (p. 3)

Only the first four "facts" were used. The last two emphasize the differences between countries or economies and are thus not general enough.

Although we have seen that equilibrium business-cycle modelers aim to model from invariants (Hoover 1995), the choice to take these stylized facts as empirical facts of growth is dubious. Solow already remarked that "There is no doubt that they are stylized, though it is possible to question whether they are facts" (1970, p. 2). The danger is that stylized facts may turn out to be more conventional than empirical. Hacche (1979) provided an account of the British-US evidence relating to Kaldor's six stylized facts and showed inconsistencies between economic history and Kaldor's stylized facts.

> In any event the data for the United Kingdom provide little support for the hypothesis that there is some "steady trend" or "normal" growth rate of capital or output or both running through economic history - which is what Kaldor's stylised facts suggest - unless the interpretation of the hypothesis is so liberal as to bear little meaning (p. 278).
>
> Also, the experience of other industrialized countries appears to support the conclusions drawn form UK data (p. 280).

Although in the case of the SI base units the values of some of the used physical constants are fixed by convention, their stability is not a matter of convention but an empirical fact. In contrast to this observation, whether the stylised facts of growth do have empirical counterparts is seriously doubted. Hence, whatever values are chosen for them, they will be seen as arbitrary. And so the same sense of arbitrariness will adhere to the calibration procedure and therefore ultimately to the measurement results.

## 6. THE HODRICK-PRESCOTT FILTER

The model economy is meant to measure facts about the business cycle phenomenon, therefore first the business cycle has to be made visible. In their experiment, Kydland and Prescott use a "filter", a term that is reminiscent of spectral filters in optics used to modify the spectral transmittance of an optical system. According to the *Dictionary of Statistical Terms* (Kendall and Buckland, 1960, p. 109), a filter is "any method of isolating harmonic constituents in a time-series" analogous to filtering "a ray of light or sound by removing unsystematic effects and bringing out the constituent harmonics". But one should note an essential difference between a physical filter and a mathematical one, a physical filter removes impurities by withholding them, rather than transforming them like a mathematical filter (Kendall and Stuart, 1966, p. 424).

A well-known filter in time-series analysis is the Kalman filter, first developed by Kalman (1960). Kalman filtering is a method of extracting the behavior of a signal $x(t)$ given observations $y(t)$ which are subject to error $v(t)$, so that $y(t) = x(t) + v(t)$. The term "filtering" refers to the removal as much as possible of the error term $v(t)$ to give an estimation of the true signal $x(t)$. A prerequisite to the application of the Kalman filter is that the behavior of the system under study be described through a quantity known as the system "state", which may be defined in terms of a first order differential or difference equation known as the system equation.

Given the state-space representation of a discrete linear dynamic system the problem is then to estimate the state $x_t$ from the noisy observations $y_1$, $y_2$, ..., $y_t$; in this context, three distinct problems may be distinguished:

a) *filtering*: the measurements $y_1$, $y_2$, ..., $y_t$ are used to form an estimate $\hat{x}_t$ of the state at time $t$;
b) *smoothing*: the measurements $y_1$, $y_2$, ..., $y_t$ are used to form an estimate $\hat{x}_t$ of $x_s$ at some past time point for $1 \leq s < t$;
c) *prediction*: the measurements $y_1$, $y_2$, ..., $y_t$ are used to form an estimate $\hat{x}_t$ of $x_s$ at some future time point for $s > t$.
(Kalman, 1960, p. 36)

In general, filtering in economics is motivated by one of several objectives:

1. extraction of a component such as a growth, cyclical, or seasonal component,
2. transformation to induce stationarity, or
3. mitigation of measurement error that is assumed to be particularly important at specific frequencies (King and Rebelo, 1993, p. 213).

In their experiment, Kydland and Prescott use the "Hodrick-Prescott filter" (HP-filter). Its application is motivated by objective (a), that is taking a moving average of the observations to extract one of the components. To simplify the discussion, we focus on time series containing only growth and business cycle components:

$$y_t = y^g_t + y^c_t \tag{1}$$

where $y^g_t$ is the unobserved growth component and $y^c_t$ is the unobserved business cycle component. In the subsequent discussion, filtering is understood as a way of detrending by representing the growth component as a moving average of the observed $y_t$ that permits us to extract both components. That is, we assume that

$$y_t^g = \sum_{k=-\infty}^{\infty} g_k y_{t-k} = G(B) y_t \quad (2)$$

where $B$ is the backshift operator with $B^n x_t = x_{t-n}$ for $n \geq 0$. Then, $y^c_t$ is also a moving average of $y_t$,

$$y^c_t = [1-G(B)]y_t \equiv C(B)y_t \quad (3)$$

In the language of filtering theory, both $G(B)$ and $C(B)$ are linear filters.

The functioning of linear filters is mainly discussed in terms of the frequencies extracted by taking the Fourier transform of a linear filter, also called the frequency response function of the filter. The frequency response function makes it easier to interpret the effects of filtering. For example, the Fourier transform of the growth filter is

$$G(\omega) = \sum_{k=-\infty}^{\infty} g_k e^{-ik\omega} \quad (4)$$

where $i$ denotes the imaginary number $\sqrt{-1}$ and where $\omega$ is the frequency measured in radians.

At a given frequency $\omega$, the frequency response $G(\omega)$ is simply a complex number, so that it may be written in polar form as $G(\omega) = |G(\omega)|\exp(-i\Psi(\omega))$. The "gain" of the linear filter, $|G(\omega)|$ , yields a measure - at the specified frequency $\omega$ - of the increase in the amplitude of the filtered series over the original series. The phase, $\Psi(\omega)$, yields a measure of the time displacement attributable to the linear filter. Symmetric filters, filters that possess the symmetry property that $g_k = g_{-k}$, have the property that they do not induce a phase shift, because their Fourier transform is real so their gain function is equal to the frequency response.

The specific linear filter suggested by Hodrick and Prescott (1997)[1] determines the trend component series $\{y^g_t, t = 1, ..., T\}$ by minimizing

$$\sum_{t=1}^{T} (y_t - y_t^g)^2 + \lambda \sum_{t=2}^{T-1} [(y_{t+1}^g - y_t^g) - (y_t^g - y_{t-1}^g)]^2 \quad (5)$$

where $\lambda$ is a fixed parameter. "The maintained hypothesis, based upon growth theory considerations, is that the growth component of aggregate economic time series varies smoothly over time" (Hodrick and Prescott,

[1] The Hodrick-Prescott filter was originally introduced in a 1981 working paper. The 1997 paper is "substantially the same" as their 1981 one.

1997, p. 2). The measure of the smoothness of the growth path is the sum of the squares of its second difference. The parameter $\lambda$ can be interpreted as a penalty on variability in the growth component series. The larger the value of $\lambda$, the smoother is the solution series. If $\lambda$ goes to infinity, the growth component approaches a linear deterministic time trend. If $\lambda = 0$, the growth component series coincides with the observed series and the cyclical component is zero. Hodrick and Prescott propose a value of $\lambda = 1600$ for quarterly time series data as reasonable, and their recommendation has been widely followed in the literature applying the HP-filter.

King and Rebelo (1993) show that the HP-filter takes the form

$$C_{HP}(B) = \frac{\lambda(1-B)^2(1-B^{-1})^2}{1+\lambda(1-B)^2(1-B^{-1})^2} \tag{6}$$

Thus, the HP-filter is a two-sided symmetric filter and it depends only on the value of the parameter $\lambda$.

The Fourier transform of the cyclical component filter takes the form:

$$C_{HP}(\omega) = \frac{4\lambda[1-\cos\omega]^2}{1+4\lambda[1-\cos\omega]^2} \tag{7}$$

The cyclical component filter places zero weight on the zero frequency and close to unit weight on high frequencies, see Figure 1. Increasing $\lambda$ shifts the gain function upward, moving a given frequency's gain closer to unity. The larger $\lambda$, the more the HP filter looks like an ideal high pass filter. A high pass filter removes the low frequencies or long-cycle components and allows the high frequencies or short-cycle components to pass through. An ideal high pass cyclical filter is a filter for which the frequency's gains are defined as: $C_{high}(\omega) = 0$ for $\omega < \omega^*$ and $C_{high}(\omega) = 1$ for $\omega > \omega^*$.

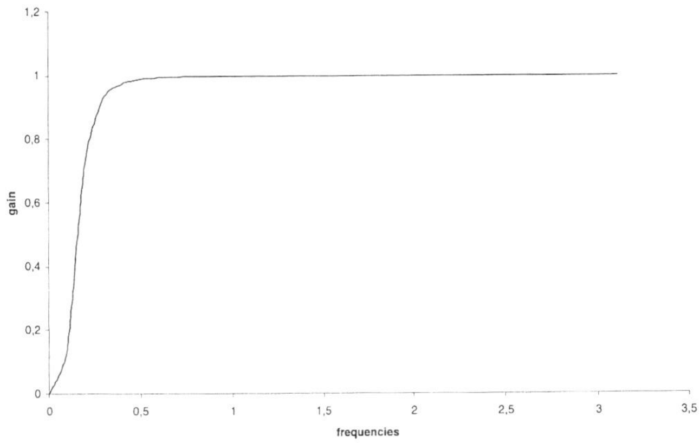


*Figure 1.* Hodrick-Prescott filter.

The effect of filtering can be analysed with a formula linking the spectrum of a stationary process, $x_t$, with the spectrum of the filtered process, $y_t = F(B)x_t$:

$$g_y(\omega) = |F(\omega)|^2 g_x(\omega) \tag{8}$$

where $g$ denotes the spectrum and $|F|$ the gain of the filter $F$. A stochastic process is said to be stationary if its first moment is independent of $t$ and if the covariance $\sigma_{t,s}$ depends only on $t$-$s$.

Before we continue discussing the HP-filter, let us first examine the spectra of two simple processes. First consider a white-noise process, $y_t = \varepsilon_t$. As its name indicates its spectrum is flat, that is all frequencies are equally important and equal $\sigma_\varepsilon^2$:

$$g_{white}(\omega) = \sigma_\varepsilon^2 \tag{9}$$

Secondly, we consider a random walk which is a first-order autoregressive process $y_t = y_{t-1} + \varepsilon_t$, $\varepsilon_t$ white noise. Then

$$y_t = \frac{1}{1-B}\varepsilon_t \tag{10}$$

Therefore, the spectrum of a random walk is

$$g_{walk}(\omega) = \frac{1}{2-2\cos\omega}\sigma_\varepsilon^2 \tag{11}$$

See figure 2, for the graphs of both spectra, spectrum values are in fractions of $\sigma_\varepsilon^2$.

The effect of the HP-filter applied to these simple stationary processes can now easily be shown. Substituting equation (7) and (9) into equation (8) yields the spectrum of filtered white noise:

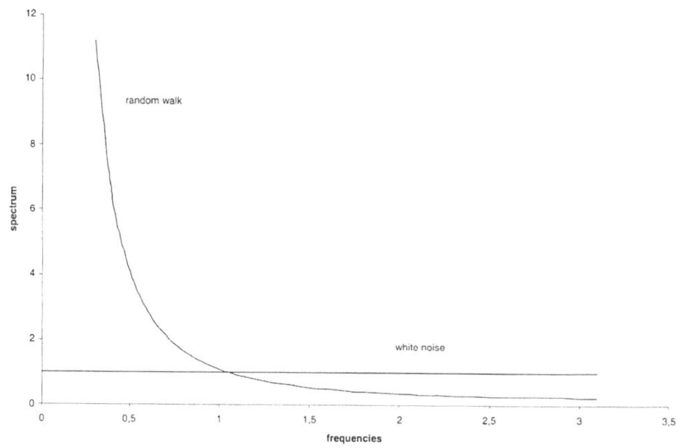


*Figure 2.* Spectra of two simple processes.

$$g_{filterwhite}(\omega) = |C_{HP}(\omega)|^2 g_{white}(\omega) = \left[\frac{4\lambda[1-\cos\omega]^2}{1+4\lambda[1-\cos\omega]^2}\right]^2 \sigma_\varepsilon^2 \tag{12}$$

And substituting equation (7) and (11) into (8) yields the spectrum of a filtered random walk:

$$g_{filterwalk}(\omega) = |C_{HP}(\omega)|^2 g_{walk}(\omega) = \left[\frac{4\lambda[1-\cos\omega]^2}{1+4\lambda[1-\cos\omega]^2}\right]^2 \frac{1}{2-2\cos\omega}\sigma_\varepsilon^2 \tag{13}$$

See Figure 3.

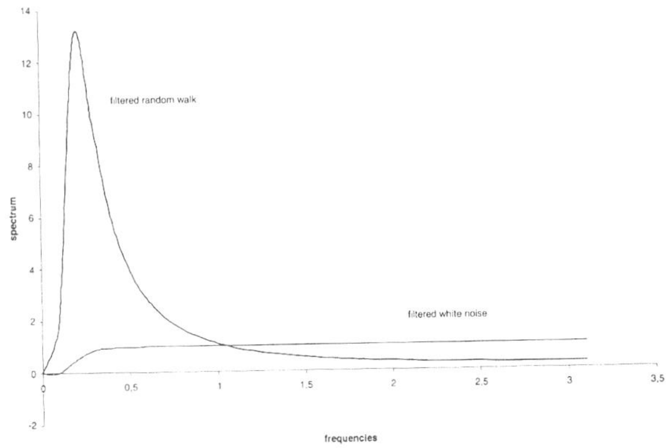


*Figure 3.* Two filtered processes.

The Hodrick-Prescott filter has been widely adopted in the business cycle literature to make the business cycle visible. Beside this growing interest in the use of this filter, several studies have appeared in which the HP-filter has been critically examined. One of the worries is whether the filter does not in fact lead to spurious cyclical behavior and distorts the measure of persistence, variability and co-movement - that is, distorts Lucas's stylized facts of the business cycle phenomenon. Cogley and Nason (1995) assert that "the filter can generate business cycle periodicity and comovement even if none are present in the original data" (p. 254). Thus, "it is not clear whether the results should be interpreted as facts or artifacts" (p. 255). Harvey and Jaeger (1993) claim to have shown that "the uncritical use of mechanical detrending can lead investigators to report spurious cyclical behavior" (p. 231). In a note, Jaeger (1994) asserts that applying the HP-filter to time series with stochastic trends "may extract cyclical movements which are entirely spurious" (p. 493).

These papers share a common argument in their claim that the HP-filter may extract spurious cycles. This argument in its most simple form goes as follows. If one of the components of the time series being investigated is the above-described random walk, the filtered process contains a peak at a frequency which for $\lambda = 1600$ corresponds to a period of about 30 (see Figure 3). Because we deal with quarterly data, this period of about 7.5 years equals the business cycle period.

> Thus applying the standard HP filter to a random walk produces detrended observations which have the characteristics of a business cycle for quarterly observations. Such cyclical behavior is spurious and is a classic example of the Yule-Slutzky effect. (Harvey and Jaeger, 1993, p. 234; see also Jaeger, 1994, p. 497).

> [T]he power spectrum for an HP filtered random walk [...] has a peak at 7.6 years per cycle. Hence there is business cycle periodicity in the elements of $c(t)$ even though the elements of $y(t)$ are random walks. (Cogley and Nason, 1995, p. 259)

Many textbooks of statistics (e.g. Kendall and Stuart (1966) and Malinvaud (1966)) warn of the use of filters or moving averages because they might produce artificial oscillations due solely to the statistical treatment of the data. This is the so-called (Yule-)Slutzky effect, after the two statisticians who studied it in detail. In particular, Slutzky (1937) considered the effects of starting with a white noise $\varepsilon_t$, taking a two-period moving sum $n$ times, and then taking first differences $m$ times. If $m/n$ is held constant and $n \rightarrow \infty$, then a single frequency is very much emphasized by this process and a cycle results.

The question now is whether the application of the HP-filter also leads to a Slutzky effect. The HP-filter on its own does not enlarge the amplitude of any frequency, so it does not emphasize a specific frequency. It only de-emphasizes frequencies below $2\pi/30 = 0.21$. The problem is that the analysis is not conclusive, spectral analysis can only be applied to stationary time series. It has only been shown that the HP-filter operates like a high pass filter when applied to stationary time series. Only a stationary time series can be decomposed into orthogonal periodic components. Hence it is not clear yet what the effect is of the HP-filter when applied to nonstationary time series.

However, macroeconomic time series often have an upward trend which makes them nonstationary and one of the objectives of filtering is "transformation to induce stationarity" (see above). To analyse the HP-filter effect for these cases, Cogley and Nason (1995), Jaeger (1994) and Harvey and Jaeger (1993) split the HP-filter into two parts. One part is chosen to make the time series stationary so that subsequently the resulting part can be analyzed to see its effect on the stationary data. Again the most simple case will be taken to discuss the results: Cogley and Nason analyze the HP filter effect on difference-stationary time series, that is a time series $y_t$ of which the differenced time series, $\Delta y_t$, is assumed to be stationary. In this way the HP filter operates like a "two-step linear filter": "Difference $y(t)$ to make it stationary, and smooth $\Delta y(t)$ with an asymmetric moving average filter" (Cogley and Nason, 1995, p. 258):

$$y^c_t = S(B)\Delta y_t \tag{14}$$

where $S(B) = C_{HP}(B)/(1-B)$. The squared gain of $S(B)$ is equal to the spectrum of the filtered random walk (when $\sigma_\varepsilon^2$ is taken as unit; see equation 13), so that Figure 3 also pictures the squared gain of $S(B)$. If one considers $S(B)$ as a filter, one can easily see that it leads to a Slutzky effect, namely very much emphasizing a period of 7.5 years. But one cannot infer from this that the HP-filter, $C_{HP}(B)$, has this effect, too. Properties of parts of the filter do not necessarily sum up to as the properties of the complete filter, they may cancel each other out. The Slutzky-effect of the $S(B)$ filter is nullified by the factor (1-$B$).

Whether filters lead to spurious cycles or not, the above account of filters shows that calibration did not work for immaterial filters. The fact that the HP-filter functions as a high pass filter for stationary data, did not strengthen the belief that the filter is working properly for nonstationary data.

## 7. CONCLUSION

How do we assess our scientific tools? To "hit the target", the tools "must mirror the target in a number of relevant aspects. Whether the mirroring should be purely formal or material in character, depends on the kind of question we are asking, and on the amount and quality of background knowledge we have accumulated regarding the target system itself" (Guala, 2001, p. 17). The questions the models of the "Computational Experiment" had to answer imposed that the mirroring should be material in character to gain confidence in the experimental results and because the experiment failed to fulfil this requirement of materiality it did not hit the target. Because of the questions the two models had to answer they had to be positioned on the theory-world axis between phenomena and data. Macroeconomic models located on that spot are assessed by calibration. However, as the above case study of a virtual experiment shows us, calibration strategies assume material intervention. Two calibration strategies were used and both failed because of the lack of materiality. The attempt to calibrate the measuring instrument did not lead to reliable evidence of the quantitative facts about the business cycle. The calibration was based on stylized facts, but they were doubted as standards because it is not clear whether they should be considered as real invariants found in "nature" or rather as conventions. Lack of their empirical basis undermined the trust in the calibration.

Also in the case of the filter, its more mathematical (nonmaterial) nature hindered the success of the calibration. Although the HP-filter works well for known phenomena, namely stationary processes, it did not bring confidence in its working for nonstationary processes. While in the case of calibration as an empirical investigation of the instruments the move from observing that the tool works properly for known domains to having confidence of its working for unknown domains is legitimized, this is apparently not the case for a mathematical investigation.

## ACKNOWLEDGEMENTS

I thank Francesco Guala and Mary Morgan and other participants of the MBR'01 for helpful remarks on the paper. All errors are mine.

## REFERENCES

Bogen, J. and Woodward, J., 1988, Saving the phenomena, *The Philosophical Review* 97:303-352.

Boumans, M. and Morgan, M.S., 2001, Ceteris paribus conditions: Materiality and the application of economic theories, *The Journal of Economic Methodology* 8:11-26.

Cogley, T. and Nason, J.M., 1995, Effects of the Hodrick-Prescott filter on trend and difference stationary time series. Implications for business cycle research, *Journal of Economic Dynamics and Control* 19:253-278.

Cooley, T.F. and Prescott, E.C., 1995, Economic growth and business cycle, in: *Frontiers of Business Cycle Research*, T.F. Cooley, ed., Princeton University Press, Princeton, pp. 1-38.

Franklin, A., 1986, *The Neglect of Experiment*, Cambridge University Press, Cambridge.

Franklin, A., 1997, Calibration, *Perspectives on Science* 5:31-80.

Guala, F., 2001, Models, simulations, and experiments, in this volume, pp. 59-74.

Hacche, G., 1979, *The Theory of Economic Growth*, Macmillan, London.

Harvey, A.C., and Jaeger, A., 1993, Detrending, stylized facts and the business cycle, *Journal of Applied Econometrics* 8:231-247.

Hodrick, R.J., and Prescott, E.C, 1997, Postwar U.S. business cycles: An empirical investigation, *Journal of Money, Credit, and Banking* 29:1-16.

Hoover, K.D., 1995, Facts and artifacts: Calibration and the empirical assessment of real-business-cycle models, *Oxford Economic Papers* 47:24-44.

Jaeger, A., 1994, Mechanical detrending by Hodrick-Precott filtering: A note, *Empirical Economics* 19:493-500.

Kaldor, N., [1958] 1978, Capital Accumulation and Economic Growth, in: *Further Essays on Economic Theory*, Duckworth, London, pp. 1-53.

Kalman, R.E., 1960, A new approach to linear filtering and prediction problems, *Journal of Basic Engineering. Transactions of the ASME* 82:35-45.

Kendall, M.G. and Buckland, W.R., 1960, *A Dictionary of Statistical Terms*, Oliver and Boyd, Edinburgh and London.

Kendall, M.G. and Stuart, A., 1966, *The Advanced Theory of Statistics, Volume 3: Design and Analysis, and Time-Series*, Charles Griffin, London.

King, R.G. and Rebelo, S.T., 1993, Low frequency filtering and real business cycles, *Journal of Economic Dynamics and Control* 17:207-231.

Kydland, F.E. and Prescott, E.C., 1982, Time to build and aggregate fluctuations, *Econometrica* 50:1345-1370.

Kydland, F.E. and Prescott, E.C., 1996, The computational experiment: An econometric tool, *Journal of Economic Perspectives* 10:69-85.

Lucas, R.E., 1977, Understanding business cycles, in: *Stabilization of the Domestic and International Economy*, K. Brunner and A.H. Meltzer, eds., North-Holland, Amsterdam, pp. 7-29.

Lucas, R.E., 1980, Methods and problems in business cycle theory, *Journal of Money, Credit, and Banking* 12:686-715.

Malinvaud, E., 1966, *Statistical Methods of Econometrics*, North-Holland, Amsterdam.

Morgan, M.S., 2000, Experiments without material intervention: Model experiments, virtual experiments and virtually experiments, University of Amsterdam Research Memoranda 00-4, Technical Paper.

Morrison, M. and Morgan, M.S., 1999, Models as mediating instruments, in: *Models as Mediators*, M.S. Morgan and M. Morrison, eds., Cambridge University Press, Cambridge.

Prescott, E.C., 1986, Theory ahead of business cycle measurement, *Federal Reserve Bank of Minneapolis Quarterly Review* 10:9-22.

Prescott, E.C., 1998, Business cycle research: Methods and problems, Working Paper Federal Reserve Bank of Minneapolis.

Slutzky, E., 1937, The summation of random causes as the source of cyclic processes, *Econometrica* 5:105-146.

Solow, R.M., 1970, *Growth Theory, An Exposition*, Clarendon Press, Oxford.

VIM, 1993, *International Vocabulary of Basic and General Terms in Metrology*, ISO, Geneva.

Woodward, J., 1989, Data and phenomena, *Synthese* 79:393-472.

# The Development of Scientific Taxonomies

Hanne Andersen
*Department of Medical Philosophy and Clinical Theory, University of Copenhagen, Copenhagen, Denmark, h.andersen@medphil.ku.dk*

Abstract: In this paper I shall compare and integrate some of the various historical, philosophical and cognitive accounts of scientific taxonomies. First, I shall present a family resemblance account of taxonomic concepts. Next, I shall give a brief overview of the research on taxonomies and family resemblance. I shall discuss the role of model-based reasoning in the development of taxonomic concepts, arguing that models may provide the means to uniting the structural and developmental analyses of scientific taxonomies. Finally, I shall illustrate the account by a historical case study on the development of nuclear physics. I shall end by arguing that we need to integrate certain social dimensions in the account of the development of scientific taxonomies.

## 1. INTRODUCTION

Within the last decade, scientific taxonomies have been the topic of several philosophical and historical analyses. Within philosophy of science, one of the key figures in the early discussions of scientific change, Thomas Kuhn, emphasized in his late works from the 1990s that his incommensurability thesis was a thesis about "the meaning of a restricted class of terms. Roughly speaking, they are the taxonomic terms, or kind terms" (Kuhn, 1991, p. 4). Likewise, Thagard (1992) in his analyses of conceptual revolutions focused on kind hierarchies and part-whole hierarchies and discussed conceptual revolutions and degrees of conceptual change in terms of hierarchical reorganization. Within history of science, a number of scholars,

*Model-Based Reasoning: Science, Technology, Values*, edited by
L. Magnani and N.J. Nersessian, Kluwer Academic/Plenum Publishers, New York, 2002

among them Peter Barker, Jed Buchwald, Xiang Chen, Daiwie Fu, and myself, have analyzed taxonomic terms in different scientific disciplines and during different historical periods, e.g. the Copernican revolution (Barker 2001), the wave theory of light (Buchwald, 1992, Chen 1995), XIX century ornithology (Chen and Barker, 2000), the Hobbes-Boyle debate (Fu, 1995), and nuclear physics (Andersen, 1996).

In the following, I shall present a family resemblance account of concepts that shows that such concepts necessarily form taxonomies, and I shall discuss how the development of such taxonomies can be analyzed.

## 1.1 History of the family resemblance account

It is an ancient discussion whether concepts are defined by necessary and sufficient conditions which hold for all instances of the concept in question, or if a concept can only be explicated by typical examples. In modern philosophy, the predominant view during most of the XX century has been that concepts are defined by a set of characteristics that are individually necessary and jointly sufficient for an object to be an instance of the defined concept. However, in his *Philosophical Investigations* (1953), Wittgenstein attacked this view, arguing instead that instances of a concept might bear no more than a family resemblance to each other.

One of the main approaches to discussions of Wittgenstein's ideas was initiated by Bambrough (1961) in his classical paper on universals and family resemblance. In this paper Bambrough introduced an interpretation of family resemblance based on sorites series which he illustrated by simple diagrams of instances of objects (a-e) and the patterns of features (A-E) that the individual features possess:

| e | d | c | b | a |
|---|---|---|---|---|
| ABCD | ABCE | ABDE | ACDE | BCDE |

These diagrams nicely illustrated Wittgenstein's point that although all instances of the concept in question share some characteristics with each of their fellows, no single characteristic or set of characteristics has to be shared by all instances.

During the 1970s, cognitive psychologists, starting with Eleanor Rosch and her collaborators, concluded from experiments that a family resemblance account of concepts seemed to reflect the actual use of concepts much better than an account in terms of necessary and sufficient conditions. Also during the late 1960s and 1970s, a family resemblance view of scientific concepts was advanced by the philosopher of science Thomas Kuhn.[1]

Contrary to the purely philosophical discussions provided by Wittgenstein and his philosophical followers, Rosch' account of concepts was developed from experiments. Rosch originally focused on everyday concepts like animals, trees, clothing and furniture and showed that such concepts have graded structures, that is, that the individual instances of a concept vary in how good an exemplar they are of the concept in question (Rosch, 1973a, 1973b). Since concepts defined by necessary and sufficient conditions should have instances that are all equally good examples - either they fulfil the definition or they do not - the existence of graded structures served as one of the main arguments for an account of concepts based on similarity rather than definitions.

Although her findings on the graded structure of concepts have been confirmed for as diverse concepts as, e.g., geometrical shapes (Rosch, 1973a), vehicles, weapons (Rosch and Mervis, 1975), and psychiatric expressions (Ekman et al. 1972), there seems to have been no empirical studies of graded structures carried out on pure science concepts such as the key concepts within classical mechanics or within electrodynamics. However, Giere (1994) has shown how the Roschian account of concepts and categories may be applied to mechanical concepts, arguing that models of classical mechanics form categories with graded structures.

Peter Barker, Xiang Chen and myself, among others, have made similar attempts to bridge the gap between the family resemblance accounts of scientific concepts advanced by philosophers, and the family resemblance accounts of everyday concepts advanced by psychologists. We have argued that Kuhn's philosophical account of family resemblance is vindicated by the empirical work of cognitive psychologists like Rosch and her collaborators, and that all of the important features of Kuhn's model for the development of science can be seen as consequences of a family resemblance account of the nature of concepts and their dynamics (Andersen et al. 1996, Chen et al. 1998).

## 1.2 Criticism of the family resemblance account

Family resemblance accounts of concepts have been met with two major objections. The first objection was originally raised by philosophers in the 1960es, and soon became adopted by psychologists. The objection is that since we can always find *some* resemblance between instances of one con-

[1] For an account of the relation between the work of Wittgenstein and the work of Rosch, see Rosch (1987). For an account of the relation between the work of Kuhn, Wittgenstein, and results within cognitive psychology, see Andersen, Barker and Chen (1996).

cept and those of another, family resemblance does not suffice to limit the extension of concepts. In this phrasing, which is due to critics of Wittgenstein from the early 1960s, the problem is referred to as the problem of wide-open texture, or the problem of the under-determination of extension (e.g. Richman, 1962; Pompa, 1967; Gupta, 1970; Bellaimey, 1990; Beardsmore, 1992). Phrasing the problem slightly differently, psychologists have usually drawn on Goodman's (1972) critical analysis of the notion of similarity, claiming that the notion of similarity is meaningless unless one can specify in what respects two things are similar (e.g. Murphy and Medin 1985, Medin, Goldstone, and Gentner, 1993).

The second objection has been raised by psychologists since the mid-1980s. This objection is that most family resemblance accounts describe categories as a division of the world according to clusters of features, but fail to explain "what the glue is that holds a concept together" (Murphy and Medin, 1985, p. 290).

In the following, I shall discuss the most important responses to these two objections: first, the suggestion to include dissimilarity on a par with similarity in the account of family resemblance in order to solve the problem of wide-open texture, and second, the suggestion that concepts, including family resemblance concepts, may be "held together" by causal models.

## 2. SIMILARITY, DISSIMILARITY, AND TAXONOMIES

As I have argued in detail elsewhere (Andersen, 2000), a way to solve the problem of wide-open texture is to extend the family resemblance account of concepts to include not only similarity between members of the same category, but also dissimilarity to members of other categories. When including dissimilarity in the account it is important to note that we are not interested in dissimilarity between members of any pair of arbitrary categories. Among philosophers, the restriction has been introduced by Thomas Kuhn in the early 1970s, but remained more or less unnoticed.[2] As Kuhn specifies the restriction, we are only interested in dissimilarity between *members of categories which can be mistaken for one another*. That they can be mistaken for one another indicates that all their instances are more similar to one another than to instances of other concepts. Hence, these concepts whose instances can be mistaken in this way also form a family resemblance class at the superordinate level. Thus, concepts constituted by similarity and dissimilarity

[2] This restriction on dissimilarity has been noted by cognitive anthropologists fairly early, see e.g. Conklin, 1969. For a similar argument within cognitive psychology, see e.g. Rosch, 1987.

relations are tied together in contrast sets. In this way, family resemblance concepts form hierarchical structures in which a general concept decomposes into more specific concepts that may again decompose into yet more specific concepts, and so forth - in other words taxonomies.[3] The inclusion of dissimilarity in the family resemblance account of concepts therefore ties concepts inextricably to taxonomies.

Within this taxonomic conceptual structure, the sorites series formed by the possible features of each family resemblance concept are therefore stopped by collision with the sorites series of contrasting concepts. However, this approach opens for the usual question concerning sorites series with contrasting ends: are the intermediate cases instances of one or the other of the contrasting ends? If the sorites series develop gradually and continuously, it is necessary to define where the extension of one concept ends and the extension of the contrasting concept begins.

This problem can only be solved if the sorites series do *not* develop gradually and continuously, that is, if there is an empty perceptual space between the families to be discriminated. However, it is important to note that there may be empty perceptual spaces along many different dimensions in perceptual space, and although some of them may be compatible, they need not all be so. Instead they may separate between different, overlapping families along different dimensions in perceptual space.[4]

Although the possibility of empty perceptual spaces along different dimensions in perceptual space implies that there may be many different categorizations of the world, this does not mean that any arbitrary categorization can be established. If attempted, anomalies will appear, that is, objects which do not fit into the categorization, but seem to belong to several contrasting categories simultaneously, or to no category at all. When this happens, the contrast set(s) in question must be changed to provide a consistent classification of all objects known, both the previously known objects which could all be consistently classified within the unchanged contrast sets and the newly discovered anomaly. Hence, although a taxonomy has proved to be a fully consistent division of the object domain for all previously known objects, a new object may still reveal it as inconsistent and as a consequence it will be changed to provide a new division which is consistent for the new, enlarged group of known objects.

Thus, taxonomies are *dynamic* entities which under certain circumstances may be subjected to change. This suggests a developmental perspective, ex-

[3] Not all scientific concepts need form taxonomies in the way indicated here. On the contrary, Nersessian and I have argued that some theoretical concepts do not form contrast sets (and thus simple taxonomies) directly, but only as part of complex problem situations (Andersen and Nersessian, 2000).

[4] For similar arguments, see e.g. Bambrough, 1961, Rosch, 1978.

plaining not only how a taxonomy is *structured synchronously*, but also how it is *changed diachronically*.

## 2.1 Triggering taxonomic change

The analysis of taxonomic conceptual structures in terms of underlying relations of similarity and dissimilarity may help us understand the dynamics of conceptual change. Taxonomies obey the following three principles:

1. *The no-overlap principle*: no extensions of the concepts in a contrast set may overlap
2. *The exhaustion principle*: the extensions of all concepts in a contrast set together exhaust the extension of their superordinate
3. *The inclusion principle*: the extension of each concept in a contrast set is a true subset to the extension of the superordinate

Therefore, instances that violate these principles are anomalous and may call for changes in the conceptual structure so as to make the altered structure comply again with the principles. For example, an object that is judged from some features is an instance of one concept, but when judged from other features it is an instance of a contrasting concept, violating the no-overlap principle and therefore calls for a change in the taxonomy.[5]

This connection between similarity/dissimilarity relations and taxonomies may provide a useful underpinning for the analysis of conceptual hierarchies. Thus, much of the standard concept mapping has covered kind-hierarchies and part-hierarchies, and the links relating concepts in such conceptual hierarchies: kind links, instance links, rule links, property links, and part links, but providing only little information on how these links and structures may depend on each other. Adopting a family resemblance approach, property links and part links reflect the similarity and dissimilarity relations which are constitutive of the taxonomy and therefore establish the kind links and instance links.

Another important implication of the family resemblance approach to analysis of science is related to the phenomenon that concepts have graded structures. Graded structures may influence the judgement of anomalies. If an object is encountered which is judged from different features is a good

[5] Taxonomic changes in response to violations of the three taxonomic principles are special instances of the anomaly resolution strategies described by Darden, basically with the same set of possible changes as Darden lists for her general account, adding new component, deleting component, specialising, and tweaking (Darden, 1995).

example of two different concepts in a contrast set, this will be a severe anomaly, as it clearly questions the adequacy of the conceptual structure. On the contrary, if an object is encountered that judged from different features is a bad example of two different concepts in a contrast set it may not call the conceptual structure in question, but just suggests that further research may be necessary before the object can be properly assigned to one of the known, contrasting concepts. Since different members of the scientific community may develop different graded structures, they may not necessarily agree on which anomalies are severe and which are not. What one scientist considers to be such a severe anomaly that the conceptual structure has to be changed, another scientist may just ignore.

Hence, a similarity based account of taxonomies explains how individual differences between members of the same scientific community can exist in a latent form that is not apparent in the usual linguistic practice, but only becomes manifest when triggered by anomalies. In addressing the question of taxonomic change, two aspects therefore have to be distinguished: changes in the structure that is shared by the community, and changes in the graded structures possessed by individuals. Thus, a developmental account of taxonomies that addresses the synchronous structure of the taxonomy as well as its diachronic development that it may be required to explain both the development of structural aspects that is shared by the whole scientific community, and the development of latent differences.

## 3. COHERENCE OF TAXONOMIES

On the family resemblance account just described, categories divide the world according to clusters of features. But features may have complex relationships to one another. To represent family resemblance concepts and the taxonomic structures they form, mere feature lists or sorites series are therefore insufficient. To capture the complex relations between features, various researchers have suggested to represent concepts by *frames* (Barsalou, 1992, Minsky, 1975), *schema* (Rumelhart, 1980), or *schemata* (Brewer, 1999).

On the basis of Barsalou's work, Peter Barker, Xiang Chen and I have used the frame representation to analyze various developments from the history of science such as the Copernican revolution (Barker, 2001), XIX century ornithology (Chen and Barker, 2000), and nuclear physics (Nersessian and Andersen, 1997). However, this approach mainly discloses the underlying problem, *viz.* why specific features are bundled together to form a concept.

## 3.1 Causal models

Within the psychological literature, Medin and collaborators have argued that concepts are coherent to the extent that they fit people's theories about the world (e.g. Murphy and Medin, 1985). Murphy and Medin argue that people are not only sensitive to feature correlations, but that they deduce *reasons* for those correlations. Thus, the connection between features seems to be not just a simple link but a causal explanation for how the two are related (Murphy and Medin 1985, p. 300).

From the mid-1980s, various psychologists developed theory-theories of concepts, arguing that underlying theories (or models) provide the causal and explanatory links that hold individual concepts together and establish connections to other concepts.[6]

Similar views have been explored from early on within cognitive history of science. For example, Nersessian in her analysis of the development of the field concept adopted a schema representation to explain how a scientific concept develops over time. On Nersessian's view, a schema contains for each stage of the conceptual development a summery of features selected by their explanatory role. Across the different historical stages the schema contains the chains of reasoning that connect one developmental stage of the concept with its later alterations (Nersessian, 1984, section 7.2).

Much of the recent cognitive-historical work has focused on conceptual change and especially the problem-solving heuristics that are used to create new representations in science. Many scholars now agree that the reasoning processes underlying conceptual change is model-based, that is, that scientists "reason by constructing an internal model of the situations, events and processes that in dynamic cases provide the basis for simulative reasoning" (Nersessian, 1999, p. 12). Much cognitive-historical work has focused on analysing the different kinds of model-based reasoning involved in conceptual change, e.g. analogical reasoning, imagistic reasoning, thought experiments, and limiting case analyses (see Nersessian, 1992, 1999 for overviews of this research).

However, in focusing so strongly on conceptual change, that is, on the diachronic development of concepts and conceptual structures, some aspects of the synchronous constitution of concepts and conceptual structures may have been overlooked. In particular, it is necessary to examine how the accounts of conceptual change as the result of model-based reasoning can be refined to include graded structures, providing an explanation of how the individual differences in concepts' graded structures among scientists working within the same domain may arise. To investigate these issues, I

[6] For an account of this development and an overview of various theory-theories, see e.g. Laurence and Margolis, 1999.

start from a brief analysis of the development of nuclear physics leading up to the discovery of nuclear fission, and conclude by pointing to a new area of research that has to be opened to achieve an integral account of conceptual structures and conceptual development.

## 4. A HISTORICAL CASE

One of the primary taxonomies in nuclear physics in the 1930s was the taxonomy of nuclear disintegration processes, that is, the taxonomy of radioactivity.[7] Initially, the background for this taxonomy was the taxonomy of radiation types which arose from the work with the Geissler tube, that is, cathode rays and X-rays. X-rays were emitted when the cathode rays reached the end of the Geissler tube and made the glass fluoresce, and this phenomenon made the French physicist Henri Becquerel wonder if other fluorescent or phosphorescent materials would emit radiation like the X-rays emitted from the fluorescent glass of the Geissler tube. There was yet no model that explained *why* the features of fluorescence and radiation were correlated, there was only the observation of the correlation itself. Hence, what Becquerel examined was under which circumstances the correlation was to hold.

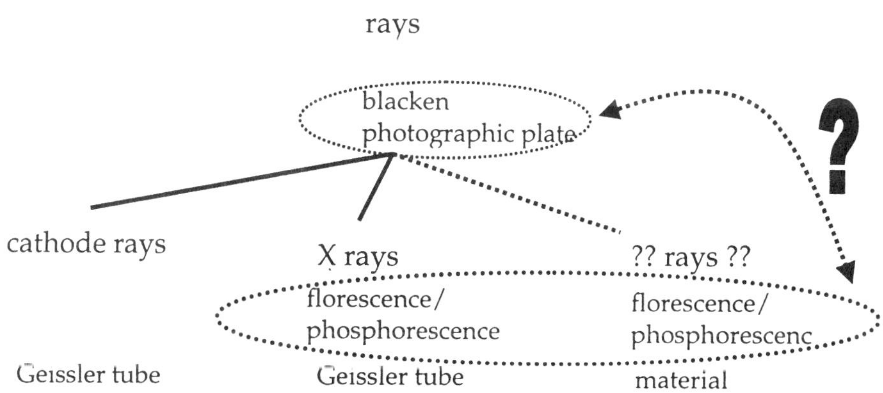


*Figure 1.* Becquerel hypothesizing new correlations of features within a taxonomic conceptual structure.

[7] For lack of space, this presentation concentrates on the major principles in the development. For detailed historical accounts, see e.g. Andersen, 1996, Stuewer, 1994, Weart, 1983.

Examining various luminescent minerals, Becquerel discovered that minerals containing uranium emitted some sort of radiation, even if the mineral was *not* fluorescent. Hence, originally hypothesized by analogy to concepts within a well-known taxonomy, a new concept - by Becquerel termed *Uranium rays* - was being formed. But Uranium rays were not triggered by or dependent upon something, such as fluorescence, high voltage or a good vacuum, they were simply intrinsic to the material that emitted them, and the new concept therefore formed a whole new branch in the taxonomy.

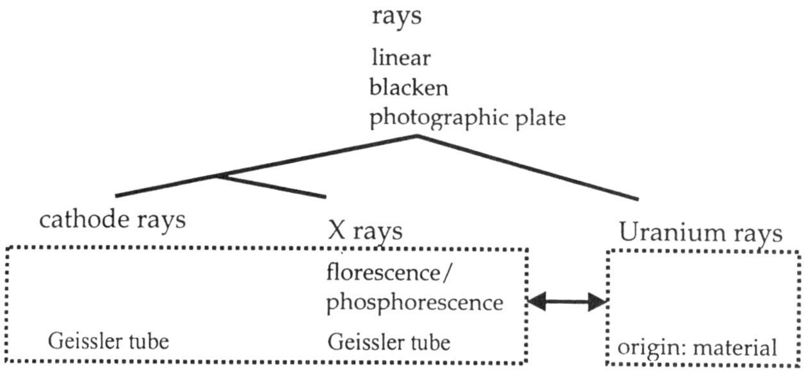


*Figure 2.* Branch splitting. Uranium rays forming a new branch in the taxonomy of different rays.

Two years later, on the background of a series of penetration experiments, Rutherford distinguished between two kinds of radiation, ∀ and ∃. Yet two years later, Villard found a third kind of radiation, (radiation. It was further discovered that the three kinds of radiation did not only differ with respect to their penetrating power, but also with respect to the deflection in a magnetic field. This stable correlation between additional differentiating features further stabilized the taxonomy.[8]

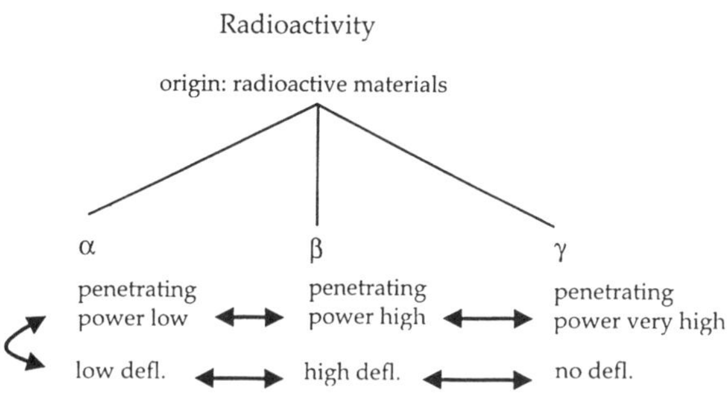


*Figure 3.* Correlation between differentiating features in the taxonomy of radioactivity.

[8] As Buchwald has argued, a "taxonomy's strength [...] depends to a very large extent on its device-independence", that is, on the possibility of distinguishing between the concepts in a contrast set in several, different ways" (Buchwald, 1992, p. 44).

At this point, it was still open whether the taxonomy was now exhaustive, or if still further differentiations were to come. Blondlot postulated the existence of N-rays, but after careful experiments their existence was rejected by the majority of physicists.

With the development of nuclear physics it became clear that the nucleus was not an elementary particle, and that ∀ and ∃ radiation was the result of the spontaneous disintegration of the nucleus, whereas (-radiation was the result of a transition of the nucleus from an excited state to a state of lower energy.

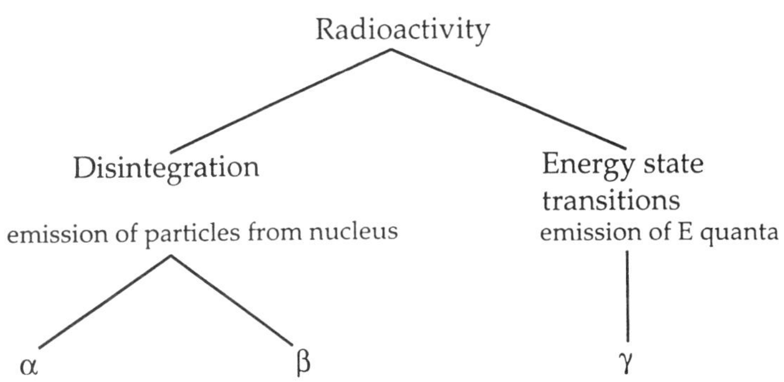


*Figure 4.* Branch splitting. Nuclear disintegrations and energy state transitions forming the main branches in the taxonomy of radioactivity.

Models of the ∀ and ∃ particles explained the correlations of features - both were particles with specific weights and specific charges and this explained both their deflection in a magnetic field and their penetrating power in various materials. Hence, these models explained the correlations of features for each concept in the contrast set. But what about the range of the contrast set? Why did these kinds of disintegrations of the nucleus exist, but not others?

By 1928/29, Gamow suggested that the nucleus should be seen as a droplet which consisted of individual ∀ particles, protons and electrons (Gamow, 1929a, 1929b). Like a droplet, the particles were held together by surface tension. Thus, it was the surface tension that determined which particles or conglomerate of particles that could escape the nucleus. This representational model was used to create a mathematical model of nuclear decay. The core of the mathematical model was the potential well created by the short-ranged attractive forces between the particles inside the nucleus. From the energy function, the probability of a particle tunneling through the potential barrier could be computed, and these computations showed that only particles up to the size of the ∀-particle could be emitted from the nucleus. In this way, the model could explain the range of the taxonomy: the taxonomy was exhaustive because particles bigger than the ∀-particle could not tunnel through the potential barrier.

Gamow's liquid drop model became very popular, and the result that only particles up to the size of the ∀-particle could be emitted lead to checker-board like diagrams depicting the possible disintegrations. These diagrams were purely descriptive; they did not serve any explanatory purposes, but they represented the scope of the taxonomy very well. Using checker-board like diagrams to represent nuclear disintegrations, only processes in which one nucleus transformed into another nuclear nearby in the period table by the emission of a small particle could be represented. These diagrams could not represent the idea of one nucleus splitting into two much smaller nuclei. Hence, as a simple tool, the diagrams helped consolidate the taxonomy.

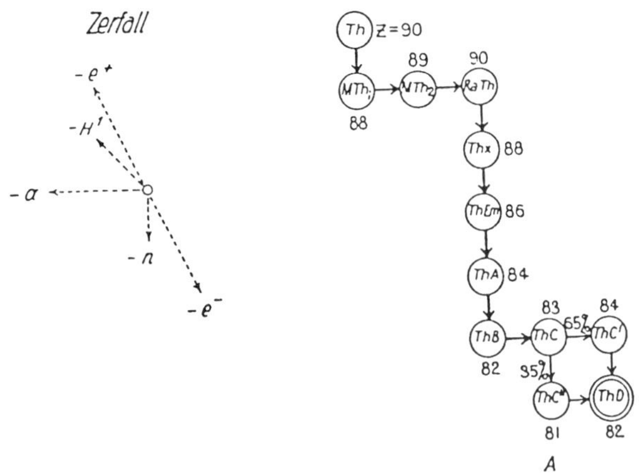


*Figure 5.* Checker-board like diagrams of possible nuclear disintegrations. From Meitner and Delbrück (1935) and Gamow (1931).

When induced radioactivity was discovered, Gamow's liquid drop model still delimited the range of the taxonomy of nuclear disintegrations produced artificially. The processes that were theoretically possible were the two naturally occurring processes, emission of ∀ particles and emission of ∃ particles, plus some additional processes, emission of protons, emission of deuterons, and emission of neutrons. In the case of bombardment with neutrons, a further possible process was neutron capture. The model showed which processes were possible, and experiments showed that only some of these actually took place. Thus, von Weizsäcker in his book *Die Atomkerne* (1937) discussed all possible induced radioactive processes in the form of a list of all possible permutations of protons, neutrons, deuterons, ∀ particles ∃ particles, respectively. For neutron bombardment of heavy nuclei the possible decay processes had been reported to be ∀ emission, proton emission, neutron capture and ∃ emission. Again, these lists of all possible permutations of projectiles and decay products helped consolidate the taxonomy. The list showed which processes existed and left out what did not exist. In this way the list clarified the ontology which all nuclear physicists agreed upon.

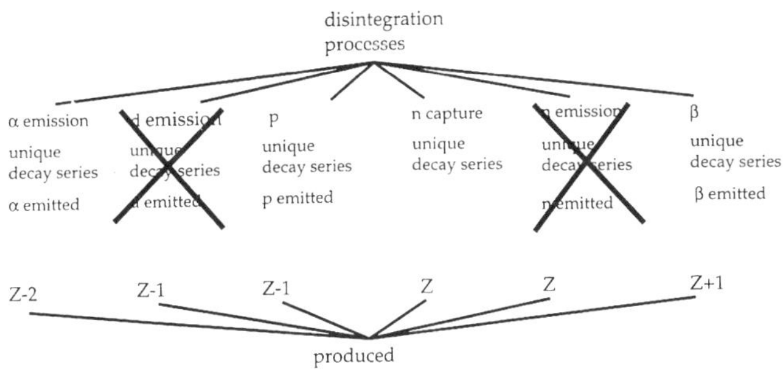


*Figure 6.* The taxonomy of disintegration processes. Gamow's model determined which processes were theoretically possible. Experiments showed which of these that actually occured.

However, a few years after Gamow had developed his theory of decay on the basis of a droplet model that emphasized surface tension, a different way of using the droplet model for discussing artificially induced radioactivity was gradually developed by Bohr. The idea of a potential barrier which the emitted particles had to overcome was visualized by a mechanical model of billiard balls in a shallow basin. But Bohr used this model to argue for a compound nucleus. In bombarding the nucleus with a neutron, the projectile would divide its energy first with one of the balls, these two would share their energy with others, and so on until the kinetic energy of the projectile was divided among all balls.

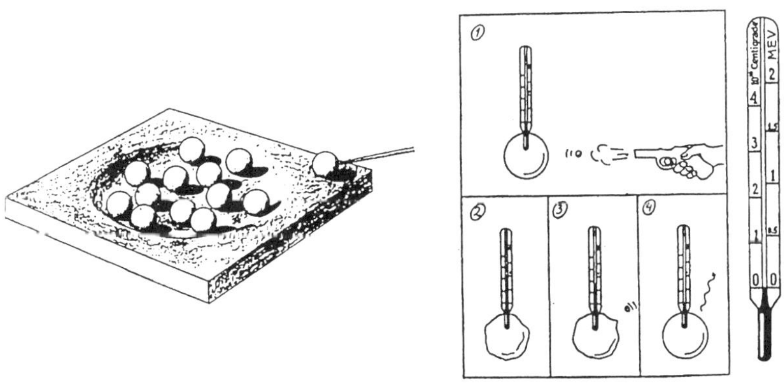

*Figure 7.* Bohr's mechanical (left) and thermodynamical (right) models of the compound nucleus. From Bohr (1937).

This led back to the droplet model, now as a thermodynamical model about temperature, rather than Gamow's fluid dynamic model about surface tension. On this model, after the nucleus was hit by the projectile particle, the temperature of the droplet rose and the droplet oscillated correspondingly. A particle might evaporate from the oscillating droplet, causing the temperature to fall, and finally the last part of the energy might be emitted in the form of electromagnetic radiation. On this model, there was no objection to the idea of several particles being emitted in induced radioactivity, provided the energy of the projectile particle was sufficiently high. Bohr even noted that for extremely violent impacts, an explosion of the whole nucleus was conceivable.

But Bohr did not specify what the end product of such an explosion would be, and here it seems like the previous scope of a well-sustained taxonomy together with the visual images from previous models served as restrictions. Contrary to Gamow's model, on Bohr's model large decay products were theoretically possible - but this theoretically possible extension of the taxonomy was not realized.[9]

## 5. CONCLUSION

The development we see here is a continuous interplay between, on the one hand categorizations and recategorizations of observed phenomena on the basis of characteristics that make them seem similar or dissimilar, and on the other hand the development of models. Early models of the nucleus explained ∀ and ∃ radiation as the emission of charged particles and as different from the emission of energy quanta in the form of ( radiation. In this case the model explains how the branches of the taxonomy are organized. Likewise, Gamow's later droplet model explained the scope of the taxonomy. In

[9] Further, although anomalies had been encountered in the investigation of induced radioactivity, none of these attempted were resolved by extending the taxonomy to encompass large decay products. Instead, all attempted were resolved within the range of decay products determined from Gamow's mode. An early attempt to suggest nuclear fission that had been dismissed as simply inconceivable was not reevaluated in the light of the new model that made the phenomenon conceivable. The taxonomy retained its scope and was still considered exhaustive although the new model could in fact have removed its claim for exhaustiveness. Not until a highly anomalous experimental result made two scientists suggest the existence of nuclear fission was the taxonomy extended to encompass this new concept. When that happened, physicists very quickly realised that this new phenomenon which had previously been dismissed as inconceivable was not as revolutionary a finding as it might seem; it could easily be explained from Bohr's model. See Andersen, 1996 for a detailed account of the anomalies that were discovered during this period as well as for a discussion of the early suggestion of nuclear fission.

both cases, the models explain the correlation of specific characteristics and thus the occurrence and non-occurrence of categories, and they make predictions about new characteristics or about new categories. In this respect, the models provide the causal and explanatory links that hold individual concepts together and establish taxonomic relations to other concepts.

However, models are rarely all-encompassing in the sense that one model explains all aspects of a taxonomy. Gamow's model focused on the energetic aspects of decay and explained the range of the taxonomy. Bohr's model focused on excitations and explained various characteristics of different types of decay. The two models were useful for different kinds of research questions - Gamow's model was useful for examining static characteristics like nuclear mass defects; Bohr's model was useful for examining dynamic characteristics like nuclear reactions.[10] Consequently, some research groups primarily worked with the one model, other research groups with the other model.

The interplay between taxonomy and the various models is therefore clearly contextualized. In working with the same taxonomy, scientists adopt different models to explain different things. Although they agree on the overall ontology and thus on the range of the taxonomy, still the different models they employ may lead them to develop differences that may not, at first, be apparent.

Hence, combining analysis of the synchronic constitution of conceptual hierarchies with analysis of the diachronic change of conceptual hierarchies illuminates a tension which has so far received only little attention within cognitive history of science: the tension between the cognitive and the social. Some aspects of taxonomies relate to their status as the possession of a scientific community. There is an agreement within a community on the ontology of a given subject area and thus on the range of concepts used to describe particular phenomena. Other aspects of taxonomies relate to their status as part of the mental operations of the individual. Different models may lead individual scientists to develop the same taxonomy in different, inconsistent ways, although this may not, at first, be apparent.

Both aspects are closely connected to models - models that explain the coherence of concepts and taxonomies and that are used in scientists' reasoning processes to develop the taxonomies further. But models are partial and contextural: no model explains all aspects of a taxonomy, instead, scientists adopt different models to explain different things. In some cases, these different models employed to solve different questions may turn out to be incompatible or inconsistent. Gamow's and Bohr's droplet models were employed to solve different research questions, but turned out to imply different

[10] See Stuewer 1994 for a comparison of the two research traditions building on Gamow's and Bohr's models, respectively.

ontologies. This opens for a whole range of new research questions within cognitive history of science: How does a community divide its effort in working with different models? How are inconsistent models handled? When do inconsistencies become a problem? How are they resolved? These are some of the questions we need to address in order to integrate the accounts of the cognitive and the social dimensions of science.

## REFERENCES

Andersen, H., 1996, Categorization, anomalies and the discovery of nuclear fission, *Stud. Hist. Phil. Mod. Phys.* 27:463-492.

Andersen, H., 2000, Kuhn's account of family resemblance: a solution to the problem of wide-open texture, *Erkenntnis* 52:313-337.

Andersen, H., Barker, P. and Chen, X., 1996, Kuhn's mature philosophy of science and cognitive psychology, *Philosophical Psychology* 9:347-363.

Andersen, H. and Nersessian, N., 2000, Nomic concepts, frames, and conceptual change, in: *Proceedings of the Philosphy of Science Association*, D.A. Howard, ed., University of Chicago Press, Chicago, 67, pp. 224-241.

Bambrough, R, 1961, Universals and family resemblances, *Proc. Aris. Soc.* 61:207-222.

Barker, P., 2001, Incommensurability and conceptual change during the Copernican revolution, in: *Incommensurability and Related Matters*, H. Sankey and P. Hoyningen-Huene, eds., Kluwer Academic Publishers, Dordrecht.

Barsalou, L., 1992, Frames, concepts, and conceptual fields, in: *Frames, Fields and Contrasts: New Essays in Semantical and Lexical Organization*, A. Lehrer and E. Kittay, eds., Erlbaum, Hillsdale, NJ, pp. 21-74.

Beardsmore, R.W., 1992, The theory of family resemblance, *Philosophical Investigations* 15:131-146.

Bellaimey, J.E., 1990, Family resemblances and the problem of the underdetermination of extension, *Philosophical Investigations* 13:31-43.

Bohr, N., 1937, Transmutations of atomic nuclei, *Science* 86:161-165.

Brewer, B., 1999, Schemata, in: *The MIT Encyclopedia of the Cognitive Sciences*, MIT Press, Cambridge MA, pp. 729-730.

Buchwald, J.Z., 1992, Kinds and the wave theory of light, *Stud. Hist. Phil. Sci.* 23:39-74.

Chen, X., 1995, Taxonomic changes and the particle-wave debate in early nineteenth-century Britain, *Stud. Hist. Phil. Sci.* 26:251-271.

Chen, X., Andersen, H., and Barker, P., 1998, Kuhn's theory of scientific revolutions and cognitive psychology, *Philosophical Psychology* 11: 5-28.

Chen, X. and Barker, P., 2000, Continuity through revolutions: A frame-based account of conceptual change during scientific revolutions, in: *Proceedings of the Philosophy of Science Association*, D.A. Howard, ed., University of Chicago Press, Chicago, 67, pp. 208-223.

Conklin, H.C., 1969, Lexicographical treatment of folk taxonomies, in: *Cognitive Anthropology*, S.A. Tyler, ed., Holt, Rinehart and Winston, New York, pp. 41-59.

Darden, L., 1995, Exemplars, abstractions, and anomalies: Representations and theory change in Mendelian and molecular genetics, in: *Concepts, theories, and rationality in the biological sciences*, G. Wolters and J.G. Lennox, eds., University of Pittsburgh Press, Pittsburgh, pp. 137-158.

Ekman, P., Friesen, W.V. and Ellsworth, P., 1972, *Emotion in the Human Face*, Pergamon, Elmsford, NY.

Fu, D., 1995, Higher taxonomy and higher incommensurability, *Stud. Hist. Phil. Sci.* 26:273-294.
Gamow, G., 1929a, Über die struktur des atomkerns, *Physikalische Zeitschrift* 30:717-720.
Gamow, G., 1929b, Zur quantentheorie der atomzertrümmerung, *Z. Phys.* 52:510-516.
Gamow, G., 1931, *Constitution of Atomic Nuclei and Radioactivity*, Clarendon Press, Oxford.
Giere, R., 1994, The cognitive structure of scientific theories, *Phil. Sci.* 61:276-296.
Goodman, N., 1972, Seven strictures on similarity, in: *Problems and Projects*, N. Goodman, ed., Bobbs-Merrill, Indianapolis, pp. 437-446.
Gupta, R.K., 1970, Wittgenstein's theory of "family resemblance" in his philosophical investigations, *Philosophia Naturalis* 12:282-286.
Kuhn, T.S., 1991, The road since structure, in: *PSA 1990, Vol. 2*, A. Fine, M. Forbes and L. Wessels, eds., Philosophy of Science Association, East Lansing, pp. 3-13.
Laurence, E. and Margolis, S., eds., 1999, *Concepts. Core Reasings*, MIT Press, Cambridge MA.
Meitner, L. and Delbrück, M., 1935, *Der Aufbau der Atomkerne. Natürliche und künstliche Kernumwandlungen*, Springer, Berlin.
Minsky, M., 1975, A framework for representing knowledge, in: *The Psychology of Computer Vision*, P.H. Winston, ed., New York, McGraw-Hill, pp. 211-277
Murphy, G. and Medin D., 1985, The role of theories in conceptual coherence, *Psychological Review* 92:289-316.
Nersessian, N., 1984, *From Faraday to Einstein: Constructing Meaning in Scientific Theories*, Martinus Nijhoff, Dordrecht.
Nersessian, N., 1992, How do scientists think? Capturing the dynamics of conceptual change in science, in: *Cognitive Models of Science, Minnesota Studies in the Philosophy of Science, vol. XV*, University of Minnesota Press, Minneapolis, pp. 3-44.
Nersessian, N., 1999, Model-based reasoning in conceptual change, in: *Model-Based Reasoning in Scientific Discovery*, L. Magnani, N.J. Nersessian and P. Thagard, eds., Kluwer Academic/Plenum Publishers, New York, pp. 5-22.
Nersessian, N. and Andersen, H., 1997, Conceptual change and incommensurability: A cognitive-historical view, *Danish Yearbook of Philosophy* 32:111-152.
Pompa, L., 1967, Family resemblance, *Philosophical Quarterly* 17:63-69.
Richman, R.J., 1962, Something Common, *Journal of Philosophy* LIX:821-830.
Rosch, E., 1973a, Natural Categories, *Cognitive Psychology* 4:328-350.
Rosch, E., 1973b On the internal structure of perceptual and semantic categories, in: *Cognitive Development and the Acquisition of Language*, T.E. Moore, ed., Academic Press, New York, pp. 111-144.
Rosch, E., 1987, Wittgenstein and categorization research in cognitive psychology, in: *Meaning and the Growth of Understanding: Wittgenstein's Significance for Developmental Psychology*, M. Chapman and R.A. Dixon, eds., Springer, Berlin.
Rosch, E. and Mervis, C.B., 1975, Family resemblances: Studies in the internal structure of categories, *Cognitove Psychology* 7:573-605.
Rumelhart, D.E., 1980, Schemata: The building blocks of cognition, in: *Theoretical Issues in Reading Comprehension*, R.J. Spiro, ed., Erlbaum, Hillsdal, NJ, pp. 33-58.
Stuewer, R.H., 1994, The origin of the liquid-drop model and the interpretation of nuclear fission, *Perspectives on Science* 2:76-129.
Thagard, P., 1992, *Conceptual Revolutions*, Princeton University Press, Princeton.
Weart, S.R., 1983, The discovery of fission and a nuclear physics paradigm, in: *Otto Hahn and the Rise of Nuclear Physics*, W.R. Shea, ed., Reidel, Dordrecht, pp. 91-133.
Weizsäcker, C.F. von, 1937, *Die Atomkerne*, Springer, Berlin.
Wittgenstein, L., 1953, *Philosophical Investigations*, Macmillan, New York.

# Production, Science and Epistemology. An Overview on New Models and Scenarios.

Simone Turchetti[*], Mauro Capocci[#], Elena Gagliasso[†]
[*]*CHSTM, University of Manchester, Machester, UK, sturchetti@stud.man.ac.uk;* [#]*Department of Animal Biology and Genetics, University of Florence, Florence, Italy, capocci@libero.it;* [†]*Department of Epistemological and Philosophical Studies, University of Rome La Sapienza, Rome, Italy, elena.gagliasso@mail.uniroma1.it*

Abstract: During the XX century, the industrial and economic organization of developed countries has radically changed. First, the development of new technologies has moved production from linear mechanisms such as the Fordist *assembly line* to complex industrial networks in which information and communication play a central role. Moreover, this development has deeply modified scientific and technological R&D and more generally it has had an impact on the structure of scientific and technological research establishments and their epistemological framework. This new scenario opens still unquestioned problems. Has the end of the *Fordist factory* something to do with the end of obsolete *Big Science institutions*? Does the birth of new research establishments fit within a general change of production models? And finally, to what extent does this change affect the epistemological framework? The changes involving the context of production, science and epistemology are crucial to understand the development of model-based reasoning. Therefore we will try here to compare and contrast them.

## 1. INTRODUCTION

During the XX century, the industrial and economic organization of developed countries radically changed. First, the development of new technologies moved production from linear mechanisms such as the Fordist *assembly line* to complex industrial networks in which information and communication played a central role. Moreover, this development deeply modi-

*Model-Based Reasoning: Science, Technology, Values*, edited by
L. Magnani and N.J. Nersessian, Kluwer Academic/Plenum Publishers, New York, 2002

fied scientific and technological R&D and more generally it impacted on the structure of scientific and technological research establishments. Finally, these structural changes in research establishments were some how correlated with the epistemological shift that was defined as a transition from "physics" models to "biology" (Cini, 1994). If such a shift was the result of a complex of intertwined factors, then one of these factors was the change of science production procedures.

This new scenario opens still unquestioned problems. Did the end of the old Fordist factory relate to the end of old Big Science institutions? To what extent did the birth of new research establishments fit within a general change of production models? Did this birth impact on science modeling and its epistemological content?

Here we will try to link three categories: production, science and epistemology in order to see whether the transformation of the scientific and industrial complex was one of the factors somehow involved in the changes of epistemological categories of thought. So, first we will observe how the changes in methods of production influenced the changes in scientific production. Second, we will see how these changes may be influential in the definition of a new epistemology.

## 2. THE TRANSITION FROM FORDISM TO POST-FORDISM IN PRODUCTION

During the second half of the XX century, a dramatic change in planning and functioning of production systems took place. The so-called *Fordist factory* transformed into something new. Historians, sociologists and experts of production management have not yet universally defined the new kinds of production systems that emerged. Nevertheless, the term *post-Fordism* has been broadly used to describe new kinds of production systems that present novelties previously unseen in the Fordist factory (Amin, 1994; Zanini, Fadini, 2000). The transition from Fordism to post-Fordism can be observed throughout four main categories: *organization* and *architecture, production policy, State's role* and *production subjects.*

In the Fordist factory, *organization* and *architecture* played a complementary role as the *assembly line* created a new space dimension for production and also organized the human and non-human factors within the factory. During the 70s, the introduction of Asian production systems defined major changes both in the architecture and the organization of factories. As far as the architecture was concerned, the production of spare parts was made in "industrial districts" and then assembled in central premises. Becattini has described this process as the constitution of a *fractal* structure of production. The large-scale organization of the Fordist factory became a complex of small-scale organizations in which the relationships in each small-scale facility was the same "in scale" of the overall organization (Becattini, 1998). Thus, it resembled a *fractal*, the geometric figure that reproduces macroscopically shapes emerging at microscopic level and vice versa. Furthermore, the introduction of new principles for the management of production such as the *zero-stock* and the *just in time* deeply changed the structure of the assembly line. Throughout the introduction of communication and feedback devices, the post-Fordist factory started to depend heavily upon communication technologies. This change was associated with a new production policy in which *mass production* was no longer taken into account as a goal. Mass production was one of the ideals of Fordism in the sense that shaping tastes throughout the mass production of some goods was perceived as a target of production systems. This ideal was substituted by the mass production of diversified goods, in which a range of options could be outlined for the same kind of commodities.

Fordism was shaped by planned economy and central States. Throughout the mechanism of welfare and warfare, the State ruled, organized, invested and shaped production. During post-Fordism, the State was no longer protagonist in production planning and left companies free to play in the global market.

Finally, in the transition from Fordism to post-Fordism, *manpower* changed. Fordism created a working environment in which machines would embody job qualifications and skills of manpower, leaving to the latter the only function of machine control and feeding. *Alienation* was the kind of feeling that characterised the people working in factories as a consequence of their dependency from machines (Panzieri, 1976). Post-Fordism maintained this dependency, but it changed places, qualifications and requirements of individuals in the factory and more generally in working places. *Flexibility* was an element thoroughly required within any kind of specific qualified or not qualified job. As shown by Bologna, Fumagalli and Rocco (1997), in *hi-tech* productions, manpower worked in a new productive dimension that was freer from physical constraints, but equally alienating (as far as dependency from machines went).

These radical changes characterised the dimension of production in the XX century, but they also deeply influenced the structure of scientific production. There are two main reasons for that: first, scientific and technological innovation was linked with production through Research and Development institutions; second, even when scientific and technological production was not immediately linked with production factories, it influenced the process of decision-making and the patterns of technological development in scientific institutions.

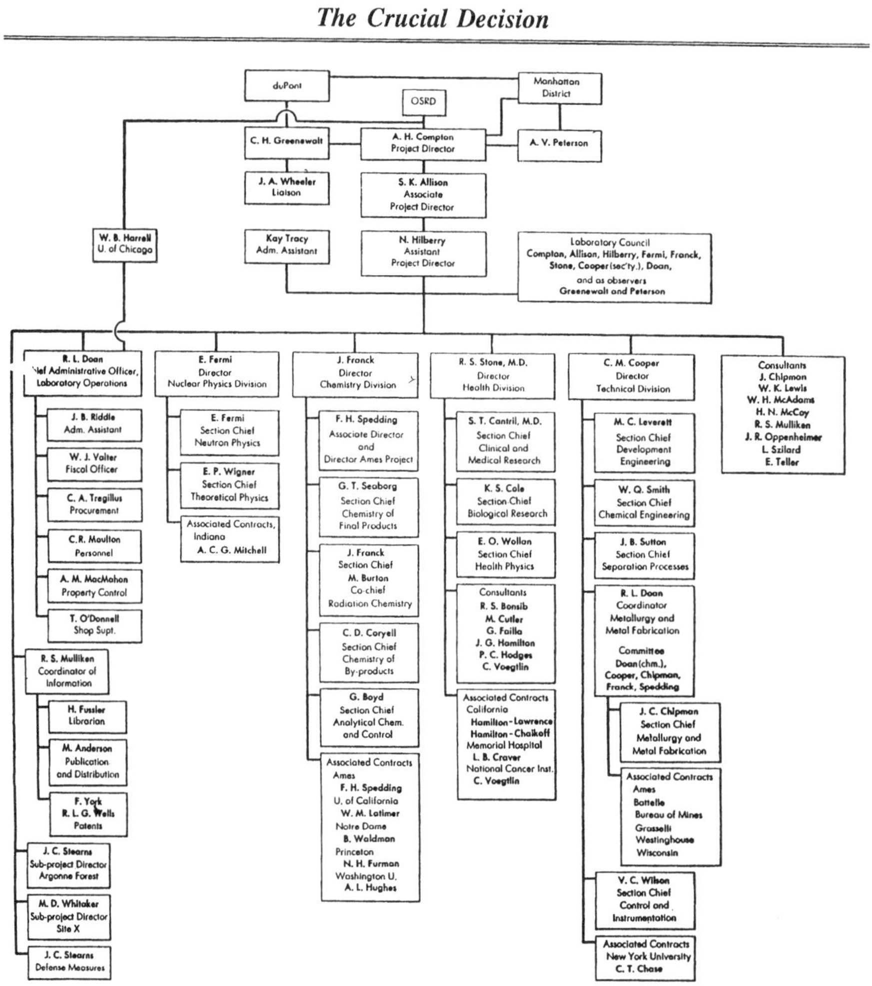


*Figure 1. Organizational Chart* of the Metallurgical Laboratory, Manhattan Project, January 1943. R.L. Kathren, J.B. Gough, G.T. Benefiel, eds., *The Plutonium Story: The Journals of Professor Glenn T. Seaborg 1939-1946* (Columbus, Richland: Battelle Press,1994), p. 237 (Courtesy of Battelle Press).

## 3. "FORDIST" SCIENCE

At the end of WW2, the chemist Bertrand Goldschmidt made one of the most celebrated comparisons between production systems and science production. Goldschmidt observed that the Manhattan Engineering District (the secret scientific research establishment that led to the production of the first atomic bomb) was "the astonishing American creation of a formidable array of factories and laboratories as large as the entire automobile industry of the United States at that date" (Rhodes, 1986, p. 605). The Manhattan project was a prototype for institutions devoted to science production in the Fordist era. From the Manhattan Project onwards, gigantic facilities devoted to scientific production were built in order to improve it. The sociologist of science Derek de Solla Price (1963) used the term *Big Science* to describe a number of changes in science production that occurred in the decades following WW2. His analysis focused on the statistical changes occurring in science production at that date. But a closer look to the *Big Science* shows how much it was largely shaped by the spreading of Fordist systems of production in the scientific context.

The organization models chosen by XX century major scientific establishments clearly presented an impressive Fordist outlook. If the assembly line played a central role in factory organization, then the *knowledge factory* was organized throughout means such as the *organizational chart* (see Figure 1), a metaphoric assembly line installed just in the core of the scientific production.

The organizational chart, as the assembly line, fragmented scientific production. It also assigned specific roles and tasks to members of a scientific community. Then, it assigned hierarchies within the "knowledge factory". Finally, it indicated the direction that research projects had to follow, from theoretical development to application. As Sharon Traweek has shown, along organizational charts are established long-term relationships within a scientific community. For instance, the organizational charts clarified in a big research establishment such as Stanford Linear Accelerator (SLAC), the path dependency between theoretical physicists, experimental physicists, technicians, chemists and postgraduates (Traweek, 1988).

The *big laboratory* or *big research facility* was the *architectural feature* of the Fordist knowledge factory. It was indeed a large-scale facility in which the whole scientific production was scheduled. Scientists were supposed to use the research facilities provided as part of the overall gigantic structure. Big laboratories like the atomic research establishments of Los Alamos or Livermore were created as secret facilities, usually with a forbidden access for the public. Major developments in crucial scientific sectors such as *computational research* were indeed produced in this kind of establishments (MacKenzie, 1991).

As far as the *production policy* was concerned, the Fordist knowledge factory had strict timetables in order to meet the scientific and technological needs of mass production. As David Noble has highlighted, science played a central role in the organization of mass production. It gave to this organization a scientific outlook throughout the introduction of *numeric control* and *automation.* As Noble (1984; 1987) observed, the military played a crucial role in mixing the Fordist ways of production, traditional criteria of command and control and scientific means to produce them automatically.

*Control* was indeed the goal of Fordist science. Every scientific object had an intrinsic need of control. Electronic systems or "green revolution" crop produced for the development of the third world countries were the result of research developed within a controlled framework of mass production. Furthermore, the Cold War needs added to this control dimension a specific character. As Paul Edwards has pointed out, the *Doomsday machine*, that is to say the nuclear complex for prompt retaliation in the case of nuclear attack, was fed with the systematic control of all the possible scenarios following a nuclear conflict. Each scenario was part of a specific research program committed to electronic machinery and scientists supposed to serve their needs. As shown by Edwards (1996):

> Both the engineering and the politics of the closed-world discourse centred around problems of *human-machine integration*: building weapons, systems, and strategies whose human and machine component could function as a seamless web, even on the global scale and in the vastly compressed time frames of a superpower nuclear war.

The Fordist science was also deeply committed to *state intervention*, as the State was supposed to provide funding for research facilities linked with military projects. Indeed, the Cold War pushed forward such economic dimension and strengthened it with the developments in aeronautics, ballistic, nuclear and telecommunication research. State allocation of funding in science had also an economic function. Since WW2, the economists John M. Keynes and William Beveridge observed that the state intervention was a means to foster private economy in moments of economic crisis. This hypothesis justified the improvement of public funded scientific research in both warfare and welfare.[1]

[1] In other terms the state intervention in science was due to strategic as well as economic reasons. Both welfare and warfare contributed to improve economy and overcome cyclical crisis. The idea of warfare as a means to overcome crisis emerged for the first time during WW2, when William Beveridge (1944) observed that through the defeat of the "tyrant" Hitler, it would have been possible to defeat another "tyrant": the unemployment.

Finally, scientific manpower changed and was embodied within the Fordist system of scientific production. After WW2, for the first time scientists worked in an environment causing *alienation*. Scientific machinery became too complicated to be manipulated by individuals. This dimension soon changed the relationship between scientist and experimental device. Until the 30s the scientist was a clever craftsman and toolmaker.[2] The innovation constituted by the construction of big laboratories with particles accelerators or huge computational devices made of the scientist a "machine's guardian" that keeps controlling and feeding with data the machine.[3] The machine became a complex of several parts and each scientist was devoted to the control of a single element. Thus, the overall complex became a "black box" (Latour, 1987). Furthermore, the object of scientific research became morally controversial, as scientists could not control the final output of their research and the results could be against moral or ethical ideals.

## 4. "POST-FORDIST" SCIENCE

During the last quarter of the XX century, the Fordist frame in which science was produced radically changed. Here, we will try to see whether these changes were related to the development of post-Fordism as complex of production systems.

At the end of the 80s, the US project for the construction of a Superconducting Super Collider (SSC), a big scale facility for the study of nuclear particles, was abandoned. In the same period, large funding was provided for a complex of smaller and disperse laboratories leading the Human Genome project, the research on human genetic set (Daniel Kevles, 1997). This change was representative of a shift from an old form of *Big Science* to a new *Big Science* with different characteristics and agenda. As Kevles (1997, p. 297) significantly has observed:

> Both were called Big Science, yet while each was a kind of Big Science enterprise, they were quite different expressions of the genre. That difference figured significantly in the ultimate fate of each, and so did the circumstances of their respective fields of science as the Cold War came to an end and a new agenda for science took hold in the United States.

[2] See for example how the cosmic ray physicist Patrick Blackett described the "string and wax" method to J.Huxley (1934). The 'string and wax' method was constantly used at the Cavendish Laboratory in Cambridge before WW2.

[3] For instance, that meant for a number of physicists to have a clear feeling of being "chased by machines". Pierre Leprince-Ringuet commented: "We must go quickly, we must run without slowing our rhythm: we are being chased [...] we are being chased by the machines!" in Peter Galison (1997, p. 227).

Kevles' analysis reflects the shift from the scientific production in the Fordist era to the new scientific production emerging in post-Fordism.

First, a new path embraced the organization of laboratories. Organizational charts were no longer representatives of the laboratory structure as they put too many constraints to science production. For instance, nowadays Genentech scientists are increasingly unwilling to present themselves as "cogs" in the science production machinery. They prefer to present themselves as independent from fixed hierarchies but linked as subject at the same level co-operating in the scientific endeavor.

In post-Fordist science, large-scale facilities were no longer the only place devoted to pinning science productions. Indeed small-scale laboratories linked each other and provided with technologically advanced experimental apparatus replaced big laboratories. The image of a 'fractal' structure, introduced in the first paragraph, is still indeed valuable to describe this new *architectural dimension* of science production in post-Fordism. For instance, the Human Genome Project was born as the enterprise of a laboratories network. Each laboratory had a similar task: pick up a fragment of DNA and decode it. They used similar procedures in order to achieve this goal and worked as a group of small-scale research facilities, provided with instruments for data processing and transmission.[4]

More generally, the idea of a "fractal" structure brings about a change in science production. As in a "fractal" type of production, scientific research moves to a complex of similar small-scale facilities, whose properties emerge at macroscopic level and are reproduced at microscopic level. The relationship that links those research facilities is still proper of a fractal structure of production. First, it shows that several small research facilities can work together in a *web*. The web can be a computer connection made with Internet or similar technologies and fit to process scientific data. Those data can be analyzed or collected even from locations very far from the places where they have been gathered for the first time and spread in all the small-scale research facilities in need of those data. Second, it shows that the big laboratory, the gigantic research facility project-driven is replaced by *science & technology parks*, a multifarious structure in which a number of research oriented small companies deal with other companies and with main research establishments like universities.

In this second example, the State provides links with knowledge institutions and the initial investments while the small research companies are

[4] "The big subject of the genome project is, of course the map and sequence of the human genome. It proceeded largely by awarding grants for research to many small groups of scientists across the country for work initiated by them on problems and organisms relevant to the overall goal. In 1991, for example, NIH funded some 175 different genome projects, at an average grant size of $ 312,000 a year [...]". (Kevles, 1997, p. 294).

willing to develop their activity in the network provided by a park facility or within other networks (Massey, 1992). This shows the different role played by the State in post-Fordist science. The State is unwilling to provide large funding for large scale research facilities, but it still produce an effort to foster scientific research as long as it proves to be self-sufficient and capable to meet market needs. That diminishes public investments in large-scale facilities and increases cases of co-operation at several levels between public and private. For instance, an interesting case of merging between private and public within the post-Fordist context is represented by the Frimorfo case. Frimorfo is a small company specialized in analysis of mice genetic modifications. Frimorfo works in co-operation with Fribourg University, which offers room and electronic microscopes (interview with Frimorfo director, LA.S.E.R., forthcoming).

This new dimension deeply transforms scientific production policy and targets. Science and technology institutions increasingly become goods providers in the free market of the global economy. Scientific and technological innovation needs long-term funding that the State is not able to provide. Therefore the scientific enterprise has to rely upon a new solid network of financial institutions ready to support and to sustain research. New scientific ventures in fields such as biotechnology have brought into science production a new link with financial institution and brought together the stages of patent licensing, stock investing and industrial production of living beings (Gagliasso, 1996). Furthermore, a number of new scientific ventures are likely to explore the potentialities of the technological transfer that is to say the application of an innovation obtained in one scientific discipline in other fields. In other words, a scientific innovation is not a piece of knowledge within a *closed knowledge factory*. Indeed, it is an *open source of transferable technology* within the context of a new networked web devoted to knowledge production. A new software based upon mathematical or physical algorithms can be used has instrument in the context of economics (as in the new field of econophysics). For example, software originally devised for the analysis of stress in nuclear devices could be applied with success in normal biophysics procedures in order to understand the stress in molecules structure (Fabio Sterpone, PhD student, CEA Paris, private interview, in LA.S.E.R., forthcoming). Soluzioni Informatiche (S.IN.), a small private company made of 4 employees provided software and hardware for "computer chemistry". S.IN. provided software for a number of different tasks in chemistry: molecules modeling, chemical metrics and database (LA.S.E.R., cit.).

This big transformation of the context of science production involves also a big transformation for the scientific and technological expertise. *Flexibility* is a word thoroughly used in scientific environments nowadays and reflects changing attitudes of scientists towards their research. The hierarchical pyramid linking individuals in Fordist science fell down when the process of decision making was completely restructured both for the sense of "claustrophobia" of researchers in those institutions and the needs of new research facilities. Within these, scientists are now relatively free to organize their timetables, thus changing dramatically the perception of lifetime and work-time. Moreover, the decision to invest time, money and energy in research is taken by scientists themselves in the first place. Every scientist becomes a fund-raiser and a manager. This happened in two ways. First, scientist provides the expertise for deciding if a project was to be funded by a private enterprise (Kornberg, 1995). Second, the scientist becomes a private entrepreneur ready to challenge the market with his own innovations. These changes may result also in the increase of working hours with respect to the past, and the endangering of job securities (in terms of long-term careers).

Post-Fordist science was born reproducing clear patterns emerging at the level of production systems and gave to these patterns its own typical character. At the levels of organization, policy production, subjects and State's role, these novelties can be seen still nowadays and will continue to structure scientific research in the years following.

## 5. NEW EPISTEMOLOGICAL FEATURES WITHIN THE POST-FORDIST SCIENCE

The new patterns of scientific production define a new environment in which the work of scientists takes place. This new environment (new technologies, knowledge networks, new social hierarchies) affects (even if strictly speaking, *it does not determine*) the development of scientific thinking and modeling. So, in this last paragraph we will try to find an answer to the question: does the shift from Fordism to Post-Fordism in science affect the epistemological frame in which science is produced?

Generally speaking, a significant epistemological shift takes place in the last decades: scientific thinking moves from knowledge as a form of *factory production* to knowledge as a form of *networked production*. The knowledge factory based on law modeling was localized within one methodological system, considered as self-sustained because it contained all the knowledge in the given domain.

Axiomatic structures, uniformation patterns and deductive reasoning shaped that scientific thinking. The solution of specific tasks would take place in a *virtual assembly line of reasoning* in which the significant problems were specific, but the methods of solution universal.

On the contrary, the network of knowledge production sets up a frame in which "a common background allows forms of *distributed reasoning in which communication plays an essential role*" (Tagliagambe, 1997, p. 155).

For instance, deduction is not anymore dominant in reasoning, and a wide range of approaches to scientific thinking emerge. Among others, abduction, analogy and metaphor play a significant role, allowing the emergence of narrative explanations, particularly in process-based scientific reasoning. Some case studies introduced during this conference have shown the importance of this methodological choice in epistemology.

As a consequence of this change in scientific thinking, the epistemological complex becomes an "horizontal" and multidisciplinary set of hybrids rather than a pyramid of knowledge organizing hierarchically scientific disciplines. Some disciplines gain their autonomy but, at the same time, they co-operate more strictly with other disciplines in the solution of specific tasks. Thus, scientific endeavor moves towards a process of *fractalization* as observed in both industrial and scientific production. In other words, from the methodological point of view the merging of several disciplinary sets increasingly shapes scientific research. For example, in the case of epidemic diseases, a number of disciplines converge in order to find successful solutions to a clearly defined problem: epidemiology, immunology, molecular genetics, population genetics, pharmacology, and environmental ecology.

We can also observe a shift from an ideal of a standard-science, rigid and context-independent, formulated throughout analytical statements (or *protocols*), to discursive practices founded on blurred concepts whose meaning depends both on a wider context and on their fruitful migration from discipline to discipline by means of metaphors. Concepts such as "selection", "environment", "information", "feed-back" find a proper meaning in the transfer among disciplines. Their *flexibility* emerges as very helpful tool in the birth of new theories, particularly in life sciences:

> Instead of formulating laws, biologists usually organize their generalizations into a framework of concepts. The claim has been made that laws concepts is only a formal difference, since every concept can be translated into one or several laws. Even if this were formally true [...] such a translation would not be helpful in the actual performance of biological research. Laws lack the flexibility and heuristic usefulness of concepts (Mayr, 1982, p. 43).

Finally, new technologies provide tools like computer modeling which change the perception of what science can be. Scientists' work with simulation procedures leads to a true form of *reification* of analogies and metaphors. Indeed, computer modeling of specific scientific theories and phenomena pushes to think about the objects of scientific enquiry as something highly abstract and de-materialized.

This promotes a new phenomenon of metaphor transfer from computer-based sciences to other disciplines, such as social sciences. It is interesting to underline that the reification of metaphors takes place in the disciplines which receive the metaphors. For instance, the use of complex systems dynamics comes to be a form of reality in social studies.

## 6. CONCLUSIONS

We have tried to prove that production and science are intertwined categories, and that progressive shift in one lead to the changing of the other. As a third element, we added epistemology, showing that the shift from Fordism to post-Fordism let new forms of reasoning emerge. Thus, a virtuous circle is set between these three categories, in which the feedback from one to the others may configure new forms of industrial production, scientific production, and scientific reasoning.

However, we do not intend these relationships as a form of *direct social determinism* of science and epistemology. We are fully aware that science and society are in a complex relation which is not straightforward, but still it can not be ignored that such a correlation exists. So, we suggest a way of relating science and production to an epistemological model which is progressively moving to the centre of the stage in philosophy of science. Further research could then help in clarifying which is the role of those three elements in shaping the others, then helping the setting up of a framework in which further developments of both science and society could be better understood.

## ACKNOWLEDGEMENTS

We would like to thank LA.S.E.R. research network for the very helpful interviews and ideas.

## REFERENCES

Amin, A., ed., 1994, *Post-Fordism: A Reader*, Blackwell, Oxford.

Becattini, G., 1998, *Distretti industriali e made in Italy. Le basi socioculturali del nostro sviluppo economico*, Bollati Boringhieri, Turin.

Beveridge, W.H., 1944, *Full Employment in a Free Society*, Allen & Unwin, London.

Bologna, S. and Fumagalli A., 1997, *Il lavoro autonomo di seconda generazione*, Feltrinelli, Milan.

Cini, M., 1994, *Un paradiso perduto: dall'universo delle leggi naturali al mondo dei processi evolutivi*, Feltrinelli, Milan.

De Solla Price, D., 1963, *Little Science, Big Science*, Columbia University Press, New York.

Edwards, P., 1996, *The Closed World*, MIT Press, Cambridge, MA.

Gagliasso, E., 1996, L'uniformazione e la varianza: una critica delle monoculture disciplinari, in: *Biotecnologie: le frontiere nello sfruttamento della natura*, LA.S.E.R., ed., University "La Sapienza", Rome.

Galison, P., 1997, *Image and Logic: A Material Culture of Microphysics*, Chicago University Press, Chicago.

Huxley, J., 1934, *Scientific Research and Social Needs*, Watts and Co., London.

Kevles, D., 1997, Big science and big politics in the United States: Reflections on the death of the SSC and the life of the human genome project, *Historical Studies in the Physical Sciences* 27:269-297.

Kornberg, A., 1995, *The Golden Helix: Inside the Biotech Ventures*, University Science Books, Sausalito (CA).

LA.S.E.R., forthcoming, *La scienza e il post-fordismo*, DeriveApprodi, Rome.

Latour, B., 1987, *Science in Action: How to Follow Scientists and Engineers through Society*, Open University Press, Milton Keynes.

Mackenzie, D., 1991, The influence of the Los Alamos and Livermore National Laboratories on the development of supercomputing, *Annals of History of Computing* 13:179-201.

Massey, D., 1992, *High-Tech Fantasies: Science Parks in Society, Science and Space*, Routledge, London.

Mayr, E., 1982, *The Growth of Biological Thought. Diversity, Evolution, and Inheritance*, Belknap press at Harvard University Press, Cambridge, MA.

Noble, D., 1987, Command performance: A perspective on the social and economic consequences of military enterprise, in: *Military Enterprise and Technological Change*, Smith, M.R., ed., MIT Press, Cambridge, MA, pp.330-346.

Noble, D., 1984, *Forces of Production. A Social History of Industrial Automation*, Knopf, New York.

Panzieri, R., 1976, Sull'uso capitalistico delle macchine, in: *Lotte operaie nello sviluppo capitalistico*, Panzieri, R., ed., Einaudi, Turin.

Rhodes, R., 1986, *The Making of the Atomic Bomb*, Penguin, London, New York.

Rocco, A., 1997, Il mondo racchiuso in un computer. Intervista con Paulina Boorsock, *Il Manifesto*, February, 26, 1997.

Tagliagambe, S., 1997, *L'epistemologia del confine*, Il Saggiatore, Milano.

Traweek, S., 1988, *Beamtimes and Lifetimes. The World of High Energy Physicists*, Harvard University Press, Cambridge, MA.

Zanini, A. and Fadini, U., eds., 2001, *Lessico Postfordista. Dizionario di idee della mutazione*, Feltrinelli, Milan.

# Modeling Practices and "Tradition"

Elke Kurz-Milcke* and Laura Martignon#
*Department of Psychology, University of Tübingen, Tübingen, Germany, elke.kurz@uni-tuebingen.de; #Max Planck Institute for Human Development, Berlin, Germany, lauram@mpib-berlin.mpg.de

Abstract: The question is posed whether the concept of tradition can enhance our in-depth understanding of scientific modeling and models. We present a progression of uses that the concept has found in the study of science, broadly conceived. From there we take a fresh look at our case study of linear models in judgment and decision making research. We distinguish two traditions in this research area, a ratiomorphic tradition and a paramorphic tradition. We show how the linear regression model was crucial to the inception of these traditions but not to their continuation. Following Galison's example, we also search out connective areas between the two traditions. These "trading zones" can be found in close proximity to computer and (Monte Carlo) simulations, in particular. Finally, a tradition is also an abstraction, which in the end opens the possibility of using our understanding of modeling to enhance our understanding of the concept of tradition.

## 1. INTRODUCTION

The question that we are posing with this chapter is simply this: can the concept of tradition be used to enhance our in-depth understanding of modeling and models in science? Tradition, it would seem, is a poor candidate for fostering insight into the processes and practices of modeling and the use of models, and thereby into a central aspect of modern science (see Giere, 1999, on the prevalence of model-based reasoning). After all, it is a vague category brimming with meaning and value. Indeed, from an everyday un-

*Model-Based Reasoning: Science, Technology, Values*, edited by
L. Magnani and N.J. Nersessian, Kluwer Academic/Plenum Publishers, New York, 2002

derstanding of the concept, tradition connotes food, dance, dress, family, education, and so on. In these contexts the concept seems precariously close to an exceedingly conservative appreciation of historical continuity, and it would seem advisable to stay away from the concept when it comes to the study of science. Some, however have given different advice, among them Karl Popper, Thomas Kuhn, and recently, Peter Galison.

For us, the topic of tradition emerged from our interest in psychologists' use of models in the area of judgment and decision making research. Linear regression models have been widely used in this area and, as we found, this use has followed different traditions. Most interestingly, and we will also elaborate this point, critical voices to the use of regression models in this area have not broken with these traditions but rather continued them on a different note (see our case study below). In still a different context, the concept of tradition caught the interest of the first author in a study on how scientists use differential calculus to represent a physical process (Kurz, 1997; Kurz and Tweney, 1998).

With this chapter we are setting out to analyze in a more systematic fashion how the concept of tradition can contribute to the study of scientific modeling practices. Part of our project has been to examine how others have used the concept in the context of the study of science, broadly conceived. Based on this survey we have taken a fresh look at our case study, the uses of (regression) models in the area of judgment and decision making research. We present aspects of this case study in Section 3: *Ratiomorphic and paramorphic: traditions in regression and beyond.* Many questions have emerged for us from the more refined notion of tradition that we have acquired, and we feel that its value has not yet been fully cashed out.

We present the uses that the concept of tradition has found in the following progression: first, we analyze a pattern in which the concept is frequently used. Next we introduce Karl Popper's argument in favor of tradition, then Kuhn's uses of the concept. We proceed with the ways in which those studying investigative practices in a historical perspective, namely, Kurt Danziger and Peter Galison, have used the concept. The latter has also studied the tradition of instruments. We like to think of this progression as one that spans from text to technology and from the lofty heights of theory to the material basis of science - and modeling.

# 2. FROM TEXT TO TECHNOLOGY

## 2.1 A frequent pattern

Since we have been pricking up when the term tradition is used we have found that the term is used quite frequently in the scientific literature. Most often the term is used as a clue to the author's position with respect to the study of a particular subject. Take, for instance, an essay by the sociologist Norbert Elias (1984) in which he analyzed the instrumental character of time and how a naturalistic conception of time came to eclipse the awareness that time foremost has been a means for orientation in the social world. In this essay Elias (1984, p. 11) argued that "traditional philosophical epistemology" conceptualizes time as a given, something nature is endowed with, either in the objective sense, sensu Newton, or in the subjective sense, sensu Kant. For Elias, a sociologist of knowledge, this has been the traditional way to conceive of the problem of time. His, Elias', understanding of time differed from this "traditional" view. In his essay Elias presented his non-traditional, that is, better way of thinking about time. The pattern is obvious. Very often the term traditional signals or rather emphasizes that something new and distinct is presented. This is a very common pattern (you may think of other such categories, such as "traditional history" or "traditional AI", etc.) We do not wish to criticize this pattern per se. However, what we would like to emphasize is, of course, that there is more to the concept of tradition than this particular literary figure.

## 2.2 Karl Popper

In a talk given in 1948 Popper (1963) developed the idea of *a rational theory of tradition*. Such a theory would need to be sociological, a theory that would deal with the function of tradition and also with the emotional aspects that lead humans to continue the traditions that they encounter in their communities. Such a theory, suggested Popper, would also need to take into account that traditions are usually not the product of willful planning. With respect to the role that tradition plays in the advancement of science Popper argued that the scientist has to build on what others have already achieved and by picking up a certain line of research the entire background of its earlier development is implicated.

But most importantly, science as such has only become possible, according to Popper, with the inception of a tradition; the tradition of critical discussion that he believed originated with the pre-Socratics. This tradition of critical discussion enabled *the* scientific tradition, writ large. Science,

Popper argued, is different from myth only because it was merged with a tradition of second order, the tradition to discuss myth critically. Before that time the only traditions that existed were traditions of first order, those that carry myths from one generation to the next.

## 2.3 Thomas Kuhn I

The concept of tradition was used in prominent ways by Thomas Kuhn, especially in the collection of papers published in 1977 under the title *The Essential Tension: Selected Studies in Scientific Tradition and Change*. We concentrate on two earlier papers that were reprinted in this collection, one that pre-dates the first edition of Kuhn's *Structure of Scientific Revolutions* of 1962 and the second that was first published in 1976 (see Kuhn II, below).

Kuhn's paper "The essential tension", which gave the volume its title, carried as its subtitle "Tradition and innovation in scientific research". There is, according to Kuhn (1977a, p. 227), an essential tension implicit in scientific research that results from the fact that "only investigations that are firmly rooted in the contemporary scientific tradition are likely to break that tradition and give rise to a new one". Innovation, continued his argument Kuhn, 1977a, p. 234), is usually the result of "tradition-bound work". Therefore the successful scientist, or rather the professional group as a whole, "must simultaneously display the characteristics of the traditionalist and of the iconoclast" (Kuhn, 1977a, p. 227).

Kuhn read this paper at a conference on the identification of scientific talent. He introduced himself to the audience, many of them psychologists, by saying that he had nothing much to say about the meeting's central concern of early identification of the creative personality. Rather, he expressed his concern about the image of the scientific process and of the scientist that he found implicit in some of the working papers for the conference. In this image Kuhn saw an over-emphasis on the requirement of flexibility and open-mindedness in the basic scientist, on the one hand, and underestimation of the requirement of a thorough-going commitment to a scientific tradition on the other: All scientists at most times aim to elucidate the scientific tradition in which they were raised rather than change it. Yet, and this was Kuhn's point (1977a, p. 234), "the ultimate effect of this tradition-bound work has invariably been to change the tradition". Therefore, it was only an apparent paradox that the disciplines most noted for the persistent production of novel ideas and techniques have also been relying on a rigid, highly textbook-bound education reminiscent of "an educational initiation" (Kuhn, 1977a, p. 230).

This was Kuhn speaking mostly as philosopher of science, experienced scientist and teacher, positing an implicit and "essential tension" in scientific research. We cannot but wonder how the audience reacted to the "tension" that Kuhn's argument must have induced.

## 2.4 Thomas Kuhn II

Another persona, Kuhn, the historian of science, made good use of the concept of tradition, as well. In a paper titled "Mathematical and experimental traditions in the development of physical science" Kuhn (1977b, p. 32) sought a balance between two "long-traditional and generally noncommunicating historiographic modes", the one treating science "as at most a loose-linked congeries of separate sciences", the other as "a single enterprise". The problem that Kuhn diagnosed with the first mode was that the contemporary subdivisions of subject matter and the contemporary disciplinary organization of the sciences cannot simply be projected into the past but are themselves subject to historical change. The drawback of the second is its general tendency to avoid dealing with, as Kuhn (1977b, p. 33) put it, the "evolving content" of science. The use of the concept of tradition allowed him to consider developments that spanned several centuries without having to assume that the subject matter of the physical sciences had remained unchanged.

Kuhn conceptualized the origins of modern science and the historical development of the physical sciences, in particular, in terms of two traditions: a classical, mathematical tradition and a Baconian, experimental tradition. Kuhn sought to understand the nature of the seventeenth century *Scientific Revolution* as well as the present state of the physical sciences in terms of these traditions. We suggest that it is also and especially his interest in this present state that led him to write a history in terms of traditions.

## 2.5 Studying investigative practices

Recently, the concept of tradition has also been used by those who have undertaken detailed studies of the history of investigative or experimental practices: Kurt Danziger in his book *Constructing the Subject* and Peter Galison in *How Experiments End*. Both accounts are mostly concerned with twentieth century science, Danziger's with psychology, Galison's with microphysics. Both focus on investigative practices in the respective fields, and both give fine-grained accounts of the interplay between theory and practice in the laboratory.

### 2.5.1 Kurt Danziger

During the first half-century of its existence, the research practices of the new discipline, psychology, were still in flux and diverged widely from each other. However, soon "certain traditions of investigative practice" formed in the new field (Danziger, 1990, p. 16). One such tradition, for instance, stemmed from Francis Galton's anthropometric approach and was soon to eclipse other traditions, for instance, the one continuing the research practices found in Wilhelm Wundt's laboratory at Leipzig. These and other traditions could be identified *within* the new discipline. Other traditions relevant to psychological research practices were external to psychology, for instance, "the Prussian traditions about what characterized a good officer" (Danziger, 1990, p. 172) that shaped the investigative practices of German army psychologists. The pattern of influences has also taken the reverse direction, "traditions and conventions of psychological experimentation" (Danziger, 1990, p. 9) were projected outside the field of psychology: "In those societies in which it is practiced, the psychological experiment has become a social institution recognized by most people with a certain level of education".

Social institutions, according to Danziger, have a certain social structure that can be analyzed but also a history that can be traced. Tracing the history of the social structure of psychological experimentation, Danziger (1990, p. 49ff) identified "three models of practice": 1) the Leipzig style of experimentation in which the participants were fellow investigators and in which the roles of experimenter and of subject were interchangeable, 2) the clinical experiment in which these roles where rigidly segregated, essentially extending the medical context into the experimental situation, and 3) Galton's mental testing services, which provided "applicants" (Danziger, 1990, p. 55) with information on their relative performance, linking the performances of individuals in a statistical fashion. These models implied not only particular social structures of experimentation but entailed distinctive knowledge goals - be they about elementary processes in the generalized human mind, or knowledge of pathological states, or of individual differences in performance. And as Danziger (1990, p. 59) has emphasized, "the knowledge goals that these investigators pursued were as much rooted in certain traditions as were the practices they adopted". In addition, these models supported particular modes of handling and reporting data and results - whether individual or group data were reported - as well as differing rhetoric when it came to the identification of "subjects" (Danziger, 1990, p. 88ff).

In many places throughout *Constructing the Subject*, Danziger utilized the concept of tradition. However, the concept appears neither in the headings nor in the subject index of the book. Instead, it seems to us, in key places Danziger opted for the term "model" instead, as in the headings "Alternative models of psychological investigations" and "Three models of practice compared". Model, in the sense of a pattern of practices that is emulated by others in the field, very nearly means tradition. In fact, the uses of the term tradition in other places in the book suggest that in a number of contexts these terms were used interchangeably.

Finally, a "footnote" on the use of tradition in Danziger's book: As mentioned above, searching the index in the back of the book one finds that "tradition" is not among its entries. There are two entries in the index that imply tradition: "Galtonian investigative practice" and "Lewinian investigative practice". The first refers to the above-mentioned anthropometric practice of Francis Galton, the second to the style of experimentation practiced by Kurt Lewin and his research group in Berlin. One may count this relative non-appearance of tradition in the index as an omission or not; we are here just noting that the concept disappears at a particular point from explicit consideration. Our hunch is that this may be a more general pattern in the science studies literature.

So far, in a nutshell, Kurt Danziger's use of tradition in a landmark book on the history of modern psychology.

### 2.5.2 Peter Galison I

Galison (1987, p. 248) has diagnosed two research traditions in XX century microphysics, an "image tradition" and a "'logic' or electronic tradition". The image tradition "embraced the visual detectors, such as cloud chambers and the bubble chamber, which etched onto film the fine details of individual events", whereas the "logic" tradition "formed around electronic detectors". Associated with these traditions were differing beliefs and attitudes towards demonstrations. The first, the image tradition, sought convincing evidence in the study of "golden events", the second, the "logic" tradition, in the analysis of large numbers of events (Galison, 1987, p. 131). The two traditions merged in the early 1980s "when electronic detectors become capable of producing computer-constructed images that were so well resolved that individual events acquire[d] significance" (Galison, 1987, p. 249).

Inspired by Fernand Braudel's history of the Mediterranean civilization, Galison (1987, p. 246f) distinguished three levels of historical constraints, long-term, middle-term, and short-term. In essence, neither of these constraints need be theoretical. Twentieth-century microphysics knew two contrasting styles of experimentation that constituted "long-term constraints" under which experimentalists worked. Galison has referred to these as the "logic" and the image traditions. Thus, Galison (1987, p. 248) argued: "Experimental work does not change radically with each theoretical innovation – very frequently the same people using similar apparatus continue their work across the theoretical divide". There is "a hidden history of physics, one that is not periodized by theoretical divisions". And further he argued that" for a proper long-term history of experimental culture we must descend to the instrumental practices conveyed through apprenticeship from one scientific generation to the next" (Galison, 1987, p. 249). Traditions rather than theoretical divisions appear as the proper framework for a history of experimental culture.

Specific models - whether quantitative, visual, heuristic, or formal - Galison (1987, p. 254) posited, constitute "short-term theoretical constraints". However, in this context Galison (1987, p. 255) also noted that "it is not inevitable that when the over-arching theoretical structures are abandoned, logically lower-level models must be discarded as well". There is, we think, a broader issue implied, namely, that models, especially quantitative models, can acquire a relative independence from the content area in which they originated or, for that matter, from any particular content area. Artificial neural networks, simulated annealing, and genetic algorithms are some of the most prominent data analysis techniques that have turned from "scientific model" to "data analysis model" (Bailer-Jones, this volume). Phenomenon-unspecific computational models of this sort can constitute medium- and long-term constraints on investigative practices. Our case study, below, is concerned with another data analysis technique, regression analysis. Interestingly, its emergence marked the gestation of an entire discipline, statistics, that has sought to deal in phenomenon-unspecific currency.

## 2.6 Peter Galison II

Galison organized his 1997 work *How Experiments End* (Galison 1997, p. xix) around experimental procedures and focussed on how the standards of demonstration shifted over time. For his extensive follow-up account of the development of microphysics, *Image and Logic: a Material Culture of Microphysics*, Galison (1997, p. xix) "inverted the perspective", the experimental demonstrations of particular effects receded and the instruments became central. Under this perspective tradition came to refer to the "handing down", or migration of objects. These migrations, Galison (1997, p. 435) proposed, can take on various forms, for instance, the passing on of laboratory equipment from war work to universities, or the incorporation of instrument parts from one experiment to another. It can also refer to the manifold ways in which computers, televisions, and other technology are combined and re-combined into entirely "new" instruments.

Considering how religious practices involve the handing down of sacral objects, such as, for example, the paten or chalice, Galison (1997, p. 435) argued that

> objects travel clothed in culture and human interactions. Objects (like the paten and chalice) are encumbered, covered with meanings, symbolism,

> power, and the ability to represent but also to preserve specific elements of continuity. Precisely, because things come dressed with meaning, it is essential not to picture the handing down as occurring without alteration. There are no purely neutral exchanges or donations, no "technological transfer" that is isolable from the contexts of origin and destination.

Further, a central argument was that "there is a partial peeling away, an (incomplete) disencumbrance", as Galison (1997, p. 436) said, "of meaning that is associated with the transfer of objects". This argument, we agree, can apply to objects of various kinds, including models.

However, it is not sufficient to think of the trading of objects and instruments only in historical time, from one experiment to the next, from one generation to the next. Trading occurs also, prominently, across various subcultures of physics, and across various professional groups involved in the large-scale research projects characteristic of twentieth century microphysics. To capture trading in this broader sense, Galison (1997, p. 803) introduced the concept of a "trading zone".

"Tradition", used in the plural, *differentiates* patterns of continuity. Galison's "trading zone" points in the opposite direction (Galison, 1997, p. 803): "I intend the term 'trading zone' to be taken seriously, as a social, material, and intellectual mortar binding together the disunified traditions of experimentation, theorizing, and instrument building". Galison's transformation of tradition into "trading zone" can also be read as a spatialization of historical time. And, to extend this analogy, Galison has already proven for the case of microphysics that parallel lines can meet; this alternative "geometry" emerged from the study of the "inner laboratory" (Galison, 1997, p. 4), the microenvironment of the working physicist including material culture and experimental practices.

## 3. RATIOMORPHIC AND PARAMORPHIC: TRADITIONS IN REGRESSION AND BEYOND

In the area of judgment and decision making research the mathematical model of regression analysis has shaped two contrasting traditions. Although regression analysis has been central to the inception of these traditions, their historical continuation was not restricted to the use of this particular model.

### 3.1 Ratiomorphic versus paramorphic traditions

In 1952 the Austrian-American psychologist Egon Brunswik proposed a model that he considered constituting a unifying basis for the disparate research interests in psychology. Figure 1 shows an adaptation of Brunswik's *generalized lens model*. The lens in the middle is built of a number of cues (or possibly habits, psychoanalytic substitution mechanisms, representations, etc.). The lens mediates between an initial state and a state that is achieved after the cues or habits have been in play.

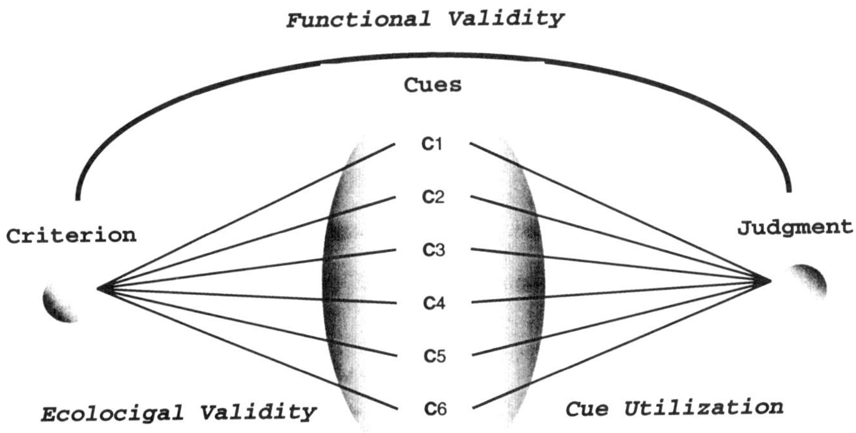


*Figure 1.* An adaptation of Brunswik's *generalized lens model* to judgment.

Brunswik initially developed his lens analogy in connection with his work on perceptual constancies. Take, for instance, size constancy: the size of the projection of an object placed at various distances from an observer varies with distance, nevertheless it retains, within certain limits, its apparent size. In the case of size constancy Brunswik's lens consisted of the cue "retinal size" as well as of a number of cues indicating depth. These cues were thought to function vicariously.

Vicarious functioning in Brunswik's sense meant both sides of the lens model, the side that relates criterion (actual size) and cues (retinal size and depth) and the side that relates cues and judgment (estimated size). The corresponding "rays" on the left-hand side of the "lens" were termed *ecological validities*; those on the right-hand side represented *cue utilization*. In the context of his research on size constancy Brunswik had proposed to measure ecological validity and cue utilization in terms of correlations. The "wide arch", relating criterion and judgment was labeled *functional validity* and also measured in terms of a correlation.

Brunswik's generalized lens model became the touchstone of a research tradition originating with a paper by the psychologist Kenneth Hammond, published in 1955. Hammond, a former student of Brunswik's, adapted the lens model, including its correlation statistical interpretation, to the study of clinical judgment, predicting clinicians' judgments of patients' IQ from Rorschach factors. Hammond's version of the model captured the two aspects of vicarious functioning: ecological validity as the correlations between Rorschach factors and patients' IQ measured by an intelligence test, and cue utilization as the correlations between Rorschach factors and the clinician's judgment of IQ. In addition, Hammond and a doctoral student, F. J. Todd, proposed to merge the generalized lens model with regression analysis. Hammond and Todd used multiple regression on the cues to fit the judgment of the individual clinician; that is, the regression model was applied to the right-hand side (cue utilization) of the lens model. They found that the responses predicted for the clinicians by the regression model correlated quite highly with the clinicians' actual responses. With regression analysis it became possible to model vicarious functioning *and* to predict, in the sense of, assess the models fit to data. In this conjunction we locate the beginning of the Brunswikian tradition in the study of judgment and decision making.

A second tradition in the area of judgment and decision making emerged in 1960 with a paper by Paul Hoffman (see Doherty and Brehmer, 1997 for a review of citations). Hoffman divorced, so to speak, the use of the linear regression model from the lens model and thereby from its relation with Brunswik's psychology. Hoffman's (1960, p. 119) aim was to provide a different "rationale" for the use of the linear regression model in modeling judgment. For this purpose he coined the term "paramorphic representation" (Hoffman, 1960, p. 124).

In chemistry *paramorph* is used to refer to different crystalline structures. For, instance, $CaCO_3$, may represent two crystals, calcite or aragonite. In mineralogy, calcite is commonly described as a paramorph of aragonite. To Hoffman (1960, p. 125) "the mathematical representation of the judgment process (was) a level of description that approach[ed] the chemical description of minerals". The mathematical model captured some but not all properties of the mental process and the "formula" allowed one to make predictions concerning the outcomes of certain tests. "Paramorphic representation" then referred to the following "state of affairs" (Hoffman, 1960, p. 125):

> [A]s with chemical analysis, the mathematical description of judgment is inevitably incomplete, for there are other properties of judgment still undescribed, and it is not known how completely or how accurately the underlying process has been represented.

Hoffman sought to describe the mental process in clinical judgment in terms of "a functional relationship". To that effect, the mental process was conceived of as "a set of intervening variables" (Hoffman, 1960, p. 117), intervening between observables. Thus, Hoffman was looking for a prediction mechanism, equating modeling of the mental process with data fitting. He considered this modeling approach to be the best alternative to "understanding the mental process qua subjective experience" or physiological techniques. In his 1960 paper he even showed that more complex models, *configurational models* derived from protocol data of clinicians, were not able to predict the clinicians judgments better than did the linear regression model.

Back for a moment, to the other tradition and Hammond's paper of 1955 in which a similar conclusion had been reached (p. 261):

> The responses predicted for the clinicians correlated quite highly with their actual responses. [...] Evidently the multiple correlation model which predicts that the clinician combines data from the Rorschach in a linear, additive fashion is a good one - it predicts quite successfully in comparison with most psychological efforts.

Later, others standing in the ratiomorphic tradition, for instance, Brehmer (see Doherty and Brehmer, 1997, p. 546) pointed out that he and others "do not believe that the judgment process works according to the principles of multiple regression analysis"; instead regression analysis and other methods such as the analysis of variance were used to "test a series of hypotheses about the nature of the judgment process". In either case, in the ratiomorphic tradition the study of cognitive processes was grafted onto the lens model, which in its original formulation had been a schema to capture achievement (see also Hammond, 1996). Thus, the ratiomorphic tradition worked with two models the "lens" and the mathematical model; the paramorphic tradition, by contrast, knew only one, the latter.

We derived the label "paramorphic tradition" from Hoffman's (1960) proposal of "The paramorphic representation of clinical judgment". The label "ratiomorphic tradition" goes back to Brunswik's neologism "ratiomorphism" and the corresponding "Ratiomorphic explication of the cognitive process" (Brunswik, 1964, p. 23). Brunswik coined the term ratiomorphism at about the same time that he introduced the famous metaphor of the perceptual system as "intuitive statistician" (Brunswik, 1956). Both concepts, the intuitive statistician and ratiomorphism, relate to the central point of Brunswik's psychology, namely, that perception is inference from uncertain cues. In this context, ratiomorphism referred to his conviction that reasoning as well as perception should be described in terms of "lenslike patterns" (Brunswik, 1964, p. 22).

Since the early 1960s there have existed two traditions in judgment and decision making research that have been dominated by the regression model. The ratiomorphic tradition has also been referred to as the tradition of "Social judgment theory" (see Doherty and Evans, 1996) and those working within this tradition even formed their own scientific society (*The Brunswik Society*). In both traditions, the linear model was legitimated by the goodness of its fit. However, when Hoffman in 1960 (p. 123) "recognized that two or more models may be capable of accounting for judgment variance with equal efficiency" he had already put his finger on a weak spot in this modeling practice.

## 3.2 Traditions continued

A major challenge to regression models of judgment was put forth by Dawes and Corrigan in 1974. Starting out in the paramorphic tradition these authors pointed out that unit weighting, that is, employing unit weights with appropriate signs, can lead to models that correlate highly with optimal linear models and, thus, with linear regression models. Even models with random weights were shown to correlate as highly with optimal linear models as did models fitted to actual judgment. The success of unit weighting, in particular, led Dawes and Corrigan (1974, p. 105) to the conclusion that "the whole trick is to decide what variables to look at and then to know how to add". In the late 1970s Dawes (1979, p. 571) even proclaimed the "Robust beauty of improper linear models in decision making" where "improper" meant especially unit weighting, and generally weighting by "some non-optimal method;" "robustness" was the property of a model to extrapolate well from a set of known data to a set of yet unseen data (i.e., cross-validation).

In the paramorphic tradition data fitting had been synonymous to modeling of the mental process. Consequently, everything in this tradition revolved around fit of the regression model to data. Being part of this movement, Dawes and his collaborators recognized two drawbacks in the tradition's emphasis on the regression model: first, the good fit of regression models to judgment data could be achieved by simpler models. Second, generalization (i.e., robustness) had been a blind spot in the assessment of predictive accuracy that had been largely restricted to the assessment of fit to known data. This is a case in which, in the Kuhnian (1977, p. 234) sense, "tradition-bound work" led to a significant change in that same tradition: Everything continued to revolve around fit, now of non-optimizing models; and the central concept, fit, expanded its meaning to cover both fitting the known and predicting the unknown.

A comparable impulse to that of Dawes and collaborators in the paramorphic tradition came to the ratiomorphic tradition through the work of Gigerenzer and his students in the 1990s. In 1991 Gigerenzer, Hoffrage and Kleinbölting published a paper titled "Probabilistic mental models: A Brunswikian theory of confidence" in which these authors defined Brunswik's concept of ecological validity in terms of frequencies instead of correlations. This alternative treatment of cue validities allowed them to make the so-called *over-confidence phenomenon* disappear. Further research along those lines led to a new outlook on the lens model, left and right (Gigerenzer and Goldstein, 1996): on the left-hand side ecological validity acquired a frequentistic interpretation; on the right-hand side, simple inference strategies replaced regression. These simple strategies, or *heuristics*, worked on the level of individual judgments. So far, modeling of judgment, in both traditions, had worked on the level of the individual clinician.

Gigerenzer and his coworkers defined inference heuristics in terms of building blocks, namely, rules for information search, for stopping that search, and for making a decision. Vicarious functioning was also reconsidered as being based on cue hierarchies (Gigerenzer and Goldstein, 1996; Gigerenzer and Kurz, in press). This led to the proposal that individual judgments could be based on just one cue (one-reason decision making). The performance of the simple heuristics in terms of their predictive accuracy held up surprisingly well against that of optimizing benchmarks such as regression analysis and Bayesian networks. The subsequent analyses of the ways in which the structure of the (known) information in the environment matches that of the studied heuristics answered Brunswik's (1956) call for an ecological orientation of psychological research in a new fashion (Martignon and Hoffrage, in Gigerenzer et al., 1999).

The differences between the two traditions can also be fixed to their respective attitudes towards the psychological experiment. The paramorphic tradition followed the classical experimental scheme of prediction and control, involving the manipulation of independent variables and recording their effects on the dependent variable(s) (see the Introduction of Woodworth, 1938, for a prominent explication of this experimental scheme). This scheme matched the mathematical model *qua* equation in that it also related dependent and independent variables. The mental process remained an intervening variable between observables. The ratiomorphic tradition, by contrast, was wed to Brunswik's critique of that experimental scheme (see Doherty and Kurz, 1996; Gigerenzer et al.. 1991 for an instance of Brunswik's representative design).

Brunswik (1956) was convinced that the environment could not be segregated in orthogonal variables, even for the purpose of the psychological experiment. This break with the dominant tradition of psychological experimentation held deep implications for research design, including its social structure (see Danziger, 1990, for an analysis of the social structure of experimentation). The alternative experimental design that Brunswik proposed as the *representative design* of the psychological experiment was compatible with a different analysis of environments, an analysis that worked in terms of ecological validities of one kind or another.

## 3.3 A fresh look: trading

Once we have distinguished the paramorphic and the ratiomorphic tradition we are in a good position to start looking at their "trading zones" (Galison, 1997, p. 781). In fact, as Tweney (in press), has argued, the rapid rise of statistical "instruments" in psychology after WWII marked itself a trading zone which was created by the requirements of war research. What we have to offer in this section are the beginnings of an analysis along the lines of "trading", that is, of connective areas between the traditions.

### 3.3.1 Trading zone: computer simulation and model selection

Researchers working within the ratiomorphic and the paramorphic traditions recognized each other's work by habitual cross-referencing. At times the references to the work belonging to the *other* tradition seemed to be not much more than a general acknowledgement of the others' existence. Yet, it came to trading as well. Next we examine a trading zone, namely, the use of computer simulation including Monte Carlo simulations, as used in the later development of the two traditions.

In the paramorphic tradition model selection (regression models vs. models with random weight or unit weights) had been treated computationally in the work of Dawes and collaborators. Tackling this issue became possible thanks to the advent of the computer, which enabled psychologists to perform simulations. With the use of simulations "experiment" acquired a new meaning, as it had in other sciences (Galison, 1996, p. 735); it could be performed in the laboratory as well as on the computer. This expansion of experimentation to new means became a commonality between the two traditions. In fact, as Tweney (in press), the use of statistical instruments has argued

In the ratiomorphic tradition, model selection was systematically pursued in the work of Gigerenzer and Goldstein (1996) and in subsequent work on simple strategies for inference (Gigerenzer, Todd, and the ABC Research Group, 1999). The models that Dawes and coworkers had analyzed in the 1970s became important competitors in the "model selection races" that this research group performed; for instance, the linear strategy employing unit weights entered the races under the label *Dawes's rule*. However, it was also the case that Dawes's rule was in its psychological-ecological implications foreign to the ratiomorphic tradition, which emphasized hierarchies of ecological validities (Brunswik, 1956, p. 50). Therefore, the prototype of a model, sensu simple heuristic, was one that went under the name *Take The Best* and constituted a case of one-reason decision making by relying on the cue with the highest ecological validity. In part then, the model selection races performed by Gigerenzer and his coworkers also constituted a competition between the two traditions.

As Galison (1997; see quote above) has emphasized, there are no neutral exchanges that are isolable from the contexts of origin and destination. When computational simulations became central in the ratiomorphic tradition they sparked new kinds of ecological analyses that ended model selection in a Solomonic sense, assigning, depending on the structure of environment, a time for "cue hierarchies" and a time for "unit weights" (Martignon and Hoffrage, in Gigerenzer et al. 1999).

### 3.3.2 Trading continued: the experimental study of strategy selection

In 1988 Payne, Bettman, and Johnson specified and investigated a number of decision strategies, relying on evidence generated by (Monte Carlo) simulations as well as in the psychological laboratory. Their work shared and continued the paramorphic tradition's emphasis on model comparison and selection. What was more, Payne, Bettman and Johnson (1988) researched the corresponding and knotty issue of strategy selection in humans.

These authors expressed decision strategies in terms of production rules, that is, in terms of "condition $\rightarrow$ action pairs" (Payne, Bettman, and Johnson, 1988, p. 535). The strategy corresponding (in terms of computational outcome) to a linear regression model was designated as "weighted additive" strategy or *WADD rule* (Payne, Bettman, and Johnson, 1988, p. 536):

> The strategy considers the values of each alternative on all of the relevant attributes (outcomes) and all of the relevant importances (weights or probabilities) of the different attributes (outcomes) to the decision maker. The rule develops a weighted value for each attribute by multiplying the weight (probability) by the value and sums over all attributes to arrive at an overall evaluation of an alternative.

The WADD rule was the "most information intensive strategy" that these authors examined. Other strategies relied on less information. For instance, the *equal weight rule* which corresponds to the unit weight model of Dawes and Corrigan, examined all alternatives on all attributes for each alternative but "ignore[d] information on relative importance (probability) of each attribute", or the *lexicographic rule* considered only one attribute across all alternatives (Payne, Bettman, and Johnson, 1988, p. 536). This way of specifying choice heuristics allowed the authors to investigate potential trade-offs between accuracy and effort for the different strategies in particular contexts (e.g., degrees of dispersion in probabilities). The results of their simulations were used to generate hypotheses about adaptive strategy choice in humans, especially also under conditions of time pressure.

In fact, the WADD rule as specified by Payne, Bettman, and Johnson (1988, p. 536) was a "version of expected value maximization" which constitutes still another way (besides those advanced by the ratiomorphic and paramorphic traditions) of looking at linear regression analysis *qua* linear decision rule. Payne et al.'s approach amalgamated aspects stemming from utility theory (e.g., cost-benefit analyses) with aspects stemming from information processing accounts of cognition developed by Alan Newell and Herbert Simon.

Work continuing along the lines of the ratiomorphic tradition approached the same question, namely, what kind of decision strategies humans use under particular conditions, such as limited time and knowledge (Rieskamp and Hoffrage, in Gigerenzer et al., 1999). To monitor information search of the participants in their studies, Rieskamp und Hoffrage relied on a software system developed by Johnson, Payne, Schkade, and Bettman. As was the case with Payne et al.'s approach, the approach in terms of the "simple heuristics" (Gigerenzer at al., 1999) was an amalgamation, in this case of an information processing approach, especially informed by Herbert Simon's notion of bounded rationality, and of a particular continuation of the Brunswikian tradition, emphasizing ecological analysis (in distinction to Payne et al.'s cost-benefit analyses).

No final word is attempted on the recent history of research in the area of judgment and decision making. However, it seems that the two traditions that were sparked by the use of regression analysis have met over computer simulations (including Monte Carlo simulations) and the study of human strategy choice. The differences that have remained between approaches, for instance, concerning the question what "adaptation" means in the context of strategy choice, are still related to the differences between the two traditions and the particular amalgamations that have occurred in the meantime.

## 4. CONCLUSION

Our aim in this chapter has been to explore and reflect upon the use of the concept of tradition for the study of modeling. Although the concept has had its advocates, it has continued to have a shadowy existence. Possibly by being overshadowed by more luminous concepts, such as *paradigm*, it has not been given the attention that we think it deserves in the study of science, including the study of modeling practices. With our case study we have benefited much from Galison's example. We have distinguished between traditions but we have also started to search out the respective "trading zones". Interestingly, we found that coordination between the traditions has occurred in close proximity to the computer, the site of a *versatile lingua franca* that goes beyond "variables" (see Danziger, 1997) and thus beyond the idea of "model equals equation".

In our case study we have focussed on traditions in fairly recent developments in psychology. However, we see these traditions embedded in a larger historical context, especially the history of probability and statistics. Most notably, the terminology that is often used in connection with regression models is informed by the mechanistic metaphor of summing and weighting. For instance, the regression coefficients are often referred to as "weights". Optimization in terms of least squares analysis was originally (in Legendre's 1805 exposition of the method) understood in this same mechanistic sense, as a balancing of errors. Much of the history of regression analysis in the area of judgment and decision making research is related to the question of what these "weights", now in a probabilistic sense, are supposed to mean. However, we find that this story can only be told by also considering traditions in probability theory and statistics.

Finally, a tradition is always also an abstraction. We can think of this abstraction in terms of Galison's (1997, p. 436) "partial peeling away", or "(incomplete) disencumbrance of meaning that is associated with the transfer of objects", and also as a *generic abstraction* that represents features common to a class of phenomena (see Nersessian, 1999). Not unlike abstraction in the context of modeling, the (historical) construction of a tradition goes beyond and falls short of the given. At the end of this chapter, an inversion of perspective suggests itself: from using "tradition" to enhance our understanding of modeling to using our understanding of modeling to understand "tradition".

## ACKNOWLEDGEMENT

This chapter was written while the first author was funded by a Schloessmann Fellowship of The Max Planck Society. The first author appreciates the conversations on various aspects of this paper with Hanne Andersen, David Gooding, Kenneth Knoespel, and Ryan Tweney. We thank Anita Todd and Christoph Wassner for their assistance with preparing the manuscript for this chapter.

## REFERENCES

Brunswik, E., 1952, The conceptual framework of psychology, in: *International Encyclopedia of Unified Science: Vol. 1, No. 10*, O. Neurath, R. Carnap, and C. Morris, eds., University of Chicago.

Brunswik, E., 1956, *Perception and the Representative Design of Psychological Experiments*, University of California Press, Berkeley, CA.

Brunswik, E., 1964, Scope and aspects of the cognitive problem, in: *Contemporary Approaches to Cognition: A Symposium Held at the University of Colorado*, H. Gruber, R. Jessor, and K.H. Hammond, Harvard University Press, Cambridge, pp. 5-31.

Danziger, K., 1990, *Constructing the Subject: Historical Origins of Psychological Research*, Cambridge University Press, Cambridge.

Danziger, K., 1997, *Naming the Mind: How Psychology Found its Language*, Sage, London.

Dawes, R. M., 1979, The robust beauty of improper linear models in decision making, *American Psychologist* 34:571-582.

Dawes, R. M. and Corrigan, B., 1974, Linear models in decision making, *Psychological Bulletin* 81:95-106.

Doherty, M.E. and Brehmer, B., 1997, The paramorphic representation of clinical judgment: A thirty-year retrospective, in: *Research on Judgment and Decision Making: Currents, Connections and Controversies*, W.M. Goldstein and R.M. Hogarth, eds., Cambridge University Press, Cambridge, pp. 538-551.

Doherty, M.E. and Evans, J.St.B.T., 1996, Editorial, *Thinking and Reasoning* 2:105-107.

Doherty, M.E. and Kurz, E.M., 1996, Social Judgment Theory, *Thinking and Reasoning* 2:109-140.

Elias, N., 1984, *Über die Zeit* [An Essay on Time], Suhrkamp, Frankfurt.

Galison, P., 1987, *How Experiments End*, The University of Chicago Press, Chicago.

Galison, P., 1997, *Image and Logic: A Material Culture of Microphysics*, The University of Chicago Press, Chicago.

Giere, R. N., 1999, Using models to represent reality, in: *Model-Based Reasoning in Scientific Discovery*, L. Magnani, N.J. Nersessian and P. Thagard, eds., Kluwer Academic/Plenum Publishers, New York, pp. 41-57.

Gigerenzer, G. and Goldstein, D.G., 1996, Reasoning the fast and frugal way: models of bounded rationality, *Psychological Review*, 103:650-669.

Gigerenzer, G., Hoffrage, U., and Kleinbölting, H., 1991, Probabilistic mental models: A Brunswikian theory of confidence, *Psychological Review* 98:506-528.

Gigerenzer, G. and Kurz, E.M., in press, Vicarious functioning reconsidered: A fast and frugal lens model, in: *The Essential Brunswik: Beginnings, Explications, Applications*, K.R. Hammond and T.R. Stewart, eds., Oxford University Press, New York.

Gigerenzer, G., Todd, P., and the ABC Research Group, 1999, *Simple Heuristics that Make Us Smart*, Oxford University Press, New York.

Hammond, K.R., 1955, Probabilistic functioning and the clinical method, *Psychological Review* 62:255-262.

Hammond, K.R., 1996, Upon reflection, *Thinking and Reasoning* 2:239-248.

Hoffman, P.J., 1960, The paramorphic representation of clinical judgment, *Psychological Bulletin*, 57:116-131.

Kuhn, T.S., 1977a, The essential tension: tradition and innovation in scientific research, in: *The Essential Tension: Selected Studies in Scientific Tradition and Change*, The University of Chicago Press, Chicago, pp. 225-239.

Kuhn, T. S., 1977b, Mathematical and experimental traditions in the development of physical science, in: *The Essential Tension: Selected Studies in Scientific Tradition and Change*, The University of Chicago Press, Chicago, pp. 225-239.

Kurz, E.M. and Tweney, R.D., 1998, The practice of mathematics and science: From calculus to the clothesline problem, in: *Rational models of cognition*, M. Oaksford and N. Chater, eds., Oxford University Press, New York, pp. 415-438.

Nersessian, N.J., 1999, Model-based reasoning in conceptual change, in: *Model-Based Reasoning in Scientific Discovery*, L. Magnani, N.J. Nersessian, and P. Thagard, eds., Kluwer Academic/Plenum Publishers, New York, pp. 5-22.

Payne, J.W., Bettman, J.R., and Johnson, E.J., 1988, Adaptive strategy selection in decision making, *Journal of Experimental Psychology: Learning, Memory and Cognition* 14:534-552.

Popper, K.R., 1963, *Conjectures and Refutations: The Growth of Scientific Knowledge*, Routledge and Kegan Paul, London.

Rieskamp, J. and Hoffrage, U., 1999, When do people use simple heuristics, and how can we tell?, in G. Gigerenzer, P. Todd, and the ABC Research Group, pp. 141-167.

Tweney, R.D., in press, What happened to the brass and glass? The rise of statistical "instruments" in psychology, 1900-1950, in: *Festschrift for John Poppelstone and Marion McPherson White.*

Woodworth, R., 1938, *Experimental Psychology*, Holt, New York.

# Modeling Data: Analogies in Neural Networks, Simulated Annealing and Genetic Algorithms

Daniela M. Bailer-Jones* and Coryn A.L. Bailer-Jones#
*Department of Philosophy, University of Bonn, Bonn, Germany, bailer-jones@t-online.de;
#Max-Planck-Institute for Astronomy, Heidelberg, Germany, calj@mpia-hd.mpg.de

Abstract: This paper examines several analogies employed in computational data analysis techniques: the analogy to the brain for artificial neural networks, the analogy to statistical mechanics for simulated annealing and the analogy to evolution for genetic algorithms. After exploring these analogies, we compare them to analogies in scientific models and highlight that scientific models address specific empirical phenomena, whereas data analysis models are application-neutral: they can be used whenever a set of data meets certain formal requirements, regardless of what phenomenon these data pertain to. Through the analogy, computational data analysis techniques inherit a conceptual idea from which the principle of the technique is developed. In all cases of computational data analysis techniques, the analogies used - and the metaphors generated by them - help us to understand the technique by providing a more concrete framework for understanding what is otherwise an abstract method. In the different examples, however, the significance of the analogies varies. Analogy can, though need not, be indispensable for a technique.

## 1. THE TOPIC

Doing science is about studying empirical phenomena and hence involves collecting data about these phenomena. Today this often involves assembling very large amounts of data, made possible through the automation of technical processes and the availability of computational support. Large

*Model-Based Reasoning: Science, Technology, Values*, edited by
L. Magnani and N.J. Nersessian, Kluwer Academic/Plenum Publishers, New York, 2002

amounts of data are in principle beneficial for the purpose of testing empirical claims. Large amounts of data can, however, present a major difficulty for analysis. The larger a data set gets the bigger the problem of how to interpret it meaningfully and lucidly. This is especially the case when the data are multidimensional, i.e. represented by many different variables. Once the data exist in more than three dimensions (i.e. variables), it is no longer possible simply to draw a plot and inspect it visually. On the other hand, once a pattern is found in the data and features of a phenomenon are discerned, then this finding is all the more significant when extracted from a large set of data. The increase in our ability to collect large amounts of data has a fundamental impact on the way science is done, and so considerable effort is invested in efficient methods for coping with such data.

To this end, various computational data analysis techniques have been developed over the last twenty years or so. Such techniques can be assembled into models of how to solve a certain kind of problem associated with a set of data. We call these *data analysis models.* For instance, artificial neural networks can provide a model with which to solve pattern recognition problems. Data analysis problems can arise from the study of nature, but need not (e.g. the travelling salesman problem). Examining data analysis models, we want to determine relationships to other types of models, especially *theoretical scientific models.* As in data analysis models, these primarily mathematical tools are used to deal with data that have been extracted from empirical phenomena. (Other types of scientific models, e.g. of spatial configurations, exist, but are not of immediate interest in our study.) Moreover, as in scientific models where the use of analogies is well-established, analogies are also encountered in computational data analysis techniques, which makes a comparison yet more desirable.

The point of departure for our study is our curiosity about the use of analogies in the context of computational data analysis techniques. The vocabulary used to describe these techniques is full of evocative metaphors, and some of the analogies employed have been instrumental in developing the techniques.

- *Artificial neural networks* exploit an analogy to the human brain. The idea behind artificial neural networks was to transfer the idea of parallel distributed processing, as found in the brain, to the computer in order to take advantage of the processing features of the brain.
- *Simulated annealing* is a method of optimization, for example of determining the best fit parameters of a model based on some data. The physical process of annealing is one in which a material is heated to a high temperature and then slowly cooled. Annealing provides a framework in

which to avoid local minima of energy states in order to reach the global minimum.
- *Genetic algorithms*, another optimization technique, employ operations that mimic natural evolution to search for the fittest combination of "genes", i.e. the optimal solution to a problem.

In Section 2, we describe these techniques in some more detail and spell out the analogies underlying them. This enables us to assess how deep these analogies go and how important they are conceptually for establishing the technique. The metaphors derived from such analogies could just be fancy talk, an attempt to advertise the method in question and to make it sound "sexy". Yet, we shall illustrate that there is more behind the analogies (and metaphors) typical of computational data analysis techniques.

In Section 3, we consider data analysis models in the wider context of modeling in science. We emphasize in particular that computational data analysis techniques can serve to analyze data from an enormous range of contexts or phenomena - which is why we call them *application-neutral*. This is in contrast to more conventional theoretical scientific models where the models employ theoretical concepts and physical laws that are deemed relevant for the specific physical problem in question. As we demonstrate, this fundamental difference between data analysis models and theoretical scientific models has repercussions for the role of analogies in either case.

Section 4 highlights the critical role of analogies in data analysis techniques, especially in comparison to the well-studied use of analogies in theoretical scientific models.

## 2. THE DATA ANALYSIS TECHNIQUES AND THEIR ANALOGIES

Data analysis techniques comprise a whole range of tools for solving data-related problems. One such problem which is relevant to the techniques we discuss is the search for the best fit of data to some model, a so-called optimization problem. To take a simple example, we may have some data on the variation of the temperature in the Earth's atmosphere as a function of altitude and a model for this variation which has one free parameter. The process of determining the best value for this parameter from the data can be cast as an optimization problem in which we want to minimize the error (i.e. the difference between the model's temperature predictions and the measured values) as a function of the model parameter. A schematic example of this is shown in Figure 1, where, for one parameter, the "error landscape" only stretches in one direction.

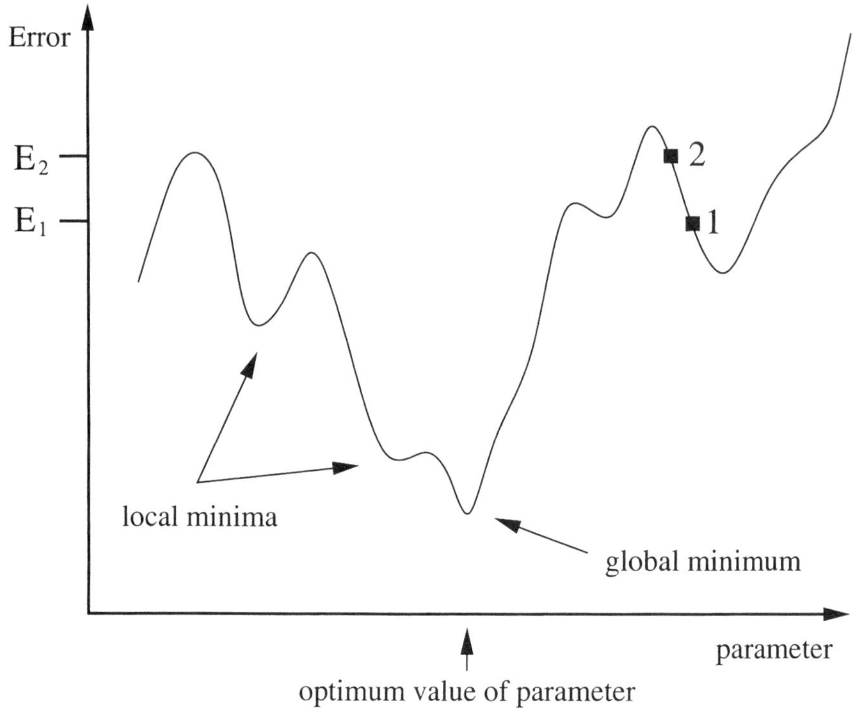


*Figure 1.* The error of a model's predictions as a function of its one free parameter. The error is determined from empirical data about the phenomenon that is being modeled.

The global minimum of the error is the smallest value of the error, and so corresponds to the optimum value of the parameter. In all but the simplest problems this optimum value cannot be found analytically. Thus we must use some kind of search strategy for locating it. In this single parameter example we could probably get very close to it simply by systematically evaluating the error at a large number of values of the parameter. However, many computational data analysis problems involve significantly more parameters, possibly many thousands. The optimization problem then occurs in a much higher dimensional space and a systematic search of all parameter permutations is out of the question simply on the grounds of computer speed.

A class of search methods exist which evaluate the gradient of the error and use this in an iterative fashion to move in the "downhill" direction of the error landscape. However, the danger of this is that the search may become stuck in local minima, which could still correspond to a large error (see Figure 1). To overcome this problem, a number of non-gradient based techniques have been developed, which frequently involve an element of random movement on the error surface. Two examples of such methods which we shall discuss are simulated annealing and genetic algorithms. Optimization techniques such as these may, for instance, be used to train an artificial neural network.

## 2.1 Artificial neural networks

### 2.1.1 Artificial neural network models

There exist many varieties of artificial neural networks (see, for example, Hertz, Krogh, and Palmer, 1991). Here we focus on just one type, albeit one of the most widely used for data modeling. This is the supervised feed-forward neural network, which gives a functional data mapping between two data domains. To illustrate this model, consider the example problem of the classification of stellar spectra, whereby one wants to determine the underlying physical conditions (temperature, gravity and composition) of a star from its optical spectrum (Bailer-Jones, 2000). The neural network is used to give a mapping between the spectrum at its input and the physical conditions at its output (see Figure 2).

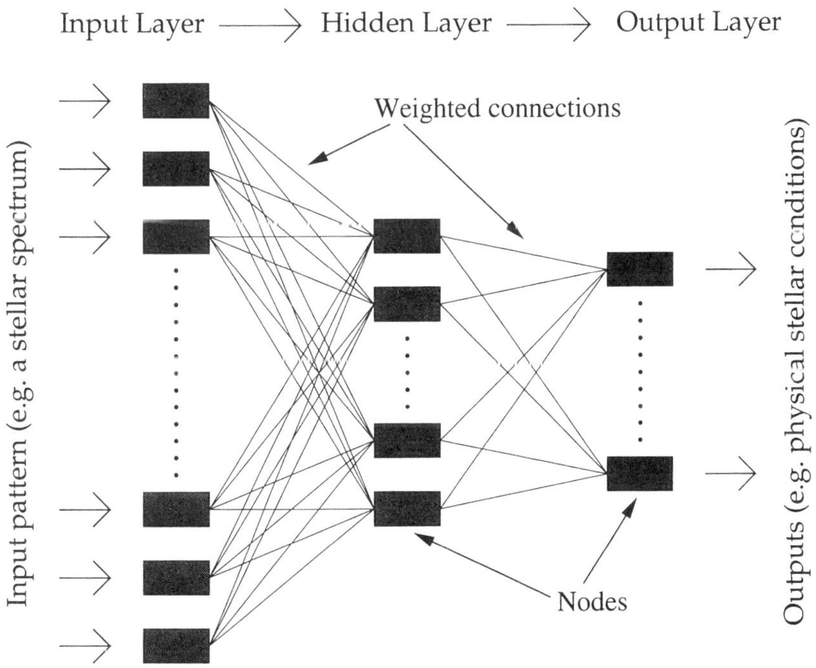


*Figure 2.* Structure of a feed-forward artificial neural network.

Between the input and output layer, there are one or more layers of hidden nodes. These nodes combine and nonlinearly process the data passed to them from the previous layer and then pass that on to the nodes in the next layer. Each connection between the nodes has a weight associated with it which modifies the value of the data passing along that connection; these weights form the free parameters of the neural network model. For the network to give the correct mapping between the spectra and their physical parameters, these weights must be set to appropriate values.

For many high-dimensional or complex problems it is difficult to quantify the relationship accurately between the two data domains solely from physical principles, so these weights cannot be assigned according to physical principles. Instead, the weights must be inferred by "training" the network on a set of pre-classified spectra, i.e. ones for which the ideal outputs (so-called targets) are already known. In this way the network "learns" the relationship from the training data. Training typically proceeds by adjusting the weights until the discrepancy between the actual network outputs and the targets is minimized; this is just a (nonlinear) multidimensional optimization problem. The network is not explicitly given any rules about which features of the spectrum are relevant for classification; it acquires this information itself from the training data. Essentially the only assumption which the network makes about the problem is that the outputs are some smooth function of the inputs.

Note that, despite any impression to the contrary that may arise from pictures such as Figure 2, the neural network is just an *algorithm*, i.e. it consists solely of equations that are used for training and applying the network. The network structure shown in Figure 2 does not really have an existence other than being a convenient way of visualizing how the network operates.

### 2.1.2 The neural network analogy

The neural network was described above in purely mathematical terms, as a nonlinear, multidimensional parametrization model. However, neural networks are often described by analogy to the brain, hence their name. The brain excels at tasks that require the processing of a multitude of information at the same time (colors, shapes, sounds) in highly variable environments, and the brain copes with such tasks by means of many processing elements that work in parallel. The idea behind developing artificial neural networks was to transfer the idea of parallel processing to the computer in order to take advantage of some of the brain's features.

The brain consists of large numbers of neurons connected to each other by synapses. The output from the neuron is a function of its inputs from many other neurons, which are "weighted" at the receiving synapses. This output is a nonlinear function of its input and the strength of the connection in the synapses can be modified by activity; in other words, the brain (the collection of neurons) learns (changes its synaptic weights) from experience. It is this behavior which an artificial neural network attempts to model algorithmically, albeit in a simplified fashion. The assumption that learning occurs in the brain when modifications are made to the effective coupling between one cell and another at a synaptic junction is simulated mathematically in artificial systems through positive or negative reinforcement of connections. This forms the basis of the analogy exploited in artificial neural networks. However, it is clear that brains are considerably more complex than artificial neural networks, just on the basis that the human brain contains of order $10^{11}$ neurons, whereas artificial neural networks are in practice restricted to a few thousand nodes. This results in an enormous difference in the degree of complexity and computational power available. Moreover, the nodes in artificial neural networks are highly simplified in comparison to the processes at work in the neurons in the brain.

In terms of the vocabulary used in the neural network analogy, there are few terms which are carried over from the biological neural network. The terms "nodes" and "connections" of course have their parallels in a biological network ("neurons" and "synapses" respectively), but it is significant that new terms have been developed for the artificial case, implying a degree of independence. The one word which is retained is "network", although this is of course a term not restricted to the brain.

Artificial neural networks were originally inspired by attempts to provide simple models of brain function and learning (McCulloch and Pitts, 1943). Although this is still an active line of research, a significant and largely separate branch of neural computation has grown out of this which is mainly concerned with taking advantage of distributed processing for the purpose of modelling multidimensional data sets. While the analogy to the brain may have been historically relevant for the conception of such artificial neural networks, it is questionable now how much the analogy really helps to comprehend the technique in a data modeling context. Similarly, although parallel distributed processing is a central idea, often the computers on which neural networks are implemented are not parallel processors. The differences between the brain and artificial neural networks are so considerable that it is probably easier to learn directly about the specific nature of this data analysis technique, rather than to take the detour via the brain. Any analogy carries disanalogies with it. This is not a problem as long as the users of the analogy are well aware of how far the analogy stretches, i.e. regarding to which issues it can be applied with benefit. On the other hand, there is always the danger of being misled by disanalogies. In the case of the type of artificial neural network we discuss, the usefulness of the analogy is doubtful considering current uses of the technique. Thus the neural network metaphor bears the risk of misleading people into thinking that artificial neural networks are *like* the brain.

## 2.2 Simulated annealing

### 2.2.1 The technique of simulated annealing

Simulated annealing (Kirkpatrick, Gelatt Jr., and Vecchi, 1983) is a method for optimization based on a controlled random walk on the *error surface* (the multidimensional generalization of the error curve shown in Figure 1). Starting at some random point (1) on this surface, the error, $E_1$, is evaluated from the model and data. A nearby point (2) is chosen at random and the error, $E_2$, evaluated (see Figure 1). If the new point has a lower error, the search moves there and the process is repeated. However, if it has a higher error (as shown), there is still a chance of moving there. The probability for this is chosen to be $p = \exp(E_1 - E_2)/kT$, as the analogy to statistical mechanics suggests (Metropolis et al., 1953). This probability ranges between 1 and 0 for very small and very large error differences respectively. In other words, "uphill" moves are permitted, albeit with decreasing probability for larger differences. This has the effect of managing to "escape" local minima, and hence permits a more comprehensive search of the parameter space. The quantity $kT$ in the equation determines exactly how probable an uphill move of a certain size is: a large $kT$ makes comparatively large uphill moves more likely. The idea behind simulated annealing is to reduce $T$ slowly (for a fixed $k$) as the search proceeds. This initially permits a large region to be searched. As time proceeds (and $T$ is reduced), large uphill moves become increasingly prohibited, thereby focusing attention on finding what is hopefully the global minimum of the parameter space.

### 2.2.2 The analogy of annealing

Simulated annealing was developed in direct analogy to physical annealing, in which a material is heated to a high temperature and melted, and then slowly cooled to encourage it to solidify into its lowest energy state. The lowest energy state corresponds to a single crystal. When in the liquid form, the molecules are constantly moving around and so are able to move into different configurations, or *microstates*, which correspond to different energies of the material. If the temperature is reduced sufficiently slowly the material will visit a very large number of microstates, corresponding to thermal equilibrium. At a lower temperature, the microstates of thermal equilibrium generally correspond to lower energy states. As the temperature is reduced, the material will converge towards a microstate which is the lowest energy state.

Simulated annealing uses exactly this approach to find the lowest value of an error function – the error function is analogous to the energy in the thermodynamic case. The formula for calculating the probability of uphill moves in simulated annealing is exactly the Boltzmann formula of statistical mechanics which describes thermal equilibrium. This means that provided we visit sufficient points (microstates) on the error surface at each temperature, a situation analogous to thermal equilibrium will be maintained. Hence, as the control parameter, T (which is called "temperature" by analogy), is slowly reduced, the lowest value of the error function will eventually be found. Of course, we would not *have* to use the Boltzmann formula to calculate the probability of uphill moves, but its choice is motivated by the analogy. If we were to take some other formula we may still be able to find the global minimum, but not in a way which could take advantage of the analogy to statistical mechanics.

The opposite of annealing is quenching, whereby a material is cooled very rapidly. In this case the material has very little opportunity to visit different configurations, so a configuration very close to that it starts in will be the one to become "frozen" in. As the material starts off at a high temperature, it is very likely that this final configuration corresponds to a metastable, high energy state, e.g. one corresponding to a polycrystalline material. In the optimization case, this is analogous to making very few moves before lowering the temperature. The result is that uphill moves are quickly discouraged and the search is forced to move to a local minimum in the vicinity of the starting point, without having explored the error surface.

The analogy between simulated annealing and the physics of true annealing is very close. In the former case, one tries to minimize an error as a function of a set of parameters (e.g. the weights in a neural network). In the latter case, one tries to minimize the energy as a function of the microstate or configuration, i.e. the set of particle positions. In both cases we can picture this as a multidimensional minimization, where each dimension corresponds to a parameter in the former case and to a particle position in the latter case. The analogy arises largely because the Boltzmann formula is used in simulated annealing to calculate the probability of uphill moves. Combined with sufficient moves at each T, this ensures that a situation analogous thermal equilibrium in a physical system is achieved, and hence that the characteristics of true annealing appear in the simulated case.

Simulated annealing is a strong analogy to physical annealing, because the underlying formalism is very similar in the two cases. This is not just a useful analogy which helps us to understand the technique, but rather a situation in which the same description and equations are used. Indeed, much of our understanding of physical systems can then be applied to simulated annealing, and even concepts such as entropy and heat capacity become meaningful in the new context. There are, nonetheless, some disanalogies. In particular, a temperature has been introduced in the simulated annealing algorithm as a convenient control parameter. While it clearly has a strong analogy to physical temperature, there is no actual temperature in an optimisation context. Similarly, in all physical systems there is a universal value for the parameter, $k$, the Boltzmann constant, which is a fundamental constant of nature. In a given computational problem, on the other hand, we are free to set this, i.e. to decide what energy difference corresponds to what probability of an uphill move at a certain temperature. Nonetheless, these disanalogies are peripheral to the extent and depth of the analogy. They by no means threaten the usefulness of the analogy, because the analogy is fundamental to the technique of simulated annealing.

## 2.3 Genetic algorithms

### 2.3.1 The technique of genetic algorithms

A genetic algorithm (Holland 1975; Mitchell, 1996) is a random, yet directed search mechanism for an optimal solution to some problem. As in biological evolution, we have a population of organisms each having a different set of genes. These genes correspond to the parameters of some model we wish to optimise, so each organism represents a potential solution to the optimisation problem. For each organism we can determine how well its parameters solve the problem in hand: this determines the "fitness" of each organism, with better (lower error) solutions corresponding to higher fitness. The idea behind a genetic algorithm is then to produce new members of the population from members of the current population using various *genetic operators*. The process of *recombination* (or *crossover*) is a means of "sexual reproduction" in which two parents interchange genes to create two new children. The operation of *mutation* is a means of "asexual reproduction" in which a child is created from a single parent by randomly changing the value of one of its genes. (There are many ways in which these operators can be implemented, but a typical example is shown in Figure 3). Thus new generations are created iteratively, and at each stage the fitness of each member is a measure of how well each solves the optimisation problem. The goal of this evolutionary approach is to produce diversity within the population and so explore various gene combinations, or solutions to the optimisation problem. However, this is not a blind search: the probability of reproduction is related to the fitness of the parents, i.e. fitter parents are generally made to produce more children. Nevertheless, not only the fittest individuals produce children: although that may seem the best strategy, it is similar to only permitting downhill moves in the search for the global minimum of the error function. As we have seen, such a strategy is prone to getting stuck in local minima. Thus, while less fit members may be produced in the interim, the goal is that, over many generations, the fitness landscape (error surface) is comprehensively explored, and that the population evolves towards a fitter state.

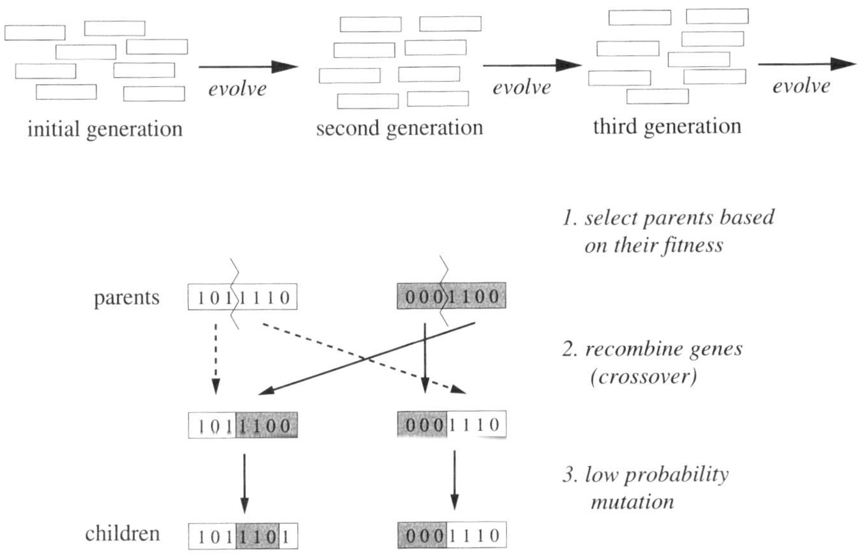


*Figure 3.* Demonstration of the major functions in a genetic algorithm. Starting from an initial population, two parents are selected with a probability depending on their fitness. The genes are split at some point and recombined to produce two new organisms, a process referred to as *crossover*. There is also a small chance that any one gene can mutate. The resulting children form members of the next generation, and the process is repeated until a second generation in size equal to the initial population is produced. The whole procedure is iterated for many generations with the expectation that fitter organisms are produced.

An important aspect of genetic algorithms is the encoding of the optimisation problem into a genetic formalism. In particular, the parameters of the problem need to be encoded as discrete genes, and the error function expressed as a fitness which can be determined from these genes. For example, a genetic algorithm could be used to determine the optimal weights of a neural network: each weight is represented as a gene, and the fitness is the negative of the error function.

### 2.3.2 The genetic analogy

The analogy with biological evolution hardly needs to be stated. It is more than evident from the vocabulary that is used in describing the method of genetic algorithms: genetics, population, organism, generation, fitness, sexual and asexual reproduction, selection, mutation. It also seems that the conception of the processes bringing about evolutionary progress are critical for the conception of this computational method: for example, the genetic operators of crossover and mutation were inspired by evolutionary processes which occur in nature. This is in contrast to the case of neural networks where the analogy to the brain may have had heuristic value, but now seems to be largely historical. The genetic algorithms technique is based on the assumption that, because in nature the analogue of these processes appear to produce an improvement in the fitness of organisms, they will do so also in a simulated environment. On the other hand, the analogy breaks down at the point of encoding. While strings of numbers may be manipulated just like genes, the way the genetic – or other – information is stored in a gene or in a string of numbers is fundamentally different. To encode information in a string of numbers involves active interference by someone who decides how best to encode some information. Another interesting disanalogy is that in biological evolution the fitness of an organism is very much dependent on the other organisms in the environment: if there are many other organisms which can eat your food (or you), then your fitness is going to be lower. This is in contrast to most optimisation problems where there is some static function which needs to be optimised, independent of the other candidate solutions.

There can be no doubt that our comprehension of the method of genetic algorithms benefits greatly from the familiarity with evolutionary concepts. This is supported by the abundance of suggestive metaphorical vocabulary. The idea for constructing genetic algorithms on the basis of the analogy to evolutionary biology requires making a considerable mental transition, in part because the encoding mechanisms are so different in the two cases. The way in which genes are manipulated, combined and expressed is very different in the biological and the genetic algorithm cases. This is unlike the case of simulated annealing where the abstract situation of the two analogues is very similar. Thus in some sense, although simulated annealing has much more in common with physical annealing than genetic algorithms do with biological evolution, simulated annealing is a much less surprising analogy precisely *because* it replicates true annealing so closely. In many ways the concepts and language carried over to the optimisation case remain the same, and are more than just an aid for dealing with the problem. With genetic algorithms, on the other hand, there is a much greater "distance" between mathematically encoded optimisation and the field of evolutionary biology from which the inspiration for the method is derived. Consequently, the language and concepts transferred are much more subject to reinterpretation. For example, a gene and a numerical encoding called a gene are not the same. Reaping the benefit of the genetic analogy first requires reinterpretation before the surprising possibilities of the analogy can be exploited.

# 3. COMPARING DATA ANALYSIS MODELS WITH THEORETICAL SCIENTIFIC MODELS

Both data analysis models and theoretical scientific models are there to solve a problem, one to solve a problem of data analysis, the other to solve a problem of describing an empirical phenomenon. Of course, solving a problem about an empirical phenomenon also involves considering data, and in principle, the same set of data could provide the basis for either a data analysis model or a scientific model. First, we discuss substantial differences between the two types of models, and then explore the impact these differences have on the use of analogies in each type of model.

## 3.1 The contrast between theoretical scientific models and data analysis models

Both theoretical scientific models and data analysis models employ mathematical strategies for solution. A theoretical scientific model is one which has been developed employing scientific theories and concepts and which applies to a specific phenomenon. An example is a model for the temperature variation in the Earth's atmosphere, where one wants to determine how the temperature in the atmosphere varies with distance from the Earth. This can be done largely on the basis of theoretical assumptions taken from theories of radiation transfer and thermodynamics and the properties of gases. The core of the model then comes from these assumptions, perhaps only taking into account a minimum of data, which then allows the modeller to solve certain equations. Whether the data set is large or small, one is interested in determining the values of some important, physically-meaningful parameter or parameters with the aim of better understanding the physics and chemistry of the atmosphere. Computational data analysis models similarly have the goal of modelling the relationship between two domains, such as stellar spectra and the physical conditions within stars. Again, the optimal values of some parameters in a model (e.g. the weights in a neural network) can be determined from some data. However, the interest here is not explicitly the values of these parameters. Instead, the primary interest is in training a model to make predictions, that is, to be able to determine the physical conditions in other stars based on their spectra, using the knowledge intrinsically present in the training data. The model parameters are a means to an end and not necessarily of physical significance themselves. This is because these models are conceived without regard to the theories and concepts of the various problems to which they can be applied. The models are sufficiently general so that they can be applied to a wide class of problems with only general requirements having to be met. In consequence, beyond the goal of accurate prediction, the scientific insight that computational data models give in a specific case may be limited. A physical understanding of exactly what aspects of the physical conditions in a star give rise to which parts of the spectrum is useful, but secondary when the main goal is accurate classification, with whatever degree of model complexity that is necessary (within feasible limits).

There is a fairly straightforward reason for the fact that artificial neural networks do not provide much physical understanding. The discussed data analysis techniques are not specific to the type of data that are modelled. The techniques are designed to be independent of specific applications – they are *application-neutral*. They are *specific to a type of data analysis problem, but not to the nature of the specific empirical problem*. For example, a neural network could be applied to almost any problem of mapping between two data domains. From which empirical phenomenon the data derive, or if they even derive from an empirical phenomenon, is secondary, just as long as the problem representation conforms to some very general requirements. A theoretical scientific model is, in contrast, specific to a type of phenomenon. The theoretical concepts and laws that give shape to the theoretical model are chosen on the basis of the physical properties of the phenomenon to be modelled. These models are therefore restricted in their range of application. For example, a general relativistic model for planetary orbits is not much use in problems in which gravity is not a central factor.

Scientific models and data analysis models do not exclude each other. It is possible to model a set of data either applying physical theories or applying a neutral data analysis technique (Figure 4). For example, in metallurgy, there exists a theoretically motivated equation for the determination of the size of microscopic crystal grains in a metal which has been mechanically processed (e.g. forged). Given knowledge of the material properties, the grain sizes can be analytically calculated from the details of how the material was mechanically processed, the processing parameters. However, it is a difficulty of this approach that some of these material properties must be determined from expensive and time-consuming electron microscopy. On the other hand, a data modelling technique such as a neural network can be trained to predict the grain sizes for given processing parameters without ever needing to explicitly know these material properties, as they are intrinsically present in the required training data (Sabin, Bailer-Jones, and Withers, 2000). Measuring the grain sizes of materials processed in a certain way to construct this training data set is cheap and quick in comparison to determining the material properties, thus offering one distinct advantage of this approach.

Optimisation techniques, such as those discussed, are application-neutral not only in the sense that they can be applied to a wide range of very different problems. In turn, a problem of a certain type can be addressed by different techniques. For example, a given optimisation problem could be tackled either with simulated annealing or a genetic algorithm. Correspondingly, a computational data analysis technique can be designed in many different ways to solve a certain problem. The choice of design is not unambiguously determined by the problem, which is why an analogy to inspire the technique can be chosen fairly freely and certain specifications within it may be arbitrary.

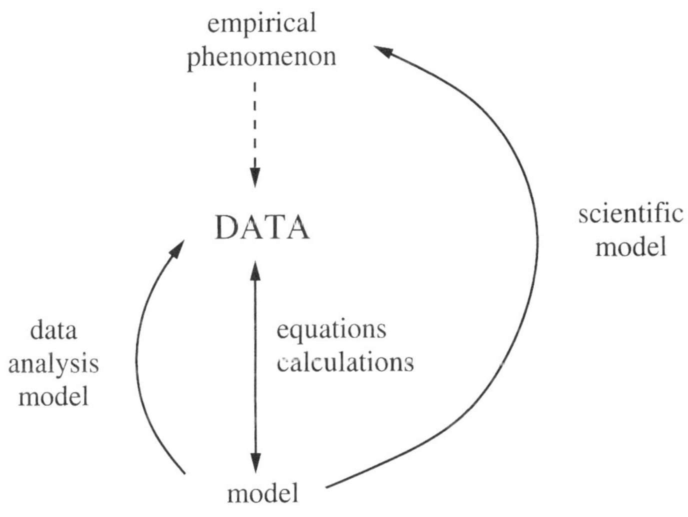


*Figure 4.* While data is derived from a phenomenon, the computational data analysis model refers to the data only, while the scientific models refers to the phenomenon. That data of a phenomenon are not the same as the phenomenon has been argued convincingly by Bogen and Woodward (1988).

On account of their application neutrality, computational data analysis techniques must *encode* or parametrize the specific problem to which the technique is to be applied. This encoding needs to be done in a way suitable for the technique. For example, with a neural network we must define appropriate inputs and outputs; with an optimisation technique we must determine a suitable energy function (in the case of simulated annealing) or fitness function (in the case of genetic algorithms). For a theoretical scientific model, this encoding is intrinsic to the very conception of the model because the model itself is motivated by the physics of the problem, so no separate encoding procedure is required to make the model match the problem. Only for an application-neutral model does the information about a *specific* application need to be encoded in the model. What is more, only the necessary minimum of information is encoded.

It is a consequence of the generality of many computational data analysis techniques that they involve an element of arbitrariness. This is exemplified by the control parameters (e.g. the number of hidden nodes in a network, the value of $k$ in simulated annealing, or the rate of mutation in genetic algorithms) which generally have to be decided on by the user. Different control parameters can be tested for efficiency, speed and reliability, but there is often little in the problem application which suggests the appropriate values. Even the free parameters of the model which are determined from the data (e.g. the weights in a neural network) retain some degree of arbitrariness, because different parameter settings may give equally good model predictions.

## 3.2 Analogies in data analysis models

Theoretical scientific models have a relationship to the phenomenon modelled that is more direct than in the case of application-neutral data analysis models. Consequently, if an analogy is used in a theoretical model, then this is an analogy between phenomena. For instance, in the 1840s, William Thomson, the later Lord Kelvin, modelled electrostatic attraction in terms of thermal conduction. He "recycled" the mathematical tools that had already been developed for conduction for the exploration of the more recent subject of electrostatics. This led to viewing thermal conduction and electrostatic attraction in analogy to each other (Harman, 1982, p. 29), a perspective which subsequently triggered further important insights into electrostatic phenomena. The mathematical analogy which Thomson exploited highlighted a *physical analogy* between the phenomena of thermal conduction and of electrostatic attraction. That analogies are important for the development of scientific models is a long-standing theme in philosophy of science (e.g. Hesse, 1966; Achinstein, 1968; for an overview see Bailer-Jones, 2001). When analogies feature in scientific modelling, then one empirical phenomenon is viewed in terms of another phenomenon, i.e. *in analogy to* that latter phenomenon. The expectation is that the interpretation of one empirical phenomenon may benefit from the interpretation of another (perhaps comparable) empirical phenomenon. If so, a transfer of insights gained from the study of the latter to the former becomes possible.

In contrast to this, the analogy that is employed in artificial neural networks - the analogy to the human brain - has nothing to do with the specific phenomenon to which that technique is applied, i.e. the stellar spectra that may be classified with the help of an artificial neural network have nothing to do with and are not analogous to the human brain. If artificial neural networks exploit the analogy to the human brain, then this by no means implies a connection to the phenomenon that is modelled by the network: a physical analogy between the analogue used in the model (e.g. the brain) and the phenomenon modelled (e.g. the formation of stellar spectra) is neither required not likely because of application-neutrality. Equally, if the optimisation technique used in the network is simulated annealing, then annealing also has no link to stellar spectra, nor, for that matter, to the way the brain works. Each of these analogies is separate in that it only illustrates or inspires a *technique*, and not a phenomenon that may be modelled with the help of that technique.

When, for the development of a theoretical scientific model (e.g. of electrostatic attraction), an analogy (e.g. to thermal conduction) is exploited, then the analogy is the source and motivation for the choice of the equations. Something similar can be said for artificial neural networks. The analogy to the brain is, at least historically, responsible for the kind of mathematical algorithm that the network employs. But the impact of an analogy on the equations is not always equally strong. Artificial neural networks only carry a vague inheritance from the true complexity of neurons, whereas in our example of simulated annealing, the impact of the analogy on the equations and concepts used in the algorithm is very significant. There, equations are transferred directly and they are employed in essentially the same way as in statistical mechanics. Thus, the impact of an analogy on the development of a computational data analysis technique can vary from example to example.

In sum, in the case of data analysis techniques, the analogies employed illuminate the technique to a greater or lesser degree, but do not illuminate the specific problem to which the technique is intended to provide a solution. However, precisely because the technique is developed independently of applications, it is important that the technique can be illustrated independently of these. Having a separate and application-independent analogy at one's disposal is likely to facilitate access to the technique for a whole range of different users with different backgrounds. Talk about annealing, temperature etc. provides common ground when applying the technique in different areas. Different users of artificial neural networks, genetic algorithms and simulated annealing may share an understanding of the analogies to the brain, to annealing and to evolutionary change, but possibly nothing more.

## 4. SO, WHY ANALOGIES?

Just as analogies are prevalent in theoretical scientific models, they also have an important - but different - role to play in computational data analysis models. There are a number of uses for a concrete analogy backing up an abstract data analysis tool. The first and foremost is to provide an idea for developing the tool, an idea for an algorithm appropriate for attacking a certain kind of problem (parallel distributed processing in the case of artificial neural networks, evolution and genetic changes in the case of genetic algorithms, the Boltzmann distribution in the case of simulated annealing).

Second, there arises the need to communicate about these abstract tools. There is a metaphorical language associated with analogies employed in models that helps to put into words what otherwise seems abstract and inaccessible for want of familiarity. Martin and Harré (1982) call this a "spin off" of the analogy when metaphors matching the analogy are introduced, but this expression may betray the conceptual importance of the vocabulary generated (Bailer-Jones, 2001): temperature in simulated annealing or fitness in genetic algorithms. Because the computational data analysis techniques themselves are application-neutral, metaphors based on analogies can be seen as constantly maintaining the link to the analogy in use. The analogy, with the metaphors attached to it, may then help to cope with the abstractness of the technique. Because the analogy only illuminates the technique and not a phenomenon to which the technique is applied, it provides an aid in overcoming the difficulty of abstractness without involving any notions about the specific application. The metaphors generated by analogies underlying computational data analysis techniques further a sense of familiarity and of concreteness in connection with the otherwise abstract computational tools. They help to attach concrete ideas to tools that are intrinsically without concrete application.

Third, there is likely to be a teaching element in the use of analogies and metaphors. Because computational data analysis techniques may be applied in a wide range of different areas, users may come to them from quite different disciplinary backgrounds. Those who enter into the area of computational data analysis may do so simply because, in their own discipline, they have a problem to solve that happens to lend itself to that type of analysis. Consequently, there will be a significant group of people who require a basic understanding of these methods, or need relatively easy access to the central ideas in order to understand better the algorithms used. Such learners depend heavily on conceptual aids linking the abstract to something more concrete.

In short, while analogies in computational data analysis techniques vary in their depth and in the extent to which they generate descriptive metaphors, analogies have an impact well beyond the original conception of computational data analysis techniques. Just like analogies in scientific models, they live on to serve not only the experts during conception and development of the technique, but perhaps as much the more casual user during acquisition and use.

## REFERENCES

Achinstein, P., 1968, *Concepts of Science*, John Hopkins Press, Baltimore, MD.

Bailer-Jones, C.A.L., 2000, Stellar parameters from very low resolution spectra and medium band filters. $T_{eff}$, log g and [M/H] using neural networks, *Astronomy & Astrophysics* 357:197-205.

Bailer-Jones, D. M., 2001 (to appear), Models, metaphors and analogies, in: *Blackwell Guide to Philosophy of Science*, P. Machamer and M. Silberstein, eds., Blackwell, Oxford.

Bogen, J. and Woodward, J., 1988, Saving the phenomena, *The Philosophical Review* 97:303-352.

Harman, P. M., 1982, *Energy, Force and Matter*, Cambridge University Press, Cambridge.

Hertz, J., Krogh, A., and Palmer, R.G., 1991, *Introduction to the Theory of Neural Computation*, Addison-Wesley, Wokingham.

Hesse, M., 1966, *Models and Analogies in Science*, University of Notre Dame Press, Notre Dame, IN.

Holland, J.H., 1975, *Adaptation in Artificial and Natural Systems*, University of Michigan Press, Ann Arbor, MI.

Kirkpatrick, S., Gelatt Jr., C.D., and Vecchi, M.P., 1983, Optimization by simulated annealing, *Science* 220:671 680.

McCulloch, W.S. and Pitts, W.S., 1943, A logical calculus of ideas immanent in nervous activity, *Bulletin of Mathematical Biophysics*, 5:115-133.

Metropolis, N., Rosenbluth, A.W., Rosenbluth, M.N., Teller, A.H., and Teller, E., 1953, Equation of state calculations for fast computing machines, *Journal of Chemical Physics* 21:1087-1092.

Mitchell, M., 1996, *An Introduction to Genetic Algorithms*, MIT Press, Cambridge, MA.

Martin, J. and Harré, R., 1982, Metaphor in science, in: *Metaphor*, Miall, D. S., ed., The Harvester Press, Sussex, pp. 89-105.

Sabin, T.J., Bailer-Jones, C.A.L., and Withers, P.J., 2000, Accelerated learning using Gaussian process models to predict static recrystallization in an Al-Mg alloy, *Modelling and Simulation in Materials Science and Engineering* 8:687-706.

# Perceptual Simulation in Analogical Problem Solving

David L. Craig*, Nancy J. Nersessian#, and Richard Catrambone†
*College of Architecture, david.craig@arch.gatech.edu; #College of Computing, nancyn@cc.gatech.edu; †School of Psychology, GA, USA, rc7@prism.gatech.edu, Georgia Institute of Technology, Atlanta

Abstract: We argue that in some situations solutions to target problems can be generated based on perceptual simulations of analogous but superficially dissimilar source problems. Unlike processes associated with perceptually neutral representations, those associated with perceptual simulations should be highly sensitive to context, whether conveyed through textual, visual or physical mediums. Consistent with this claim, we found that when the context of a source problem was manipulated to accentuate kinesthetic structure, people were more likely to generate solutions to an analogous problem. We also found that subjects who drew diagrams of solutions to source problems were most likely to transform them into solutions to an analogous problem when the contextual properties of their diagrams afforded such transformations.

## 1. INTRODUCTION

Imagine an engineer working on a cable system with an unusual set of constraints. The task requires that the engineer add a couple of tensioning wires to the system at right angles to each other. Although attaching the cables to the system is trivial, the problem is that the parts they attach to move back and forth, and when they do the two cables run into each other, causing the whole system to shut down. Ultimately, if the cables are positioned where they need to be, they will somehow have to be able to *pass through* one an-

*Model-Based Reasoning: Science, Technology, Values*, edited by
L. Magnani and N.J. Nersessian, Kluwer Academic/Plenum Publishers, New York, 2002

other without either of them disconnecting. Initially the problem seems impossible. The design engineer scratches her head and thinks if there might be a way to reposition or eliminate the tensioning cables altogether. She then thinks more carefully about the business of one thing passing through another without sacrificing its physical integrity. Suddenly she remembers the revolving door to her building which she passes through everyday day on the way to her lab. The revolving door is designed to let people pass through effortlessly while nonetheless keeping the inside of the building sealed off from the outside at all times. One part of the door, that is, always closes just as another opens and vice versa. With this insight she quickly sketches out a gate to add to one of the cables (Figure 1), which, like the revolving door, will allow the other cable to pass through it without either disconnecting, using the principle of redundancy (French, 1994).

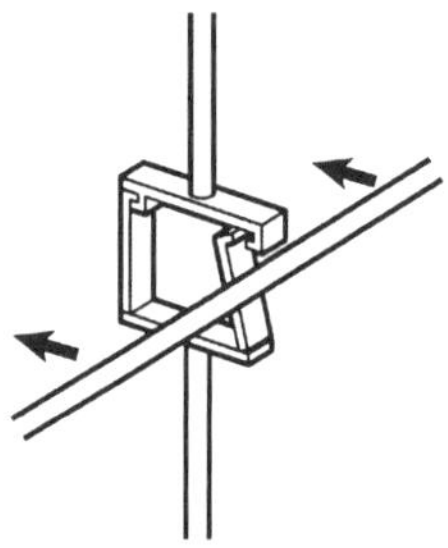

*Figure 1.* A gate mechanism, employing redundant latches, that allows one cable to pass through another at right angles.

In the scenario described above, the revolving-door exists in a context far removed from the cable problem. Yet the way the cable problem is analyzed somehow triggers the retrieval of the revolving door and the general principle behind it. The example illustrates the nontrivial process of solving a problem by constructing an analogy. Although the engineer may not have possessed explicit knowledge about how to design intersecting cable systems, the problem is recognized as being similar to one encountered elsewhere and one for which there is a known solution.

Because of their remarkable ability to transcend the limits of domain-specific knowledge, analogies play a central role in design, science, and other creative endeavours, as well as in everyday learning and problem solving. In this chapter we look at how mental representations are formed and transformed, and how differences in potential source experiences - the kinds of experiences that can lead to analogies - may result in differences in the construction process. As we look at the analogical problem solving process our primary focus is on perceptual factors. Going back to the example described above, one might ask if an engineer did in fact use her knowledge of a revolving door to solve a cable problem, would it have mattered if she imagined a particular revolving door (e.g., the one she uses every day), and if so, would the particular features of that door have mattered? Would it have mattered, for example, whether the door was big and heavy, whether it operated automatically or manually, or whether it was surrounded by a solid wall or large glass panels? Would it have mattered if the engineer had imagined walking through the door or if she instead visualized a schematic drawing of the sort produced by an architect? In most existing research on analogical problem solving these sorts of features are, strictly speaking, taken to be superficial, in that they bear no obvious relationship to the underlying principles used to solve the problem. As such, they are commonly disregarded as features that get stripped away before the real thinking gets underway.

As the chapter unfolds, we articulate a theory that explains how perceptual factors such as those described above could come into play in analogical problem solving. In particular, we present a theory that explains how analogies might be constructed via perceptual simulations. A perceptual simulation is a type of representation that likely involves the reactivation of patterns of neural activity in the perceptual and motor parts of the brain initially activated in experiencing something first hand. One possible component of perceptual simulations that happens to be fairly well understood is visual imagery. Perceptual simulations, however, can also involve other sensory modalities, including those associated with experiences of force and space. They can be richly or sparsely detailed, highly structured or flat. They can also be used to represent abstract concepts. Despite their generality, perceptual simulations contrast with the type of representation standardly used to explain high-level thought, namely predicate-based representations. Given the differences between the two, one might expect an explanation of analogical problem solving based on perceptual simulations to result in predictions somewhat different from those made by researchers employing the standard view. In particular, because perceptual simulations demand a heightened degree of physical realism, the physical setting one imagines might affect how one represents principles underlying a physical system and ultimately how one makes use of such principles in constructing analogies, even in cases where the setting is superficially distinct from that of the target problem.

Our motivation for trying to explain how analogies might be constructed via perceptual simulations stems from an interest in better understanding how external representations - including stories, pictures and even bodily reenactments - influence problem solving. In most research on analogy, external representations have been treated primarily on the assumption that they *contain* information. Reasoning with external representations, moreover, often seems limited to extracting information, in that once the information is extracted it is converted into a perceptually neutral format and at that point transformed independently of perceptual processes. To use an external representation like a diagram, for example, is to see pertinent information and discard the perceptual substrate. Alternatively, external representations might be seen as setting up perceptual simulations. The perceptual context implied in an external representation might, when integrated into a simulation, have a significant effect on how such a simulation can be transformed, particularly if the transformational processes are also perceptual.

The purpose of the chapter is not to prove that analogies are always constructed via perceptual simulations, nor to rule out all alternative explanations. We do, however, intend to show how findings produced in our lab might be explained more efficiently in terms of perceptual simulations than in terms of perceptually neutral representations. We start by explaining in more detail what perceptual simulations are. We then present a brief overview of analogical reasoning processes and describe how they might be accomplished via perceptual simulations. Finally, we describe a series of experiments designed to highlight the role of perceptual factors in analogical problem solving using perceptual simulations, both static and dynamic. In each experiment, a different type of external representation is used. In the first experiment perceptual factors are manipulated via narratives; in the second they are manipulated via physical activity; and finally, in the third they are manipulated via diagrams. We conclude the chapter by briefly discussing how our findings might impact more wide ranging studies of creative problem solving.

## 2. PERCEPTUAL SIMULATION

Representations have two significant aspects: format and content. In most research on analogy and other types of conceptual reasoning, the two have been considered independently. That is, researchers have typically employed representational forms that bear no relationship to the content they are used to represent. Representations encoded in such formats are amodal representations. Natural language is a good example of an amodal representational system in that the sound or appearance of a word typically bears no relationship to its referent - the thing it describes. Amodal representations can be used to represent nearly anything, precisely because they are analogous to natural language. They could, for instance, take the form of predicates, where a predicate is simply a statement about some thing or about a relationship between things of the type ROUND(ball) or PASS_THROUGH (ball, hoop).

A notable property of amodal representations is that they are perceptually neutral - not in the sense that they fail to represent perceptual information (ROUND(ball) is obviously perceptual), but rather in the sense that they do not preserve information about the modalities through which their referents are clearly experienced. Predicates like DIAMETER(ball, 30cm) and DIAMETER(hoop, 31cm) are perfectly useful for reasoning about whether, for instance, a ball will fit through a hoop, but they do not suggest what it is like to hold something 30cm in diameter or even how the experience of size differs from, say, the experience of weight. Some have argued that this property makes amodal representations attractive. Amodal representations, in other words, allow one to reason with information drawn from different modalities all at the same time (Palmer, 1975). The real appeal, though, of amodal representations is that they can efficiently represent things that are not perceptual. Predicates, for example, are capable of representing significant differences between things that seem otherwise perceptually similar (Smith and Medin, 1981). Oranges and orange plastic balls may look the same (and feel the same and so forth), but predicate-based representations of the two can get right to the heart of their differences, representing, for example, that one is nutritious while the other is inert, or that one is organic while the other is made in a factory.

Modal representations contrast with amodal representations in that modal representations do preserve information about the modalities through which their referents might be experienced (Barsalou, 1999; Glenberg, 1997). A modal representation is a perceptual state or a long-term copy of one. It is an analog representation, meaning that it preserves constraints inherent in the patterns of neural activation found in the perceptual and motor cortices of the brain. Reactivating a modal representation in working memory means perceptually simulating something, or, literally, reactivating a particular perceptual state. Like predicate-based representations, modal representations need not be complete. One would not, in other words, be limited to remembering things exactly as they were experienced. Modal representations could be schematic in the sense that bits and pieces of perceptual experiences could be stored separately and reactivated in novel combinations. A long-term modal representation of an orange, for example, may preserve certain properties, like shape, size, and taste, independently, allowing one to later imagine things like orange-flavored candy or orange-colored cars. That modal representations could function in such a way is supported by the fact that the perceptual system itself tends to break up stimuli (e.g., patterns on the retina) into distinct perceptual features and relations.

There are a number of reasons to believe that people do store modal representations and do make use of them, via perceptual simulations, in various reasoning tasks. One widely researched type of modal representation is mental imagery. As a number of studies have demonstrated (see Kosslyn, 1994, for a review), people who are compelled to "picture" something in their head exhibit behavior that overwhelmingly suggests they are processing short-term representations in the primary visual cortex. When, for example, people are asked to answer questions about the relative locations of points shown on a memorized spatial map, their responses tend to lag depending on the distance between the points, as if they were actually visualizing and scanning between them. If, by contrast, such a spatial map were represented using predicates, all distances would be encoded in the same way, independent of actual length, and one would expect processing speed to be independent of actual distance. There is also evidence that people use modal representations in basic conceptual tasks. One finding, for example, which Barsalou (1993) calls "linguistic vagary," is that when people are asked to list features associated with everyday objects, they provide significantly different answers at different times, even if the difference in time is just a matter of days. As Barsalou argues, such results are hard to explain using amodal representations. Predicates, for example, should produce fairly consistent output because predicates are analogous to natural language. Listing features based on modal representations, on the other hand, would be open to the vagaries of perception (physical perspective, situational context, attention, etc.), resulting in the type of behavior that is, in fact, found. None of these findings, of course, refute amodal representations outright. One could always jerry-rig an amodal representation to function like a modal representation and vice versa. Nonetheless, some types of behavior do seem more consistent with modal representations. More of this type of evidence will be presented as the chapter progresses.

The most significant limitation of modal representations is in encoding abstract conceptual information. Modal representations encode only sensory experiences and, hence, may seem inherently superficial. How, then, would one explain known differences between an orange and an orange plastic ball with modal representations? Although some perceptual differences might tip one off (oranges are tangy, for example), the most substantial differences are not superficial, such as the fact that oranges are organic and plastic balls are synthetic. Anticipating this problem, Barsalou (1999) argues that abstract concepts like organic and synthetic could be modally represented in terms of complex sequences of events in which particular objects and actions are highlighted. Organic, for example, might require a modal representation of a sequence of events in which an orange grows on a tree, while synthetic might require a representation of something being manufactured in a factory. Problematically, though, concepts like organic and synthetic are not specific to plants and factories. Rather, they seem generic and capable of entering into reasoning independent of such situations.

To represent generic, abstract concepts, we propose that people might recruit modally represented, perceptual structures to *define* them. The kinds of perceptual structures employed would be spatial and kinesthetic - in other words, structures abstracted from first-hand, bodily experiences of space and force. Consider again the concepts organic and synthetic. Organic might be represented using spatial and kinesthetic structures associated with expansion from within, something most people should be familiar with from a physical point of view. Synthetic, on the other hand, might be represented using spatial and kinesthetic structures associated with accretion. Obviously, biologists entertain more complicated representations of organic and synthetic, but these might just consist of a more complicated network of spatial and kinesthetic structures. Following Lakoff and Johnson (Johnson, 1987; Lakoff, 1987; Lakoff and Johnson, 1998) and Mandler (1992), we refer to spatial and kinesthetic structures as image schemas. Image schemas serve primarily to structure mental images, each possessing structure that can be analyzed but that nonetheless is elemental in form. An image schema for convergent force, for example, serves to structure that which is compressed in relation to that which is doing the compressing. By incorporating a convergent-force image schema in a perceptual simulation (e.g., by overlaying it on an image), one could represent things ranging from a person being compressed by a crowd to a walnut being compressed by a nutcracker. Importantly, though, image schemas need not always coexist with other modalities, visual or otherwise. An image schema for compression, for example, could be activated by itself simply as an independent representation of space and force. Although experiences of space and force typically coexist with other sense modalities, they too can be abstracted via perceptual processes and reactivated separately, in the same way that one might imagine a color or a shape as independent of an object.

To review so far, both modal and amodal representations, as theoretical entities, would seem capable of representing knowledge in schematic, composable units, ranging from minute perceptual details to abstract concepts. These units of information, however they are formatted, are, in turn, likely bound in rich associative networks in long-term memory known as schemas, scripts, frames, or mental models (Barsalou, 1992; Gentner and Stevens, 1983; Minsky, 1975; Palmer, 1975; Rumelhart and Ortony, 1977; Schank and Abelson, 1977). Despite the similarities in the types of information that modal and amodal representations might encode and the way that information might be associated in long-term memory, one significant difference between modal and amodal representations that could impact higher-level reasoning, particularly reasoning associated with analogies, is in how structural relations are encoded, specifically in how information is *encapsulated* in them. As mentioned before, in modal representations, structural relations must be represented in image schemas. A major difference between an image schema and a predicate-based representation of the same thing is that the former is considerably more condensed. The kind of information implicit in an image schema, in other words, would take a lot of predicates to represent explicitly. That image schemas automatically encapsulate a wealth of information may prove important in things like constructing analogies. An increase in representational power, that is, might increase the potential overlap between two domains and thereby increase the chance that similarities between them will be detected.

A second important difference between modal and amodal representations that we will consider has to do with how information is processed in working memory. In general, to complete any reasoning task requires that some subset of information stored in long-term memory be activated. The simplest reasoning tasks might be accomplished deductively, by tracing links in an activated part of the associative network. To tell what color ravens are, for example, one might activate raven and follow the link to color. More complex tasks, however, might require that one create new representations in working memory by restructuring units of information from long-term memory. To reason about a raven in a raven's nest, for example, might involve using the structural relation "in" to bind existing representations of raven and nest, regardless of whether such a thing has ever been witnessed or imagined in the past. The difference, in this case, between modal and amodal representations is that in the former all information included in working memory must be perceptually coherent. In addition, if one transforms a modal representation in working memory, those transformations must also be perceptually coherent. Because of the demands of perceptual coherence, context might ultimately affect how a representation is constructed and transformed over time.

## 3. ANALOGICAL PROBLEM SOLVING

Analogical problem solving is heavily dependent on mental representations. To start, a representation of a problem in working memory - a target representation - must somehow activate another representation - a source representation - stored in long-term memory. This process, known as retrieval, could be triggered by the fact that both representations share common features, parts, events, scenes or abstract concepts. However, because any representation active in working memory is likely to overlap with a large number of representations in long-term memory, additional matching criteria are typically required. One possibility, suggested by Gentner and her colleagues (e.g., Clement and Gentner, 1991; Gentner 1983), is that the retrieval process selects only those representations that yield highly systematic matches. Systematicity, in this case, refers to the degree of consistency with which high-level relations match between representations, high-level relations being those that structure other relations as opposed to features, parts or objects. That one object revolves around another, for example, is a low-level relation, because it structures two objects. That gravity *causes* one object to revolve around another, however, is a high-level relation, since the causal relation connects two other abstract relations, namely gravitational pull and revolve.

Once a source representation is retrieved (on the promise that it can help solve the problem at hand), that representation must be aligned with the target representation. In some cases, the retrieval process itself may produce partial alignment, as is required, for example, in checks on systematicity. In any case, the alignment process brings into correspondence parts of the source and target representations that match exactly, which in turn allows one to bring things that do not match into correspondence. Cigarettes and time bombs, for example, can be brought into correspondence despite their many differences, because both are known to cause eventual destruction (Gentner, 1983). Ideally, alignment is followed by a mapping process in which structural relations that exist in the source representation but not the target, and which are systematically connected to already aligned relations, are mapped onto the target. The mapping process, of course, is what in ideal situations produces a solution to the target problem.

In some cases, analogies can be constructed using knowledge from closely related domains, even the same domain as that of the target problem. Referring back to the example at the start of the chapter, if the design engineer was familiar with another cable system that satisfied constraints similar to the problem on which she was working, she may have been able to use it as a source for constructing a close-domain analogy. The closeness of two domains is ultimately a matter of the number of superficial similarities between them, superficial similarities being those that involve features that are causally unrelated to the goals and constraints of the problem. If the design engineer, for example, was familiar with an old cable problem involving 10-gauge cables and was working on a new problem that also involved 10-gauge cables, cable size would constitute a superficial similarity. In this case, size is not important in how the problem is solved, but it could help in retrieving the old experience (Gentner, Ratterman and Forbus, 1993; Holyoak and Koh, 1987). Analogies can also be drawn from unrelated domains, such as in the example where a revolving door is aligned with the cable problem. In the experiments discussed in this chapter, we focus on analogies at this end of the spectrum - those, in other words, for which few, if any, superficial similarities exist between the source domain and the problem at hand. We refer to such instances as abstract analogies.

Given the potential complexity of mental representations, the processes associated with retrieval, alignment, and mapping can themselves be complex. At a fundamental level, though, it comes down to matching units of information between representations. In the case of amodal representations, matching by itself is straightforward. Given two predicate-based representations, for example, one could just match identical predicates. In the case of modal representations, matching might also be straightforward, in that one could look for identical perceptual features, parts, objects, scenes, events or image schemas. Modal representations might in fact have an advantage over amodal representations in making superficial matches since modal representations tend to make perceptual features, such as visual patterns, easier to detect. Cars and people, for example, might not seem analogous until one visually simulates the front of a car (headlights, grill, bumper, etc.) and compares that with an image of a face (Davies, Nersessian and Goel, 2001).

The most significant difference between modal and amodal representations for generating analogies potentially lies with matching abstract structure between representations. As mentioned above, abstract structure can be aligned by pairing identical predicates in the case of amodal representations and, similarly, by pairing identical image schemas in the case of modal representations. Modal representations, however, present another possibility. An image schema in a source representation might also be matched with a pattern of constraints in a target problem. A pattern of constraints, that is, might describe properties implicit in an image schema and hence be perceived as approximating it, resulting in alignment. In this case, the image schema serves as a sort of generic, encapsulated solution schema. If a source representation includes such a schema, one might expect it to be retrieved and aligned more easily with a target problem than one that lacks it, simply because any given image schema, like any perceptual state, can be described, and hence accessed, in so many ways.

In the three sections that follow, we describe experiments in which the perceptual context of source situations is manipulated using different kinds of external representations: narratives, physical activity, and diagrams. In the first two experiments context is manipulated for the purpose of either inhibiting or fostering the activation of particular image schemas (Craig, Catrambone and Nersessian, under review). The goal is ultimately to determine whether image-schematic structure will affect the ease with which analogies are made, as perceptual simulations predict. In the third experiment we study how the perceptual context of self-generated diagrams inhibits or fosters the sort of perceptual transformations required to convert a representation of a system in one domain into a solution to a problem in another.

## 4. PERCEPTUAL SIMULATION AND NARRATIVE SOURCES

When people read narratives their behaviour suggests that they begin linking together parts of long-term representations in working memory, producing short-term representations generally referred to as situation models, although sometimes also as mental models (Johnson-Laird, 1983; van Dijk and Kintsch, 1983). If one reads about a cat taking a nap, for example, one might retrieve a schema for cats and one for napping and link them together. Readers may embellish stories by inferring prototypical information (e.g., that cats sometimes purr when sleeping) as well as information that serves to add coherent connections between things that are specified in the story (e.g., that cats nap on comfortable surfaces, like sofa cushions). The process of elaborating mental models likely serves to make the information more coherent, easier to remember, and easier to use in disambiguating subsequent references made in a text. It is so natural that readers often have a hard time remembering that which was explicitly mentioned in a text separately from that which they inferred (Bransford, Barclay and Franks, 1972).

We argue that readers may construct perceptual simulations that function like situation models with additional perceptual constraints. Evidence of additional perceptual constraints includes the fact that access to remembered details in a text often depends on spatial proximity (rather than, say, conceptual proximity) to the current focus of attention (e.g., Glenberg, Meyer and Lindem, 1987; Bower and Morrow, 1990) and on perceptual orientation (e.g., Franklin and Tversky, 1990). It also includes the fact that judgments about a text (regardless of whether the described situations are highly spatial or visual) are hampered when people are asked to hold unrelated visual images in memory, but not when they are asked to hold numerical sequences, even though the latter is considered more difficult (Fincher-Kiefer, 2001). Whether readers always generate perceptual simulations to track information presented in a text is, of course, open to speculation. Some texts - mystery novels, for example - might be too complex to track with one coherent perceptual simulation. At this point we are just considering what the consequences of constructing perceptual simulations might be for analogical problem solving, especially when contrasted with predicate-based situation models.

As it turns out, narratives have been studied extensively in analogical problem solving. In studies done by Gick and Holyoak (1980; 1983), for example, subjects were given a short story to read about a general leading an attack on a dictator's fortress. In the story, the general, knowing that he must use his entire army but also knowing that the army is too big to pass safely over the roads, splits up his men and has them converge on the fortress from different directions. Subjects are asked to study this story under the guise of a recall task. They are then given a problem to solve in which a doctor is trying to destroy a tumor using radiation, with the constraint that a beam of radiation strong enough to kill the tumor will also destroy vital tissue it passes through on the way to the tumor. Gick and Holyoak found that subjects who studied the fortress story were more likely to suggest converging weak beams of radiation on the tumor from different directions than were subjects who read an unrelated story. They argued that subjects may have induced a solution schema from the fortress story that was abstract enough to be aligned with the radiation problem. In focusing on abstract information contained the fortress story, Gick and Holyoak did not address the perceptual context itself. Interestingly, though, they found that subjects were less likely to come up with analogous solutions to the radiation problem if they read an abstract summary of the story than if they read the story with all the contextual details.

Either the abstract summary inadvertently missed a crucial piece of information, or the context of the story did, in fact, play a significant role in constructing analogies.

We hypothesize that if readers generate perceptual simulations to represent events described in a story, the image-schematic structure embedded in those simulations might depend to a large extent on how the story is written - from what perspective, for example, and in what context. Image-schematic structure could even depend on the use of metaphors in a text. Sentences like, "He blew him out of the water with a scathing rebuke," for example, might activate a simple, linear force schema, which in turn could be used to structure a simulation of two people arguing, despite the fact that the characters do not use physical force. In the case of Gick and Holyoak's (1980) fortress story, we argue that the presence of a particular image schema, namely a convergent-force image schema, will significantly increase the chance that the story is used to construct an analogical solution to the radiation problem. A convergent-force image schema, which might be learned through everyday activities like pinching and squeezing (or being pinched and squeezed for that matter), might be useful because it encapsulates information that can be aligned with constraints of the radiation problem. Specifically, the radiation problem requires that radiation be weak in all areas surrounding the tumor but strong at the site of the tumor, just like intensity grows at the point of focus in the convergent-force image schema. Incidentally, though, the fortress story used by Gick and Holyoak probably will not elicit such a schema, if only because the army in the story does not literally apply pressure to the fortress. Instead, it seems more likely that readers will represent a sequence of events in which a resource (the army) is split up and reunited at the target (the fortress) to get around an obstacle (landmines along the road). If that is the case, the fact that intensity grows at the point of focus will be left to be inferred from a sequence of maneuvers and hence not be readily available for aligning with the pattern of constraints inherent in the radiation problem.

To test our hypothesis we constructed two different stories, both derived from Gick and Holyoak's (1980) fortress story and both sharing the same predicate structure. The differences between the stories lie with the objects described and, hopefully, in the sense of force those objects elicit. One story (the "beehive story") is specifically meant to elicit a convergent-force image schema. It describes a man trying to kill a queen bee in a cell in the center of a beehive. The man first considers driving a big stick into the hive but decides not to, knowing that the big stick might set off the drones. Instead he pushes in two smaller sticks with less force from different sides so that they meet in the middle. Because the story describes the application of physical pressure in a convergent manner, we argue that it naturally elicits a convergent-force image schema. The other story (the "modified fortress story") tells of a man trying to kill a dictator in a bunker in the center of a compound. He first considers marching a big army into the compound but decides not to, knowing that the big army might set off landmines. Instead he marches in two smaller groups of men with less force from different sides so that they meet in the middle. The men in the fortress story do not literally apply pressure on the fortress. Instead their power adds up like a generic resource. As a result, a reader might construct a perceptual simulation of the events in the fortress story, but one that probably does not include a convergent-force image schema.

We started by giving 102 Georgia Tech undergraduates the radiation problem (Phase 1). After working on that problem for 4 minutes, we then gave them two stories to study, first an unrelated distracter story and second either the beehive story or the modified fortress story, depending on whether they were in the beehive or fortress condition respectively. After studying the stories, we gave them the radiation problem again (Phase 2), telling them that we were interested in seeing whether a period of incubation would help them solve the problem. Finally, we gave them the radiation problem a third time (Phase 3), along with a hint that one of the two stories they read earlier might be useful in solving it. Because the beehive story explicitly suggests the use of convergent force, we predicted that subjects in the beehive condition would perform better on the radiation problem than subjects in the fortress condition. After eliminating 9 subjects who produced a convergence solution to the radiation problem in phase 1, 20 out of 49 subjects in the beehive condition produced a convergence solution in Phase 2 as compared to 9 out of 44 subjects in the fortress condition. Of the remaining subjects, 22 out of 29 subjects in the beehive condition produced a convergence solution in Phase 3 as compared to 15 out of 35 subjects in the fortress condition. In both cases, the difference between the beehive condition and the fortress condition is reliable

The fact that significantly more subjects in the beehive condition solved the radiation problem before the hint suggests that, in general, the radiation problem was more similar to beehive story than to the fortress story. However, the fact that subjects also performed significantly better in the beehive condition *after* the hint suggests that a significant part of the advantage is due to structural similarities. Once, that is, subjects were given a hint to consider the stories, retrieval should not have been a problem, and hence superficial differences should not have given one story an advantage over the other (Holyoak and Koh, 1987). There could be other reasons why subjects who read the beehive story performed better on the radiation problem besides the presence of a convergent-force image schema. One is that sticks could have been perceived as looking like beams of radiation. Another is that people might have metaphorically represented the radiation problem itself using force. It is not uncommon, for example, for people to describe radiation as "hitting" a target. Third, and finally, there might be something about radiation and tumors that is semantically more similar to sticks and bees than to armies and dictators. The results, however, are consistent with the explanation that image-schematic structure accounts for some of the differences. Our subsequent experiments explore the image-schema hypothesis further.

## 5. PERCEPTUAL SIMULATION AND PHYSICAL ACTIVITY

Physical activity and conceptual thought have come together primarily in studies of gesture and language. Some of these studies are worth considering here to the extent that they hint about the significance of gesture *for* speech, not just as something that accompanies verbal communication. One finding in the gesture literature, for example, is that when gesturing is inhibited, speech production becomes choppy (Krauss and Hadar, 1999). Another is that blind children tend to gesture like sighted children, even when talking to other children they know are blind (Iverson and Goldin-Meadow, 1997). Gestures, in this case, appear to do more than embellish speech. They seem, rather, to be functionally tied to speech production and perhaps even to the conceptual thinking that underlies speech.

Complementing studies on gesture and language are those that directly address the impact of physical activity on conceptual thought. One relevant finding is that physical activity can prime sensibility judgements. When people, for example, are asked to briefly form a gesture with their hand and are subsequently asked to determine whether a given sentence makes sense, they tend to answer more quickly if the gesture is consistent with what the sentence describes (pinching one's fingers together, for example, would be considered consistent with a sentence like "He aimed a dart" but not with a sentence like "He rolled a ball") (Klatzky, Pellegrino, McCloskey, and Doherty, 1989). The fact that gestures can prime sensibility judgements, even when subjects are blind to the purpose of the gesture, is consistent with the idea that physical activity can actually help generate perceptual simulations. It seems unlikely, by contrast, that an unanalyzed gesture would activate specific predicates, given the lack of specificity in what a simple hand gesture might mean.

We hypothesize that physical activity can be used to embellish perceptual simulations by adding structure in the form of image schemas, since physical activity itself is always spatial and kinesthetic. Moreover, we argue that for physical activity to be useful it need not explicitly mimic events or situations under consideration. A lieutenant, for example, might wave his hands to simulate maneuvers on the battlefield, as an engineer might use her arms to simulate the physics behind cantilevered bridges. If physical activity can add image-schematic structure to perceptual simulations, then it should be possible to add a convergent-force image schema to the fortress story used in the last section, perhaps by having people physically enact the events in the story using their hands. A convergent-force image schema might, for example, be elicited by any sort of compressive activity, provided it is consciously attended to in the context of the story (just because it is acted out does not mean it will be used in a representation of the story). If a convergent-force image schema is activated, it could in turn help promote alignment and mapping with the radiation problem for the same reason we argued that the beehive story in the previous section would help.

To test our hypothesis we considered two possibilities, first that the abstraction involved in drawing a diagram of the story might invoke a convergent-force schema (such as in physics diagrams) and, second, that the activity of physically colliding two objects might also invoke it (by activating a kinesthetic sense of force). We gave 104 Georgia Tech undergraduates in three conditions the fortress story to study, telling them that we were interested in how people recalled stories in different modalities. After they studied the story, we had them verbally recall it in their own words. Subjects in the first condition (the "sketch" condition) were additionally asked to draw a simple map-like diagram depicting the action in the story during recall. Subjects in the second condition (the "gesture" condition) were additionally asked to slide wood blocks on the desktop to show what happened in the story during recall. There were three blocks total, one for the dictator's bunker measuring 1.5"x1.5"x1.5" and two for the army measuring 1"x1.5"x3.5". Subjects in both conditions were asked to carefully focus on what they were doing as they recalled the story. In the third condition (the "verbal" condition), subjects were not given any instructions in addition to verbally recalling the story. After the recall phase, all subjects were given the radiation problem under the guise of a separate problem-solving study. After working on the problem for 8 minutes, they were given it again with the hint that the fortress story might be useful.

We predicted that subjects in both the sketch and the gesture conditions would perform better on the radiation problem than subjects in the verbal condition simply because the additional activities might enhance and reinforce their memory of the story. More importantly, though, we predicted that subjects in the gesture condition would perform best overall because they would be more likely to simulate the story using a convergent-force image schema and would in turn have an easier time retrieving and aligning the story with based on the constraints of the radiation problem.

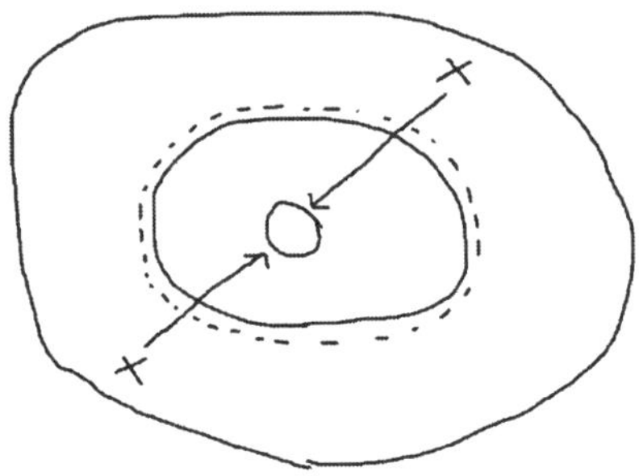

*Figure 2.* An abstract sketch illustrating the events in the fortress story drawn by a subject prior to working on the radiation problem.

Consistent with our predictions we found that significantly more subjects solved the radiation problem in the gesture condition than in the sketch condition. After eliminating subjects who already knew about treating tumors with converging beams of radiation, we found that 10 out of 33 subjects in the sketch condition solved the radiation problem before the hint to consider the story, versus 17 out of 33 in the gesture condition and 4 out of 21 in the verbal condition. The difference between the gesture condition and the sketch condition (52% vs. 30%) is reliable. The difference between the solution rates in the sketch condition and the verbal condition (30% vs. 19%), however, is not. There were also no reliable differences in solution rates among any of the conditions after the hint was given to consider the story. The lack of significant differences in solution rates after the hint could indicate that the task was not sensitive enough to reflect structural differences. Unlike in the previous experiment, no distracter story was used. Subjects thus might have been able to use a "solution-focusing" strategy based on a surface representation of the story and on pragmatic associations (e.g., between agents and goals) to construct a successful solution to the radiation problem (Gick and Holyoak, 1980; Keane, 1985).

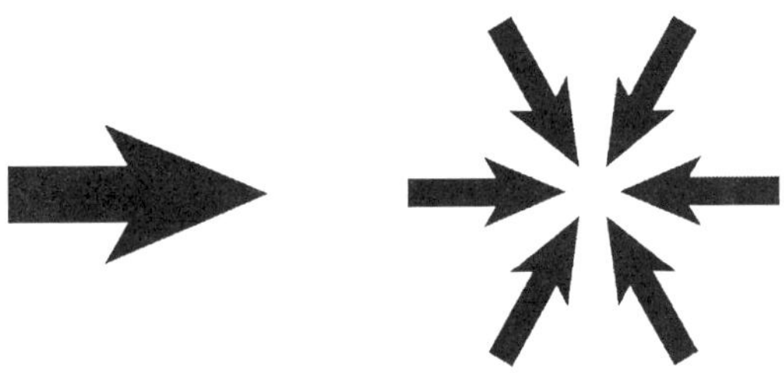

*Figure 3.* Convergence arrows similar to those used by Gick and Holyoak (1983) to help illustrate the events in the fortress story.

Before concluding this section, an additional comment on diagrams is due. Self-generated diagrams, as we found out, do not help as much as physically colliding blocks in facilitating analogous solutions to the radiation problem. We argued that although the activity of sketching is physical and although the resulting images might resemble elements of the target problem (Figure 2), sketching is not likely to activate a convergent-force image schema. One might still ask, however, why *viewing* such diagram (whether self-generated or not) does not elicit a convergent-force image schema either. The answer is that visual displays might elicit the schema, but that to do so requires visual features atypical of the sketches our subjects produced. First, viewing simple displays, like those showing converging arrows (Figure 3), is not likely to help much. Gick and Holyoak (1983), for example, found that subjects given converging-arrow diagrams in addition to the fortress story performed no better on the radiation problem than subjects who read the fortress story without the diagrams. Simple diagrams of this sort only convey a spatial schema, a pattern, in other words, similar to the convergent-force image schema but absent the force component. Without the force component, they fail to convey the fact that intensity builds at the point of focus, which we argued before might make alignment with the radiation problem significantly easier. Two experiments using modified visual displays have addressed this issue. In one study, by Beveridge and Parkins (1987), subjects were shown transparent, colored plastic strips that, when fanned out radially (in a star-like pattern), became increasingly light at the periphery where the strips did not overlap and dark at the center where they did overlap. The plastic strips show intensity growing at the center (albeit without convergent forces), and subjects who viewed this display did, in fact, perform significantly better on the radiation problem than subjects who viewed a static arrow diagram. In a separate study by Pedone, Hummel and Holyoak (2001), subjects were shown an animated version of the static converging-arrow diagram. Such a display appears ideal for eliciting a convergent-force image schema since the animated arrows literally "collide" at the center. Consistent with this analysis, subjects who viewed the animated diagram performed significantly better on the radiation problem than subjects who viewed a static diagram.

## 6. DYNAMIC PERCEPTUAL SIMULATION AND DIAGRAMMATIC SOURCES

Up to this point we have referred to perceptual simulations as if they were static. Because the perceptual system is capable of processing motion and other dynamic patterns independently of other perceptual features (such as size and shape), it is also plausible that people can store and later invoke dynamic image schemas for the purpose of structuring dynamic simulations. A simple image schema representing linear motion, for example, could be stored and then reactivated to imagine things in motion (Mandler, 1992). As with static simulations, coherence might determine when and how dynamic schemas are used. Physical properties of objects, that is, need to mesh with actions in a physically consistent way (Glenberg, 1997). An interesting finding that illustrates this possibility comes in a study by Schwartz and Black (1996). In one their experiments, Schwartz and Black asked subjects to view a diagram showing two gears meshed together, one larger than the other, and then decide whether marks on the circumferences of each gear would line up with one another if the gears were rotated. By comparing the angular disparity of the marks with response times, Schwartz and Black identified different strategies used by subjects to complete the task, one of which was mentally rotating the gears in a synchronous fashion. More interestingly, they found that when the gears were drawn to have rough surfaces rather than smooth surfaces, subjects were more likely to use the mental rotation strategy over all others, as if roughness made it easier to imagine one gear driving the other. In this case, the physical affordances implicit in the drawings might have affected the ease with which simulations based on those drawings were transformed.

In this section we consider the possibility that diagrams might help people construct perceptual simulations of source situations that can in turn be dynamically transformed into solutions to target problems. The process is similar to one of abstracting deep structure from a source situation and embedding it into a target situation. In other work we have referred to this as "generic abstraction" (see, e.g., Nersessian, 1999). In this section we focus on how perceptual simulations are transformed and specifically at how the perceptual context implicit in a diagram might influence the process. Unlike the experiments previously discussed, the experiments presented here all assume that source and target representations are available, thereby eliminating problems associated with retrieval.

We hypothesize that if subjects have access to a source solution but are not sure how to map it onto a target problem, a diagram of the former could help by setting up a perceptual simulation - static or dynamic - that can be transformed to fit the target situation. Crucially, though, we also hypothesize that the kinds of physical affordances implicit in such diagrams can have an impact, either enhancing or impeding the process of dynamical simulation.

To explore our hypothesis, we conducted an experiment in which we gave 209 Georgia Tech students two problems to work on along with a hint that the two might be solved in similar ways. The first problem, which was intended to serve as a potential source, asked subjects to design a door system that would let people in and out of a room without letting the air inside be contaminated by the air outside. The second problem, which was intended to serve as a target problem, asked subjects to design a pole capable of passing through another pole at right angles without breaking apart. We anticipated that if subjects came up with a solution to the door problem that involved redundant doors (either a vestibule system or a revolving door), they might be able to use it to come up with an analogous solution to the pole problem, namely one that incorporated redundant connections. We also predicted that although diagrams depicting a redundant-door solution might help facilitate the construction of analogous solutions, it would depend on the perceptual affordances implicit in those diagrams. Because the experiment is exploratory, we were not able to specify in advance what sort of differences might exist.

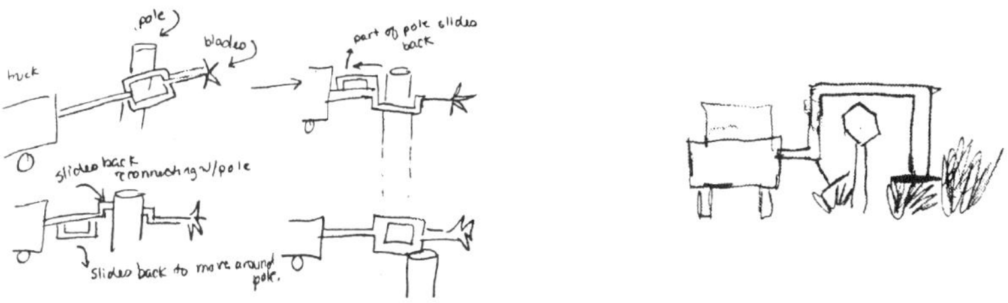


*Figure 4.* A good solution (left) to the "pole" problem that uses a redundant gate mechanism to allow one pole to pass through another, and a bad solution (right).

Of the 209 subjects who participated, 184 produced a "redundant-door" solution to the door problem. Of these subjects, 31 (17%) produced a solution to the pole problem that also involved some kind of redundant mechanism (Figure 4). In contrast, only 2 out of the 25 (8%) subjects who did not generate a redundant door solution to the door problem solved the pole problem. Drawing a diagram of a redundant door solution did not seem to help at all. Of the 131 subjects who drew diagrams, 22 (17%) solved the pole problem, compared with 9 of 53 (17%) who produced textual descriptions without diagrams.

The most interesting findings had to do with the types of diagrams people created. We found, for example, that the physical perspective of the diagram was correlated with whether subjects subsequently solved the pole problem (Figure 5). Of the 58 subjects who drew diagrams from a top view, 14 (24%) solved the pole problem, compared with only 8 of the 75 (9%) who used either used an oblique view, a side view, or an undetermined (either side or top) view. The difference in solution rates between those who drew top-view diagrams and those who drew other diagrams is reliable. We also found that the way the door system was drawn in relation to adjoining walls was correlated with whether subjects subsequently solved the pole problem. In particular, if the door system shared walls with a surrounding building – if, in other words, surrounding walls were continuous with the door system – subjects were much less likely to produce an analogous solution to the pole problem (Figure

6). Of the 104 subjects who drew physically continuous wall systems, only 10 (10%) solved the pole problem, as compared to 12 of 27 (44%) who drew discontinuous wall systems. These differences are also reliable.

The correlation between the frequency with which analogous solutions are generated and the perspective and physical continuity of source diagrams is consistent with the idea that subjects use diagrams to generate dynamic perceptual simulations and that they used these simulations, in turn, to try to adapt the source solution to the target context. In the analogy we are exploring, a significant difference in context exists. In the door problem, the object being designed (the door) is fixed relative to what passes through it (people). In the pole problem, however, the opposite is true. The object being designed (the pole) is moving relative to that which passes through it (another pole that happens to be stationary). If subjects dynamically simulate the door system and transform it into a pole solution, the motional context will have to be reversed. Subjects, in other words, will have to get the door system in motion. If this is the case, drawing the walls of the building to be continuous with the door system might cause the door to be perceived as being physically *encumbered*, and hence firmly rooted. The perspective of a diagram might produce an even more glaring perceptual discrepancy. If a door system is simulated from a side view it will be almost impossible to reconcile with the context of the pole problem, since, given the way the pole problem is stated, it is hard to see what is going on from the side. Based on our data, the physical affordances in the diagrams pertaining to the creation of a *dynamical* perceptual simulation appear to influence the transformation from source to target situation.

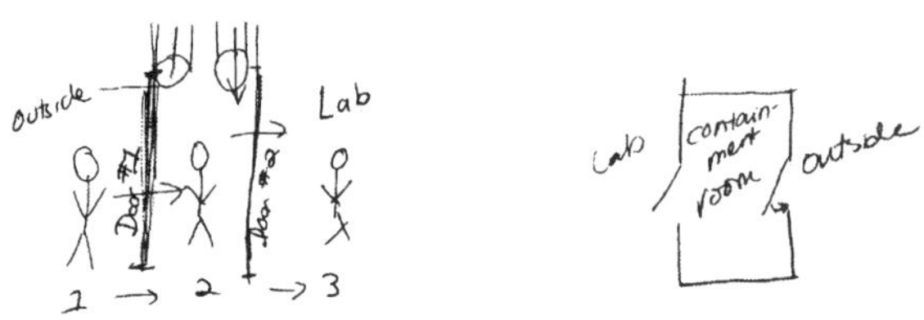


*Figure 5.* A solution to the door problem viewed from the side (left) and a solution viewed from above (right).

*Figure 6.* A solution to the door problem in which the door system shares a wall with the surrounding context (left) and one in which the door system does not share walls.

## 7. SUMMARY

The research discussed in this chapter has focused on the perceptual context of source situations - perceptual properties, in other words, that are causally disconnected from the goals and constraints of a problem as conveyed through predicate-based representations. First, we argued that different perceptual contexts could activate different image schemas, resulting in representations that differ in how they encapsulate information and thus in how easily they can be aligned with target problems. Our findings in the first two experiments are consistent with this analysis, although we acknowledge that they do not rule out other influences. We also argued that how different situations are portrayed through diagrams might affect how easily they are dynamically transformed into solutions to target problems. Our findings in the third experiment are consistent with this analysis in that contextual properties of source diagrams - notably physical perspective and physical continuity - were found to correlate with the production of analogous solutions to a target problem.

More work is needed to better understand the relationship between the perceptual context of source situations and analogies based on those situations, and to further develop the theory of perceptual simulation. Following up the research presented here, we are currently working on new studies that make a more concerted effort to control for superficial properties of source situations - both semantic and perceptual when manipulating image-schematic structure. We are also taking a closer look at the role of diagrams in analogical problem solving, this time controlling more precisely for the presence of superficial features when manipulating things like physical perspective and physical continuity.

Insights about perceptual context and perceptual simulation might also be gained reflexively, by applying the theory to real-world problem solving settings. How does an architect, for example, use drawings, physical models, embodied metaphors and fictional narratives when formulating analogies between building systems and other domains? How does a physicist make use of external representations in formulating theories? It might be found that perceptual context ultimately does more than guide interpretation or determine the choice of problem-solving strategies. Perceptual context, in fact, might be seen as a problem-solving tool itself, in the sense that problem solvers might consciously endeavour to create rich contexts for the purpose of facilitating their own thinking.

# REFERENCES


Barsalou, L.W., 1992, Frames, concepts, and conceptual fields, in: *Frames, Fields, and Contrasts: New Essays in Semantic and Lexical Organization*, E. Kittay and A. Lehrer, eds., Lawrence Erlbaum, Hillsdale, NJ, pp. 21-74.

Barsalou, L.W., 1993, Structure, flexibility, and linguistic vagary in concepts: Manifestations of a compositional system of perceptual symbols, in: *Theories of Memory*, A.C. Collins, S.E. Gathercole, and M.A. Conway, eds., Lawrence Erlbaum, London, pp. 29-101.

Barsalou, L.W., 1999, Perceptual symbol systems, *Behavioral and Brain Sciences* 22:577-609.

Beveridge, M. and Parkins, E., 1987, Visual representation in analogical problem solving, *Memory and Cognition* 15:230-237.

Bower, G.H. and Morrow, D.G., 1990, Mental models in narrative comprehension, *Science* 247:44-48.

Bransford, J.D., Barclay, J.R., and Franks, J.J., 1972, Sentence memory: A constructive versus interpretive approach, *Cognitive Psychology* 3:193-209.

Clement, C.A. and Gentner, D., 1991, Systematicity as a selection constraint in analogical mapping. *Cognitive Science* 1:89-132.

Craig, D.L., Catrambone, R., and Nersessian, N.J., under review, The role of implicit perceptual factors in analogical problem solving.

Davies, J., Nersessian, N.J., and Goel, A.K., 2001, Visual models in analogical problem solving, *Georgia Institute of Technology Cognitive Science Technical Report GIT-COGSCI-2001/03.*

Fincher-Kiefer, R., 2001, Perceptual components of situation models, *Memory and Cognition* 2:336-343.

Franklin, N. and Tversky, B., 1990, Searching imagined environments, *Journal of Experimental Psychology: General* 119:63-76.

French, M.J., 1994, *Invention and evolution: Design in nature and engineering*, Cambridge University Press, New York.

Gentner, D., 1983, Structure-mapping: A theoretical framework for analogy, *Cognitive Science* 7:155-170.

Gentner, D., Rattermann, M.J., and Forbus, K.D., 1993, The roles of similarity in transfer: Separating retrievability from inferential soundness, *Cognitive Psychology* 25:524-575.

Gentner, D. and Stevens, A.L., eds., 1983, *Mental Models,* Lawrence Erlbaum, Hillsdale, NJ.

Gick, M.L. and Holyoak, K. J., 1980, Analogical problem solving, *Cognitive Psychology* 12:306-35.

Gick, M.L. and Holyoak, K.J., 1983, Schema induction and analogical transfer, *Cognitive Psychology* 15:1-38.

Glenberg, A.M., 1997, What memory is for, *Behavioral and Brain Sciences* 20:1-19.

Glenberg, A.M., Meyer, M., and Lindem, K., 1987, Mental models contribute to foregrounding during text comprehension, *Journal of Memory and Language* 26:69-83.

Holyoak, K.J., and Koh, K., 1987, Surface and structural similarity in analogical transfer, *Memory and Cognition* 15:332-340.

Iverson, J.M. and Goldin-Meadow, S., 1997, What's communication got to do with it? Gesture in children blind from birth. *Developmental Psychology* 33:453-467.

Johnson, M., 1987, *The Body in the Mind: The Bodily Basis of Meaning, Imagination, and Reason,* University of Chicago Press, Chicago.

Johnson-Laird, P.N., 1983, *Mental Models: Towards a Cognitive Science of Language, Inference, and Consciousness*, Harvard University Press, Cambridge, MA.

Keane, M., 1985, On drawing analogies when solving problems: A theory and test of solution generation in analogical problem-solving tasks. *British Journal of Psychology* 76:449-458.

Klatzky, R.L., Pellegrino, J.W., McCloskey, B.P., and Doherty, S., 1989, Can you squeeze a tomato? The role of motor representations in semantic sensibility judgments, *Journal of Memory and Language* 28:56-77.

Kosslyn, S.M., 1994, *Image and Brain: The Resolution of the Imagery Debate.* MIT Press, Cambridge, MA.

Krauss, R.M. and Hadar, U., 1999, The role of speech-related arm/hand gestures in word retrieval, in: *Gesture, Speech, and Sign*, L.S. Messing and R. Campbell, eds., Oxford University Press, New York, pp. 93-116.

Lakoff, G., 1987, *Women, Fire, and Dangerous Things: What Categories Reveal about the Mind*, University of Chicago Press, Chicago.

Lakoff, G. and Johnson, M., 1998, *Philosophy in the Flesh*, Basic Books, New York.

Mandler, J.M., 1992, How to build a baby: II. Conceptual primitives, *Psychological Review* 4:587-604.

Minsky, M., 1975, A framework for representing knowledge, in: *The Psychology of Computer Vision*, P.H. Winston, ed., McGraw Hill, New York, pp. 211-277.

Nersessian, N.J., 1999, Model-based reasoning in conceptual change, in: *Model-Based Reasoning in Scientific Discovery*, L. Magnani, N.J. Nersessian, and P. Thagard, eds., Kluwer Academic/Plenum Publishers, New York, pp. 5-22.

Palmer, S.E., 1975, Visual perception and world knowledge: Notes on a model of sensory-cognitive interaction, in: *Explorations in Cognition*, D.A. Norman and D.E. Rumelhart, eds., Freeman, San Francisco, pp. 279-307.

Pedone, R., Hummell, J.E., and Holyoak, K. J., 2001, The use of diagrams in analogical problem solving, *Memory and Cognition* 29:214-221.

Rumelhart, D.E. and Ortony, A., 1977, The representation of knowledge in memory, in: *Schooling and the Acquisition of Knowledge*, R.C. Anderson, R.J. Spiro, and W.E. Montague, eds., Lawrence Erlbaum, Hillsdale, NJ, pp. 99-136.

Schank, R. and Abelson, R., 1977, *Scripts, Plans, Goals and Understanding: An Inquiry into Human Knowledge Structures*, Lawrence Erlbaum, Hillsdale, NJ.

Schwartz, D.L. and Black, J.B., 1996, Analog imagery in mental model reasoning: Depictive models. *Cognitive Psychology* 2:154-219.

Smith, E.S. and Medin, D.L., 1981, *Categories and Concepts*, Harvard University Press, Cambridge, MA.

van Dijk, T.A. and Kintsch, W., 1983, *Strategies of Discourse Comprehension*, Academic Press, New York.

# Building Demand Models to Improve Environmental Policy Process

Bryan G. Norton
*School of Public Policy, Georgia Institute of Technology, Atlanta, GA, USA,*
*bryan.norton@pubpolicy.gatech.edu*

Abstract: Integrated assessment modeling is gaining favor in several countries as a way of informing policy processes. Examining modeling styles and assumptions from the viewpoint of policy making, this paper, following Rotmans (1998), distinguishes between "supply" models and "demand" models, treating supply models as "curiosity motivated" and demand models as examples of "mission-oriented" science (Funtowicz and Ravetz, 1997). Models can be understood as communicative devices, similar to languages. A case study, based on research on the management of Lake Lanier in Northeastern Georgia, illustrates how failures of communication inhibit policy development. Competing interest groups employ differing mental models. A demand model might improve communication and cooperation by creating a shared model that clarifies these differences.

## 1. INTRODUCTION: USING MODELS IN MANAGEMENT PROCESS

"Model" is a very elastic term. Sometimes it is used narrowly and precisely, to refer to a set of mathematical formulas that describe the behavior of a system. But we also use the same term to refer to any simplified representation of a complex relationship. According to this broad definition, the just-mentioned mathematical equations that describe relationships are models, and so are many of our day-to-day proverbs, such as "a stitch in time saves nine," and "a rotten apple spoils the whole barrel." What models have in common, on this broad view, is that they somehow describe "how things

*Model-Based Reasoning: Science, Technology, Values*, edited by
L. Magnani and N.J. Nersessian, Kluwer Academic/Plenum Publishers, New York, 2002

work" in simplified ways - they represent a set of causal relationships that allow users of them to make sense of new information and to project, or at least speculate about, likely outcomes of actions. Models, however, are always *simplified* representations - they allow us to ignore some information as irrelevant and to consider certain possible outcomes as not worth worrying about because they are unlikely or because we are indifferent regarding them. Models, as simplified representations, can thus function to avoid information overload, and to provide a structure into which to integrate new information; if properly used, they help us to interpret and evaluate our experience.

A few groups of environmental policy researchers in several countries, sometimes referring to themselves as "Integrated Assessment Modelers", (see for example, Schneider, 1997; Rotmans', et. al., 1997; Rotmans and Van Asseltt, 1998), have undertaken to use modeling techniques to improve our understanding of environmental policy problems and solutions. These interdisciplinary teams - members of which are trained as technical, mathematical modelers - have developed many sophisticated models and model components, and some of them have experimented with using the models as tools in public management. For example, they often run simulations of their models, treating them as generators of possible scenarios that might unfold in the future. Some members of these groups have also become more reflective about their own role as modelers, and about the role of their models, in the processes of policy debate and formation. In this paper, I want to build upon some interesting developments in the European and American IAM communities, concentrating on the *motivation* and *uses* of models in actual policy processes. In particular, I want to ask the question: What kind of models would one build if one's motivation were mainly to build a model that would be *useful* in environmental policy processes?

I have defined models as a kind of simplified representation - but we must be very careful with this term, "representation"; models represent reality in a very broad sense of that term and, speaking logically, models have more in common with analogies, metaphors, similes, and simulations than they do with any strict sense of empirical representation. Empirical models based on measurable data about the real world are one kind of model, but mimicking and predicting nature is only one of many possible uses of models. This point is important because it is easy to over-emphasize the more descriptive and representational aspects of models, treating them as "predictions" or "causal sequences", despite many warnings by modelers that models offer not predictions, but at best "projections".

The shift from treating models as predictive and representational to understanding them as projections is the first step on a slippery slope on which it becomes harder and harder to maintain that models are, by their nature, representational at all, because making projections is not a descriptive activity but a goal-directed activity. As such, it must be understood by reference to a stated goal or purpose. Models are better seen as tools to fulfill various purposes; in constructing models we therefore face the question of motivation, because one can only evaluate the success of a modeling effort if one determines the extent to which the purpose of the model is fulfilled. There has been little discussion of motivations for modeling, probably because motivations are understood as invisible and "inside", and are often not discussed systematically. In this paper I take the viewpoint that, while some models are clearly designed to be predictive and descriptive, a model is always judged according to its intended use. In this paper, I look at modeling from the viewpoint of the uses to which the models will be put in an environmental policy process, such as an ecosystem management program, or a watershed management program with a citizens advisory committee and ongoing public deliberation about what to do to protect an important resource. For concreteness, I refer to such processes as "adaptive management", although one will find them referred to variably as "ecosystem management" or "community-based programs". Adaptive management, the version favored by Holling, Lee, and others, emphasizes the importance of experimental management carried out within ecologically defined complex systems (Holling, 1978; Lee 1993; Gunderson, Holling, and Light, 1995). Once one looks at environmental modeling as part of a management process, one can think of a variety of uses for models, to describe and predict, yes, but also to communicate, to persuade, to plan, and for many other purposes. In particular, I concentrate on the use of models to improve communication among stakeholders and participants in environmental policy processes.

In Section 2, I discuss important changes in the way practical philosophers think about science and its role in the policy process. Especially, I discuss why the often-implied "serial model," on which science is completed prior to addressing policy questions, is inadequate, and examine recent calls for more "mission-oriented science". This discussion will set the stage for asking the question: how can models be useful in policy processes? As a first step in this process, I articulate a distinction made by some IA modelers between "supply" and "demand" models. Then, in Section 3, I develop an analogy between models and languages, urging that we focus explicitly on the constraints placed on models by our intentions to use them to respond to real environmental problems that are faced within complex, real-world situations. Here, it will be useful to introduce, by analogy to "formal" and "semantic" systems in logic, the idea of formal, as opposed to mental models. This convention provides us with a useful analysis of the relationship between technical and public discourse, and between formal and mental models, and allows us to characterize problems of communication that can occur when people with different interests discuss what to do. In Section 4, I present a case study, a study of one area's struggle to develop a watershed management plan for an important resource, Lake Sidney Lanier. Lake Lanier is a large, multi-use impoundment in Northern Georgia; so far, discussion of an integrated management plan for the watershed has been characterized by lack of cooperation among jurisdictions and failure of communication across interest groups. This case study will illustrate how some of the conceptual distinctions drawn here can help to characterize the disagreements and lack of cooperation among parties in the dispute, and can be used to form hypotheses regarding how to increase communication and common understandings among participants in a public, adaptive approach to environmental management.

## 2. MISSION-ORIENTED SCIENCE AND THE DEMANDS OF DEMAND MODELING

The traditional view of science as being constituted by "basic" science and "applications" of basic science, is proving inadequate to understand the interplay of science production and use in complex public decision making. The problem was effectively characterized in a classic theoretical paper by Rittel and Webber (1973), who distinguished "benign" and "wicked" problems. Benign problems, such as a mathematical problem or a straightforward scientific problem, have a single answer, and once this answer is found, there is ready agreement on a definitive answer. Benign problems can be formulated in a precise way, and admit of a single, well-determined answer. Problems involving public values and diverse interests, on the other hand, are "wicked" problems (though Rittel and Webber are careful to disassociate their use of the label with moral opprobrium) in the sense that there is no single, accepted formulation of the problem, and answers are often in more-or-less terms in which planners and managers can at best find reasonable, but shifting balances among competing interests and values." Urban sprawl exemplifies a wicked problem. Another such example, to be explored in Section 4, below, is the management of a multi-purpose lake and the watershed around it. Members of different interest groups and of different scientific disciplines characterize the problem differently, as they differ in what is thought to be symptom and what is judged to be a cause; as Rittel and Webber say, the correct formulation of the problem cannot be known until a solution is accepted. They say, "The information needed to understand [a wicked] problem depends upon one's idea for *solving* it." (161)

Rittel and Webber showed - and no author known to me has provided an adequate rejoinder to their argument - that wicked problems including, all controversial public policy decisions, cannot be adequately analyzed by using optimization models. The sophisticated models of operations research or microeconomics, in other words, cannot be employed until wicked problems are made benign by the acceptance of a problem formulation, which often implies a solution in its acceptance. They say, "setting up and constraining the solution space and conducting the measures of performance is the wicked part of the problem", implying that optimization modeling is only applicable after the hard work is done. For our purposes, we can emphasize one particular characteristic of wicked problems: wicked problems have no "stopping rule". Rittel and Webber contrast policy decisions to chess, which has a stopping rule, checkmate, arguing that wicked problems must be seen as an open-ended, constant and ongoing compromise between competing interests. This characteristic suggests that dealing acceptably with wicked public policy problems must be iterative and ongoing.

Rittel and Webber's idea of social problems in modern society as having no stopping rule is also consonant with interesting developments in ethics and moral discourse. Sen's formulation of the Liberal Paradox, for example, supports the idea that political problems in a democracy cannot be modeled formally as optimization questions; negotiations among people with different values and utility functions is essential to reach rational, cooperative solutions to complex social problems (Sen, 1970a; Sen, 1970b; Sager, 1997). Similarly, the German pragmatist, Habermas has usefully noted that law in a democratic society must always be two-faced; it must on the one hand be binding, and enforced, to be law; but it must also be "legitimate" if the democracy is to even approach the ideal of government by consent (Habermas, 1998). This dual aspect is crucially tied to the idea of iteration because, according to Habermas, the losers in any particular legislation or adjudication are always capable of raising the issue again once new evidence is gathered or new affected interests are voiced. This open-ended aspect of deliberation in complex democratic processes reinforces Rittel and Webber's idea of iteration, and provides a possible connection between discourse ethics and the study of policy discourse. Details of this connection must be left for another place (Norton, in preparation.).

Interestingly, most environmental modeling violates this important policy-based requirement, as most flow diagrams of science/decision models are unidirectional and lack loops back to science to take into account new information and changes in interests. The Environmental Protection Agency's "Risk Analysis/Risk Management" model, which was used as the dominant decision tool at the Environmental Protection Agency and other regulatory agencies from 1983 (National Research Council, 1983) until recently is only the most notorious example. On this model, science is gathered, the risk is analyzed, and the risk is "characterized" - a neutral-as-possible-summary of the scientific data - and the problem is then "managed" by actors who are sensitive to economic valuation and to public opinion (usually, at EPA, political appointees). This is a perfect example of what I have elsewhere called the "serial view" of science and policy. It is as if scientists study the world without any indication of what is important to the society; then, once the best available science is gathered, it is possible to move on to consider policy. Other examples of this serial understanding of science and policy abound (Norton, 1995b). I intend, by contrast to this serial understanding of science and policy formation, to understand the use of models in policy process as an iterative process of choosing, discarding and modifying models. Science, when employed to address wicked problems, must be embedded in an ongoing process in which scientific data and policy demands are both imposed upon scientific experimentation. In such a process - what I have here called adaptive management - science must be viewed not only as an end product that feeds into the policy process, but also as a flexible tool to address open issues and public controversies about what is happening and about what should be done.

Given the firm starting point of necessary iteration, apparently supported by powerful theoretical proofs and by developments in ethics, we can build on Rittel and Webber's work by stipulating that we seek models that will work in an iterative, adaptive process in which there is ongoing, democratic deliberation about what to do - a democratic approach to what I have called adaptive management. When science is undertaken within the policy process, a new understanding of the separation between basic and applied science emerges. Under a dynamic and circular/iterative approach to science and policy, it is possible, indeed important, for policy discourse to query science, to encourage certain experiments, and to cooperate in the development of management activities that will allow testing of hypotheses and evaluation of management goals (Plough and Krimsky, 1990). Now it is possible to see how the old distinction between pure and applied scientific research has become inapplicable, given the importance of iteration. Once one recognizes the importance of both science and political process in formulating wicked problems, science must be embedded in an iterative system of problem formulation, identification of uncertainties, experimentalism and research demanded by management. The iterative process is driven by a social need to know. It is reasonable on this new understanding of science and policy for an unresolved scientific question to be subjected to the rigors of basic scientific research, because that question is important in resolving a management controversy. Scientific work, so characterized, would be "basic" science - in the sense that it must be rigorous and quantitative, designed to determine how the system in question works or responds to a given treatment - but the subject matter chosen for research is chosen to serve a dynamic that is important in producing public goods. Science in this vein is firmly embedded in praxis; it is problem-oriented and practical.

Here, it is useful to introduce another distinction, due to Funtowicz and Ravetz (1994, 1997), who divide science into "curiosity-motivated" and "mission-oriented" research. Historically, disciplinary scientists have engaged in curiosity-motivated science; problems and solutions have been defined and pursued according to standards that are articulated independently of practical applications, "internal", one might say, to the defined goals of the discipline. Important arbiters of curiosity-motivated science are professional societies reinforced by the peer review process, which are used to control access to journals, government funding for research, and disciplinary status.

Curiosity-motivated science is guided by the strictest controls in the use of the scientific method; it provides models based on the cutting-edge theories and techniques available in a scientific discipline, and is often pursued within the confines of a single academic discipline. In early forays into more interdisciplinary work, most scientists have retained this academic bias, attempting to minimize the effect of values on their work and struggling to develop cross-disciplinary science that attains the standards of disciplinary methods. "Post-disciplinary" or "post-normal" science, on the other hand, is value-laden science that is undertaken as a part of a mission. It is science of this latter sort that is most likely to be relevant in Adaptive Management processes such as Lake management (Norton, 1995, 1999; Norton and Steinemann, forthcoming). Mission-oriented science differs from traditional, disciplinary and curiosity-motivated science in that the community that reviews scientific results is expanded to include not just scientists from established disciplines, but also to include affected parties - "stakeholders - who have varied interests and viewpoints, and who express these viewpoints in an open and public process. One might say that, in mission-oriented science, proposed research must fulfill the requirements of "good science" and, in addition, it must be socially relevant. Funtowicz and Ravetz (1994, p. 206) say:

> Curiosity-motivated and mission-oriented research have complementary properties; the former produces public knowledge but within rigid disciplinary boxes; the latter is transdisciplinary, but bureaucratized and private. The issue-driven research of post normal science must combine the positive features of these other forms, and develop appropriate institutional arrangements or structures to achieve this.

Having developed this more dynamic, iterative, policy process and a social-values-motivated approach to mission-oriented science, we can now return, more specifically, to the role of models in such an iterative process.

Following Rotmans (1998), I distinguish between models that are motivated from the "supply" side and those that serve the "demand" side. In supply modeling, scientists and mathematicians survey available techniques and frameworks, attempting to push forward the frontiers of knowledge in a particular direction. In supply modeling, a model is proposed, based on best available techniques, and designed to increase our understanding in important ways, where importance is understood according to accepted disciplinary paradigms. Demand modeling, by contrast, is motivated by a social problem or issue. Professional modelers in this case bring their techniques and frameworks into a practical situation and success or failure will depend upon whether proposed models improve the decision process.

We can therefore understand Rotman's category of supply models as models that are motivated by curiosity. The goal of supply modeling is to use cutting-edge tools of science to advance our understanding, and in the process to improve those tools to allow further advances. Supply modelers apply such stringent requirements because their goal, as academic scientists, is to push the envelope of our basic understanding of the world. Here, we demand the strictest levels of proof, and we attempt to describe phenomena independently of perspectives or interests, treating value-neutrality as an ideal to be striven for, if never to be fully reached. Demand modeling, on the other hand, is often constrained by bureaucratic controls as well as the need to deal with unexpected, action-forcing events. In demand modeling, it may not always be possible to apply the normal, stringent scientific standards. When some action is forced - when inaction itself will have irreversible impacts - it may be appropriate to apply the criterion of "preponderance of evidence", rather than the academic scientists' stronger standards of proof (Norton, 1999).

Supply models - like disciplinary science - motivated by an interest in developing and using disciplinary (or super-disciplinary) tools, exercising these tools, and extending their application to new and unexplained areas of knowledge. The motivation is directed at advancing the internal goals of a discipline or super-discipline. Descriptive validity, often measured as success in predicting outcomes in experimental or other controlled situations is often the most important measure of success in supply modeling, Demand modeling, on the other hand, arises in situations of social conflict, in response to real and sometimes-bitter disagreements about what is happening and what should be done. Demand models are constructed within the maelstrom of public discussion and debate; they respond directly to problems about what to do. Adaptive managers, ever-alert to the importance of introducing knowledge that reduces uncertainty in particular situations, will have occasion to refer to both kinds of models; but the goals of management and the need to integrate information across disciplines creates its own demand for studies to resolve specific disagreements in specific action situations. This latter, mission-oriented style of science – and the demand-type models it produces - will thus play the most central role in Adaptive Management activities.

One might ask: But won't supply models be important contributors to demand modeling? Sometimes. One way to make supply models into demand models is to develop a complex and methodologically sophisticated mathematical model, which combines great power with analytic precision, in a demand context and then build an "interpretive shell" around the mathematical model, allowing users lacking specialized training to interact with the model, even though they have very little understanding of the internal workings of the model. While this can be a useful strategy in many situations, and can be one effective way for scientists to contribute to policy processes, it is difficult to gain the transparency necessary for a model to work as an educational and communicative tool. These activist goals diverge sharply from the goals that motivate the original supply-driven process of building the model, which was built under a motivation of curiosity, not issue-driven communication. It remains to be seen, then, whether the strategy of building supply-motivated models outside a problem-context and then retrofitting them to fulfill a demand in real, conflict-ridden management situations will prove more useful than building demand-oriented management models from scratch in specific demand contexts.

The attitude of demand modelers toward supply models that have been already constructed is that of a mechanic in a yard of junk cars. Supply models represent a vast array of components and spare parts that can be adapted to the demand modeling problem at hand. To salvage components from existing supply, however, requires that we judge the component not only for its scientific precision and accuracy, but also for its usefulness in solving the problem at hand. And, we must continually remind ourselves, in a demand situation, the goal should be to create a *useful* model, a model that will improve communication in a particular management context. Obviously, we assume that demand modelers will want, in the end, a model that "runs" - that generates plausible possible futures for the system; but as demand modelers we are also interested in whether use of any given components will create a model that increases communication, reduces misunderstandings, and institutes creative dialogue designed to learn from disagreements.

Figure 1 provides an abstract representation of the role of a demand model in a particular management situation that has been studied by our research team. "The Table" represents the ongoing public discussion, negotiations, and deliberations that constitute the process of participatory decision making. The demand model sits on the table, waiting to be invoked to clarify agreements and disagreements that occur in this process. The public is understood as organized into many "cells", which are more or less organized groups, centered around varied interests. Insofar as different interest groups view the problem differently, from different value perspectives, systematic miscommunication can take place. These types of communicative failures are endemic when communities face and address wicked problems.

This conception of ongoing experimental science as embedded in a larger, iterative decision-oriented process, gives new depth and meaning to the starting point of this paper, which was described as examining the motives for modeling within ecosystem and watershed-sized public decision processes. While models that will be useful in reducing confusion and failure of communication, for example, will no doubt have to be descriptive of natural systems, they must also fulfill a more practical goal of encouraging common understandings and the development of new, shared analogies and metaphors that will reduce dissonance and miscommunication.

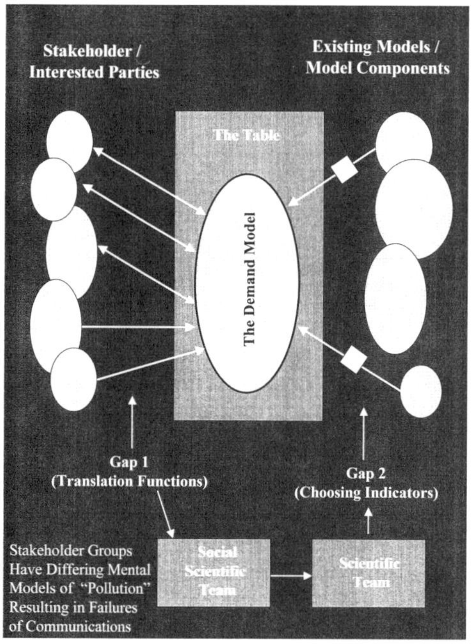


*Figure 1.* The role of a demand model.

## 3. MODELS AS LANGUAGES

As simplified representations, models can be thought of as a special category of linguistic entity; a model, once in use, conveys information by virtue of conventions, as do languages. In this Section, I examine models and their use in policy process by thinking of models as linguistic conventions; as such, they can be well suited or not so well suited to particular tasks. Once we begin thinking about models as linguistic tools, it also seems natural to expect models to be as diverse as the uses to which they are put in both everyday and in scientific situations. If we consider models to be as diverse as languages - and to have as many uses as languages - and languages to be either natural or consciously constructed, then we might liken scientific models to the formal, constructed languages of logic and some branches of mathematics. They are constituted by consciously chosen rules, rules that are constructed for particular purposes, and they are judged and chosen according to their success in achieving those purposes.

This pragmatic approach is in keeping with a conventionalist view of language. Conventionalism regarding language has largely replaced the traditional "essentialist" view of language, which saw objects and collections of objects as "given" in reality. On the essentialist view, linguistic assertions can be seen as "picturing" pre-linguistically organized elements of nature. Post-positivist philosophers have argued, however, that consistent application of empiricist principles undermines essentialism and encourages a more conventionalist view on which the objects and categories attributed to nature by language users are constituted by linguistic choices, not by pre-linguistic, self-evident essences (Quine, 1953, 1961). Once this conventionalist viewpoint is adopted, it is appropriate to be "tolerant" about choices of linguistic forms (Carnap, 1937); and, once conventionalism is adopted, it becomes possible to evaluate, and alter, linguistic forms that are ill suited to accomplish important human purposes, and to construct new linguistic forms for explicitly adopted purposes (Carnap, 1937 1950; Norton, 1977, 1978).The emergence of conventionalism, thus, inspired the study of the use of language in human communication - the "pragmatics" of language (Carnap, 1955). This paper, which aims to examine uses and motivations of models viewed as linguistic systems for communication, can then be viewed as an exercise in practical pragmatics of the language of environmental management science.

Notice that the term "language" is ambiguous in a way that parallels the narrow and the broader senses of "model" mentioned in the Introduction, above. While the term language usually refers to the everyday speech patterns of many cultures - what are called "natural" languages - logicians and philosophers also speak of "formal" or "canonical" languages. While one can imagine mixes and in-between cases, we can also think of models as falling into two main functional categories. Some models arise "naturally", and are constituted by their actual uses in real situations. Other models, "constructed models", are constituted by the rules that govern their use. The former resemble everyday proverbs - they have evolved to fulfill a function in a culture - while the latter, more formal models are constructed, built up of basic elements and transformation rules, like logicians' languages. If both models and languages are understood as conventionally formed symbolic systems, we can understand mental models as naturally forming a part of day-to-day activities, or these systems can be consciously "constructed", as in formal languages of logicians. Either way, the symbol systems can be understood as conventions; they are therefore mainly functional instruments that can only be judged by their usefulness for specified purposes. Natural languages survive if they are useful for communicating in real day-to-day tasks; they become embedded in day-to-day processes and survive, evolve, and adapt. Constructed languages, on the other hand, do not occur naturally; they are consciously contrived, constituted formally, by rules alone, to serve a particular purpose. These languages need not be embedded in any day-to-day practice at all. They are constituted by their rules, not by their usage, and they are in this sense contextless. Relying on this broad separation of models into "natural" and "constructed", it may be possible to think of most supply/curiosity-motivated mode as constructed and "formal" in the sense that careful development of a model requires that rules of use be stated carefully and definitively. A person could therefore make a mistake in using a formal model - by breaking one of its rules, or by solving equations in the wrong order, for example. If someone consciously applies a different set of rules, we conclude that a new model has been proposed and used. A formal model is thus defined by its rules.

If we, on the other hand, think of the at-best-partly conscious cognitive structures that we have described as naturally occurring models, it is interesting to think of these as "mental models". This distinction depends on the way the model is constituted and defined. A formal model is abstract in the sense that it is defined according to a set of rules, definitions, equations, etc. A formal model can be employed by a person, but the model exists independently of that person, in the form of a set of rules that determine the behavior of the model. A mental model, by contrast, is an internalized, operative model that must be studied through psychology, anthropology, and sociology. It is defined by the ways it is in fact used by persons or groups in a real situation. Mental models are simple representations of complex systems that people use to understand and predict the future, and to interpret new events and information as they encounter it. Note that someone - after working the same job for years, say - might internalize a formal model so that it "becomes" a mental model for that person. But the two remain definitionally and intentionally distinct - the mental model is behaviorally defined while the formal model is defined by its rules. Conversely, a psychologist might attempt, by eliciting behaviors or comments, attempt to "formalize" a person's mental models on a subject, by writing a set of rules that will predict the person's behavior in response to various stimuli.

This basic distinction cuts through all models; it defines two realms of modeling, one of which - mental modeling - is in some sense descriptive of actual thought processes and the other of which is prescriptive in the sense that certain rules prescribe behaviors of a system. Given the conventionalism advocated above, and thinking of models as linguistic items, we adopt a tolerant attitude in the construction of models, encouraging people to develop and explore many types of models for many purposes. This basic distinction between formal and mental models, however, encourages us to be clear at the outset whether the models we build are supposed to guide real behaviors or whether, like most formal models, they are idealizations constituted by the rules we write for them.

The purpose of introducing this meta-linguistic apparatus has been to prepare the way to characterize a class of problems that might be addressed through the construction of a demand model. In the next Section, I illustrate how the conceptual innovations here introduced allow a fruitful characterization of the problems faced in adaptive, community-based environmental management processes. Failures thus far in achieving integrated and cooperative management in the watershed are thus proposed as examples of problems that might yield to a carefully constructed demand model. In this way, I hope to illustrate the one important contribution that demand modeling can make to the development of a more effective watershed management process there.

## 4. CASE STUDY: LAKE SIDNEY LANIER

Lake Sidney Lanier, located 25-70 miles North of Atlanta in the foothills of the Southern Appalachians, is a large, multi-purpose impoundment of the Chattahoochee and the Chestatee Rivers, built and operated by the Army Corps of Engineers. Buford dam was approved by Congress in 1946; construction began in 1950 and the Lake filled by 1959. Originally intended to provide hydropower and control water flow, Lanier is today valued also as a major recreational resource, providing swimming, boating, and fishing – including an economically productive sport fishing industry. Lake Lanier receives more visitors per year than any other Army Corps lake project in the United States. The area surrounding the Lake is in the midst of a boom in second-home, retirement, and "bed-room" development on its shores and in surrounding areas. Lanier is also the main source of drinking water for many Atlanta communities, so interest in water quality is intense.

It is a major problem that Lake Lanier is in the path of Atlanta's most aggressive expansion, across the "Northern Arc", which stretches between the two main interstate highways, Interstate 75 and Interstate 85, and into the foothills of the Appalachians. A wave of residential growth in the 1970s and 1980s has been supplemented with a rapid growth in the high-tech business sector in the Northern exurbs, stimulating leap-frogging growth of Atlanta in a Northerly direction. While the legendary growth of Atlanta proceeds in most directions, the velocity of the Northward movement of the city causes special problems. The population center of the city is now five miles northeast of the center of town, and seven out of 10 residents live north of I-20, the main East-West expressway (Auchmutey, 2001). This Northward movement of the population and industrial centers of the city can only result in more congestion and inefficient transportation patterns, as more and more traffic must pass through or around the city center in order to reach Hartsfield International Airport, which anchors the city on the southern perimeter. Here, we are concerned, specifically, with the impact of this concentrated growth on Lake Lanier, which is now the center of some of the most rapid exurban growth anywhere in the United States. Rapid development stresses transportation, education, and waste-water treatment systems, and threatens the health of the Lake, as well as the watershed on which the Lake depends. With rapid growth in all the counties abutting the lake, and in the communities clustered around the Lake, the area represents a useful case study of what the US EPA calls a "rapidly urbanizing watershed."

In our social science research on the residents of the Lanier watershed, we have encountered two problems that seem to us to be illuminated by model-based and analogical thinking. We can illustrate some of the abstract concepts introduced in the last Section by applying them in a preliminary way to the Lanier case and to our attempts to build a "demand" model that will improve communication and clarify issues in public deliberation about what to do. The first problem we will examine is the lack of a "sense of place" around Lake Lanier, while the second, perhaps related, problem is a divergence in the "mental models" that members of different interest groups use in understanding the processes changing the Lake. Since these characteristics have been discussed in some detail elsewhere (Norton and Mills, forthcoming), here I will only describe the problems in enough detail to illustrate the possibilities of demand modeling.

**Problem 1:** residents of the area do not have a sense of place associated with the area as a whole. As an artificial lake, Lanier has no historic traditions and little "sense of place" attached to the Lake, as opposed to town, village, and development.

Environmental management today - especially management of surface-water and watersheds - is becoming more holistic, more adaptive, and more community-based. Management of large lakes, river systems, and estuarine systems must be holistic because lakes draw their water from a larger watershed, as non-point-sources of pollution become more and more important; the quality of a lake's water can be no better than the quality of the land and the land uses in the watershed it drains. Management must become more adaptive, because it turns out that management of lakes and watersheds requires pursuit of a moving target, an ongoing attempt to set goals, to undertake experimental projects, and then to monitor these scientifically, and to adjust goals and objectives of management accordingly (Gunderson, Holling, and Light, 1995: Lee, 1993). Finally, environmental management is becoming more community-based and more reliant upon inputs from local communities in goal-setting, in the choice of regulatory regimes, and in implementation (National Research Council, 1996; Norton and Steinemann, forthcoming). This is especially true of multi-purpose resources like Lake Lanier, the management of which almost always involves choices that affect competing uses and which involve conflicts among the multiple goods sought by local and regional interest groups.

Successful community-based, holistic management, requires cooperation, trust, and willingness to deliberate in search of acceptable solutions. To illustrate this point, consider the development, over the past thirty-odd years, of a strong and resilient public policy process associated with the management and protection of the Chesapeake Bay, the large and incredibly productive Bay that is located in one of the fastest-developing regions of the United States. In the early 1970s, there were initial outcries that the Bay was becoming dirty, polluted, etc., and a large EPA study of the causes of increased algal production and decreasing range of submerged aquatic vegetation was once undertaken. One can say of this case that the residents of the states surrounding the Bay set out to defend it as a "place" that they identified and cared about. It was this sense of place that unified groups and individuals from different jurisdictions to work together to manage the whole Chesapeake watershed (Horton, 1987).

So far, management of Lake Lanier has been fragmented by conflicts among multiple jurisdictions, hampered by lack of trust, and inflamed by frequent resort to lawsuits, which appear to be the only way differences regarding Lake management are expressed. Will it be possible for the citizens living in the area surrounding Lake Lanier to develop a more holistic approach? Will they progress toward cooperative solutions to the problems that are sure to accompany urbanization of the Chattahoochee watershed? What would be required for the areas north of Atlanta to begin managing Lake Lanier cooperatively as a community resource?

If one follows geographers and draws a distinction between "location," and "place," according to which *location* can be fully specified by physical coordinates, whereas *place* includes the richness and texture of a distinctive natural and cultural history, we can form the following explanatory hypothesis:

**Hypothesis 1:** the area around Lake Lanier has not become a "place" for residents of the area; they so not model the lake mutually as a unitary watershed. While some of the municipalities around the lake may be experienced as places, the Lake and its watershed are not so viewed.

**A possible solution to problem 1:** a demand model could, conceivably help residents to begin to see the watershed as a whole ecosystem; this might inspire a fuller sense of community around the Lake, and reduce the temptation to "go it alone" in water quality management. What is involved, here, cognitively is a shift in the "ontology" of the models residents and stakeholders use to understand Lake problems and management possibilities; this ontological shift is also a shift in scale - if the stakeholders around Lake Lanier saw the Lake and the surrounding counties as a whole watershed they would be more mentally prepared to consider management of the system as a whole system.

**Problem 2:** there is a serious lack of communication, with representatives of various interest groups often talking past each other. I hypothesize that different groups have different "mental models" of the processes that pollute the lake. As noted above, mental models are here understood as belief structures that allow individuals to interpret events, to assimilate new information, and to project current effects into the future

**Hypothesis 2:** different stakeholder groups around the lake have different "mental models" of how pollution and degradation of the lake occur.

**A possible solution to problem 2:** perhaps a neutrally developed demand model could reduce failures of communication by allowing members of interest groups to participate in the development and testing of the model. Because participants in discussions employ different representations in understanding the Lake and threats to it, there is great potential for misunderstandings and miscommunication, as is exemplified in Lanier management today. Perhaps the demand model can provide a neutral framework that will allow discussants to communicate more clearly. In this case, a demand model would fulfill the function of providing a neutral representation into which discussants can translate their beliefs and concerns, reducing miscommunication. This approach may also help us to address Problem 1 because, if a demand model could be constructed that would relate multiple perspectives and interests behind a watershed-sized model of the Lake, this more inclusive model would stand as a shared representation of the Lake, one that might relate the multiple perspectives of the different interest groups more clearly in a unified model. This understanding of the problem, however, suggests another, perhaps more radical hypothesis:

**Hypothesis 3:** the mental models employed by stakeholders are a function of the interests and values they hold and of the interest groups they identify with.

I call this hypothesis radical because it seems, contrary to the assumption of many academics and other researcher's assumption that "rational" actors first observe and describe a system, and then "evaluate" the system or changes to it. Hypotheses 3, however, suggests a reversal of the usual assumption of rational actor models that homo economicus acts only after having gathered full knowledge, implying instead that the values people hold, and the interests they pursue, determine which facts about a physical system are most likely to be believed and considered relevant.

**Conclusion:** complex management problems, such as managing complex, dynamic systems including lakes, bays, and watersheds, depend on shared understandings of problems; lacking this, members of competing interest groups will be addressing different forms of the management problem, employing different mental models, and considering different scientific evidence to be relevant. I have developed a set of meta-level concepts for talking about the process of developing demand models for the purpose of improving public processes and deliberation affecting management of complex systems. In this process, I hope I have also revealed how model-based thinking, and thinking about the motives for employing models, can direct environmental managers toward more interesting and tractable questions, questions that might, over time, create consensus regarding beliefs and management goals.

## REFERENCES

Auchmutey, J., Farther than the eye can see: It's all Atlanta, *Atlanta Journal-Constitution*, April 1, 2001: A1, A12, A13.

Carnap, R., 1937. *The Logical Syntax of Language*, translated by A. Smeaton, Degan Paul Trench, Trubner & Co, London.

Carnap, R., 1950, Empiricism, Semantics, and Ontology, *Revue Internationale de Philosophie* 4:20-40.

Carnap, R., 1955, On some concepts of pragmatics, *Philosophical Studies* 6:89-91.

Environmental Protection Agency, 1997, Community-Based Environmental Protection: A Resource Book For Protecting Ecosystems and Communities. http://www.epa.gov/cbep/pdf/coverrev.pdf.

Funtowicz, S. and Ravetz, J., 1994. The worth of a songbird: Ecological economics as a post-normal science, *Ecological Economics* 10:197-207.

Funtowicz, S. and Ravetz, J., 1997, Science for the post-normal age, in: *Perspectives on Ecological Integrity*, L. Westra, ed., Rowman and Littlefield, Lanham, MD, pp. 146-161.

Gunderson, L., Holling, C.S., and Light, A., 1955, *Barriers and Bridges*, Columbia, University Press, New York.

Habermas, J., 1998, *Between Facts and Norms: Contributions to a Discourse Theory of Law and Democracy*, translated by William Rehg, MIT Press, Cambridge, MA.

Holling, C.S., 1978, *Adaptive Monitoring and Management*, John Wiley and Sons, New York.

Horton, T., 1987, Remapping the Chesapeake, *The New American Land*, September, October, 1987.

Lee, K., 1993, *Compass and Gyroscope*, Island Press, Washington, D.C.

National Research Council, 1983, *Risk Assessment in the Federal Government*, National Academy Press, Washington, D.C.

National Research Council, 1996, *Understanding Risk*, National Academy Press, Washington, D.C.

Norton, B., 1977, On the metatheoretical nature of Carnap's philosophy, *Philosophy of Science* 44:65-85.

Norton, B., 1995a, Integration or reduction: Two approaches to environmental valuation, in: *Environmental Pragmatism*, A. Light and E. Katz, eds., Routledge Publishers, London.

Norton, B., 1995b, Ecological integrity and social values: At what scale? *Ecosystem Health* 1(4):228-241.

Norton, B., 1999, Pragmatism, adaptive management, and sustainability, *Environmental Values* 8:451-466.

Norton, B., in Preparation. *Constructing sustainability*.

Norton, B. and Hannon, B., 1998, Democracy and sense of place values, *Philosophy and Geography*, 3:119-146 (Philosophies of Place).

Norton, B. and Steinemann, A., forthcoming, Environmental values and adaptive management, *Environmental Values*.

Norton, B. and Mills, S., in preparation, Community-based evaluation and planning in rapidly urbanizing watersheds: the case of lake lanier", *Urban Environmental Ethics*, A. Light, ed.

Plough, A. and Krimsky, S., 1990, The emergence of risk commnication studies: social and political context, in: *Readings in Risk*, T.S. Glickman and M. Gough, eds., Resources for the Future, Washington, D.C.

Quine, W.V.O., 1953, 1961, Two dogmas of empiricism, in: *From A Logical Point of View*, Harper and Row, New York.

Rittel, H.W.J., and Webber, M.M., 1973. Dilemmas in a general theory of planning, *Policy Sciences* 4:155-169.

Rotmans, J., 1998, Methods for IA: The 3 challenges and opportunities ahead, *Environmental Modelling and Assessment* 3:155-179.

Rotmans, J., Dowlatabadi, H., and Parson, E. A., 1997, Integrated assessment of climate change: Evaluation of methods and strategies, in: *Human Choice and Climate Change: An International Social Science Assessment*, S. Rayner and E. Malone, eds., MIT Press, Cambridge, MA.

Rotmans, J., and Van Asselt, M.B.A., 1998, Integrated assessment: Current practices and challenges for the future, in: *Ecological Economics and Integrated Assessment: A Participatory Process for including Equity, Efficiency, and Scale in Decision-Making for Sustainability*, R. Costanza and S. Tognetti, eds., SCOPE, Paris.

Sager, T., 1997, Planning and the liberal paradox: A democratic dilemma in social choice, *Journal of Planning Literature* 12(1):16-29.

Schneider, S.H., 1997. Integrated assessment modeling of climate change: Transparent rational tool for policy making or opaque screen hiding value-laden assumptions? *Environmental Modeling and Assessment* 2:229-249.

Sen, A., 1970a, *Collective Choice and Social Welfare*, Holden-Day, San Francisco.

Sen, A., 1970b, The impossibility of a Paretian liberal, *Journal of Political Economy* 78(1):152-157.

# Toward a Computational Model of Hypothesis Formation and Model Building in Science

Joseph Phillips[*], Gary Livingston[#], Bruce Buchanan[†]
*Computer Science Department, University of Pittsburgh, Pittsburgh, PA, USA;*
[*]*josephp@cs.pitt.edu;* [#]*gary@cs.pitt.edu;* [†]*buchanan@cs.pitt.edu*

Abstract: We present our two-part approach to computational scientific discovery. The first part is a heuristic search through the space of hypotheses to find good hypotheses. The second stage is a heuristic search through the space of sets of hypotheses to find coherent models. Both of these heuristic searches strive for four criteria: empirical support, simplicity, novelty and utility. Four types of domain knowledge inform both searches: transforms, taxonomies, sequences and default models. We mention several experimental results that suggest the viability of our approach.

## 1. INTRODUCTION

We are writing computer programs to construct explanatory hypotheses and models for two reasons. First, we seek to understand model building and refinement better. Neither of these processes are well-understood (Popper, 1968), so it is difficult to instruct novices about how to do them. Building computer programs that reproduce many of the results of human scientists may provide insights into how the activities are carried out, and therefore how they might be improved.

*Model-Based Reasoning: Science, Technology, Values*, edited by
L. Magnani and N.J. Nersessian, Kluwer Academic/Plenum Publishers, New York, 2002

The second reason to write scientific discovery computer programs is to create useful tools for practicing scientists. Humans and computers have complementary strengths. Humans possess a depth and breadth of domain knowledge and heuristics that no computer program can yet match. Computers, however, surpass humans in terms of speed, accuracy and the ability to manipulate rote facts. Together, both can far exceed what either is capable of individually.

Since the terminology has not been uniform for describing some of the terms we discuss here, we offer the following working definitions:

hypothesis: a single relationship conjectured to have predictive or explanatory import
rule: a hypothesis represented as a conditional sentence (*e.g.* IF A AND B THEN C)
model: a set of hypotheses
law: a confirmed and accepted hypothesis
theory: a distinguished set of laws.

Our model of hypothesis formation and model building is based on our own and others' work in computational philosophy of science. The model is essentially heuristic search, in two stages:

- Search a Space of Hypotheses - starting with data, search a space of possible explanations to find those that constitute plausible explanations, and
- Search a Space of Models - starting with a collection of hypotheses, search a space of possible models to find plausible, coherent, and multi-statement models.

In both stages, heuristic search involves defining the syntactic form of items in the search space: hypotheses and models. But it requires more than combinatorial generation of statements from predicates, constants, variables, and relations. Heuristic search also depends critically on using background knowledge to avoid considering hypotheses and models that are not well supported by the data and to evaluate the candidates.

In the present work we discuss the generation and selection of acceptable hypotheses and coherent models using computer programs, and the uses of domain knowledge to guide the process.

Computers are unsurpassed at storing and accessing large amounts of information. Beyond that, it has been repeatedly shown that they can formulate hypotheses that explain a collection of data through a systematic search of possible explanations. Given a language and a variety of syntactic forms for expressing hypotheses, a total hypothesis space is defined. Under the exclusion paradigm (which Newton knew as "induction by enumeration"), hypotheses are generated and examined for reasons to exclude them. The combinatorics of examining all possible hypotheses are ameliorated when whole classes of hypotheses can be excluded. The DENDRAL (Lindsay, 1980), META-DENDRAL (Buchanan, 1976) and PROTEAN (Brinkley, 1988) programs are classic examples of heuristic search programs. As summarized by C.G. Hempel (1985):

> The great value and distinction of the DENDRAL program lies in the fact that its search for suitable hypotheses is guided by a set of subtle and efficient heuristic principles [so that it avoids exhaustive search].

Domain knowledge aids the process of inducing good hypothesis from data by informing this process of four aspects of good hypotheses. Those four aspects are empirical support, simplicity, novelty, and utility. We describe how we use domain knowledge to constrain the combinatorial search for good hypotheses. We take hypotheses as being the building blocks of models, so it is imperative to generate hypotheses well to help produce high quality models.

Similarly, domain knowledge aids the process of constructing coherent models from hypotheses, though model building is less well understood than hypothesis formation. Models may have considerable organizing and suggestive power for theory construction (Hesse, 1972) but have less stringent requirements than theories (Kuhn, 1962). Thus it seems appropriate to work on the intermediate step of model building, which we take to be searching a space of combinations of individually good hypotheses (and variations of them) to find a set that is more coherent than most collections of hypotheses.

This paper is organized as follows. The next section describes the types of domain knowledge that we use. Section 3 tells how we use this knowledge to create good hypotheses and gives examples of their usage. Section 4 details how we use domain knowledge to guide the formation of coherent models given hypotheses. The last section presents our conclusions.

## 2. TYPES OF DOMAIN KNOWLEDGE

Domain knowledge is knowledge specific to a particular subject of study (e.g., the motion of objects falling close to the Earth's surface) and to a particular paradigm (e.g. Newtonian mechanics and gravitation). Such knowledge describes one or a few interrelated phenomena with a coherent set of terms and relations.

The form and notation of domain knowledge is highly domain dependent. For example, physicists prefer to summarize their data as equations, while equations are less prevalent in most other disciplines.

Although the form and notation of domain knowledge is domain dependent, several broad types of knowledge enjoy widespread usage. We have identified four forms that are common in many useful domains and that may be exploited in a systematic fashion by domain independent mechanisms. The forms are: (1) transforms between attributes, (2) taxonomies and ontologies of domain objects, (3) domain specific series of domain objects, and (4) default models. We discuss each along with a summary of how they may aid hypothesis and model creation.

Transforms map one or more given data attributes into one or more new data attributes. For example, to study geophysics we transformed the locations of earthquakes from <latitude, longitude, depth> triplets to a measurement of a position along particular fault lines (Phillips 2000). Another transform from a medical domain created a length-of-stay attribute for patients of a rehabilitation ward given their admission and discharge dates. Both of these transforms add information about their fundamental phenomena from more generic raw data.

*Transforms affect heuristic search by specifying the areas in the space of input data representations that have been historically fruitful. The idea for earthquake location transform came from reading the geophysical literature and from examining practicing geophysicists. This transformation allowed us to make several geophysical rediscoveries.*

Taxonomies group domain objects in families by their degree of relatedness. For example, in some medical domains it is useful to divide the set of all strokes into left-body-affecting strokes, right-body-affecting strokes, bilateral strokes, no-paresis strokes and “other” strokes. Any particular stroke can be listed as an instance of one of these more specific classes. In the geophysics domain, earthquakes have been specifically grouped by the fault that they occurred upon and generally grouped by the fault’s type.

*Taxonomies and ontologies aid heuristic search by constraining the generalization of rules from specific instances to specific, pre-defined classes. Taxonomies give us the ability to state learned medical and geophysical data at the appropriate level of generality without having to enumerate all cases.*

Series provide a list of domain values with the understanding that there is a sequential relationship between the values. For example, knowing the season is important in analyzing many cyclical activities. The series [spring, summer, fall, winter] has been used in our studies of stock market data. Series values can also exist in taxonomies: the season values can be further subdivided into the series [January, February, March, April, May, June, July, August, September, October, November, December]. Series allow for the creation of new attributes (*successor-of* and *predecessor-of*) to be used to learn relationships over time, space, or other dimensions.

*Series help heuristic search by suggesting meaningful successor and predecessor attributes (out of the combinatorial space of possible sequence attributes) for machine learning. We anticipate using series to advance our discovery efforts on time sequence data.*

Default models allow the computation of the naïve expected behavior of a system. An example from the stock market domain is that prices will rise along with inflation. Computations derived from default models allow the identification of potentially novel predictions.

*Default models assist heuristic search by allowing the rapid and automatic identification of novel predictions. In addition to computing the default behavior of stock prices, we have used default models to identify novel and interesting hypotheses for crystal formation.*

# 3. GOOD HYPOTHESES

We view hypotheses as the building blocks of models. In this section we discuss our efforts to ensure that the component hypotheses each have high caliber. We (1) specify what we mean by "hypotheses", (2) discuss four characteristics of good hypotheses and the domain knowledge needed to find them, and (3) close with case studies and a brief conclusion.

## 3.1 Hypothesis form

For us, a hypothesis is a conditional statement that makes a prediction. Thus, in a rule-based representation, the terms "hypothesis" and "rule" are synonyms. An example hypothesis would be:

> "If a patient has had a stroke and the patient's impairment is the right-body region, then the patient's expected improvement is X".

The conditional nature of rules eases the credit assignment problem when wrong predictions are made. The fact that rules make definite predictions allows users to understand their purpose in the larger model. (Compare this to the weights of a hidden node in a neural network). This operational definition also enhances the modularity and understandability of models, which leads to increased model coherence.

## 3.2 Criteria

Many modern philosophers of science have discussed desirable criteria for evaluating theories (e.g., Hempel, 1945, Goodman, 1949, and Bunge, 1961). More recently, philosophers have attempted to map these philosophical discussions into computational terms (e.g. Buchanan, 1966, Thagard, 1988, and Darden, 1991). We have clustered these criteria into four categories: empirical support, simplicity, novelty and utility.

1. *Empirical support* - criteria relating a hypothesis to the data, which have been termed criteria of accuracy and coverage. These include various functions relating the numbers of true positive and false positive predictions and the number of items omitted correctly and incorrectly (true and false negatives).
2. *Simplicity* - which can be further subdivided into (a) syntactic considerations such as a count of the number of rules or the number of descriptive terms needed, and (b) semantic considerations, which are sometimes called "coherence conditions" including how well terms seem to "belong together".
3. *Novelty* - which include new predictions that might or might not contradict default knowledge. Our working hypothesis about novelty is that rules (and rulesets) containing novel predicates and values, rules conflicting with other rules or the background knowledge, or rules that cover anomalies are especially novel and interesting, provided they are not already in the knowledge base.
4. *Utility* - criteria encoding pragmatic considerations, which include the costs and benefits of making correct or incorrect predictions.

### 3.2.1 Empirical Support

Though empirical support (sometimes called "accuracy") is the obvious, well-studied, and most important criterion for rule acceptance, our experience has been that implementing it in a computer program is not as straightforward as counting errors in prediction. First, two aspects of empirical support may be considered: coverage (preferring more predictions to fewer) and deviation from observation (preferring accurate predictions). Second, several basic metrics exist for both coverage and accuracy. Common basic coverage metrics include the true positive and true negative counts, the positive predictive value (i.e., the percentage of correct predictions among predictions made) and the overall coverage (i.e., the fraction of items for which the model makes a prediction). Common basic deviation metrics include the false positive and false negative counts. Third, scientists often combine these basic metrics into composite statistical measures. For example, for our work we have employed frequency-based measures with a variety of potential corrections (*e.g.*, no correction, Yates' correction, asymmetric distribution correction, *etc.*).

The knowledge obtained from default models partially solves this problem. Default models can contain distributions or expectations of distributions. Distributions may be used to help choose an appropriate accuracy metric. The choice of the correct metric may, at least partially, depend on the domain. We view this as an issue to determine empirically.

### 3.2.2 Simplicity

Simplicity is another widely held characteristic of good rules, but there are two aspects: syntactic and semantic. Syntactic simplicity deals with the conciseness of the syntactic expression of the model. Operationally, it refers to any aspect of a rule that a program can count. Favoring rules with fewer conditions or predicates with fewer allowed values are good first approximations of syntactic simplicity.

Semantic simplicity addresses how well the referenced terms make sense in context. The challenge to us from the Philosophy of Science is to construct hypotheses with the semantics of "law-like statements" given the syntax of conditional rules. Specifically, the terms used in rules should make sense both individually and in context with the other terms of the rule. Goodman's example of the terms "grue" and "bleen" (Goodman, 1955) making less sense in our culture than "blue" and "green" highlights the need for terms to be individually understandable. Ryle stated that attributes mentioned in the same hypothesis should be cohere among themselves. His concept of category mistakes (*e.g.*, using incorrect attributes to describe an object) underscores the importance of coherence between terms (Ryle, 1949).

Transforms help us gain syntactic simplicity by potentially reducing the number of terms. For example the triplet <latitude, longitude, depth> that we used to describe the position of earthquakes may be compressed to the doublet <position-on-fault, depth> when we take in consideration the domain knowledge that earthquakes often occur on planar faults.

Taxonomies and default models can aid our search for in-rule coherence of terms. They allow us to see how attributes "cluster" in common usage and what types of objects they have historically been used to describe. We operationalize these concepts through reasoning with a uniform hierarchy of semantic types of concepts and with a representation of the degree of a predicate's entrenchment in a context.

### 3.2.3 Novelty

By striving for novel rules we hope to hold the interest of practicing scientists. One way that we address the question "What is completely new?" is by looking at the predictions of default models. Default models may be unable to a make a prediction or may differ from a prediction made from a new rule. Both cases hint that we have a potentially novel rule. Another way to identify potentially novel rules is by comparing their form with rules from the default models. Potentially novel rules could use unusual predicates or combinations of predicates (relative to the rules in the default model), could refer to values more than N standard deviations from average values, or could cover anomalies.

Although most of our machine discoveries are not novel, we find this is not necessarily bad. Scientists often become more confident in our approaches when they see rediscoveries. This argues for techniques for the rapid incorporation of rediscovered knowledge into default models – something that is easy to do with modular rules.

Another aspect of novelty is a "surprise factor." We are aware of this and have only recently begun to address it. Understanding the semantic types of the predicates used in a rule can let a program understand, for example, that using demographic factors alone to predict a patient's response to medical therapy would be surprising, since the conventional wisdom would hold that predicting response would also require knowing the severity of the patient's condition.

### 3.2.4 Utility

Finally, the cost or utility of rules is often an important factor in fields that concern society. Engineers, physicians and other users of scientific models often attribute differing costs to False Positive ("Type-I") and False Negative ("Type-II") errors. Similarly, there can be differing benefits to True Positive and True Negative predictions. We address this highly domain specific factor with specific information in taxonomies, ontologies and series.

One aspect of cost/utility overlaps with our treatment of novelty, namely "What is worth discovering?" Again, we use domain knowledge to infer, for example, that forward predictions in time are useful and to scan databases to rapidly identify data representing progressions in time.

We also rely upon taxonomies and ontologies to help us assign relative weights to false negative versus false positive predictions. Even here there are common themes: time and money are often very important. Taxonomies and ontologies allow our programs to know (or to guess) the dimensionality of attributes, and thus assist us in our search for useful rules.

## 3.3 Case Studies

We have developed a rule learning program named RL (Provost, 1993), and autonomous discovery programs named HAMB (Livingston, 2001) and the Scienceomatic (Phillips 2000). We briefly describe them and then present some of their discoveries and rediscoveries.

RL is a descendant and generalization of META-DENDRAL, which was developed specifically to learn rules for analyzing mass spectroscopy data in organic chemistry (Lindsay, 1980). RL is one of several AI machine learning programs that learn inductively. (For an overview of recent work, see Michalski, 1994).

RL may be briefly described as a heuristic search program that searches a space of rules to find a definition of the target concept class, where its search is guided by prior knowledge as well as by the data. Using a generator of syntactically possible rules that define a total rule space, RL explores the space heuristically. It starts from rules with a single feature and adds one feature at a time to partial rules that look most promising. Each partial rule is matched against the training cases to see how many of the positive cases are correctly predicted and how many of the negative cases are incorrectly predicted, thus providing statistical guidance to the search. In addition, semantic relationships among features that are specified in the partial domain theory are taken into account in exploring the rule space.

HAMB (*H*euristic, *A*utonomous *M*odel *B*uilder) is a prototype fully autonomous discovery program which uses heuristics to select discovery goals, select induction problems for RL, and analyze the rules induced by RL in order to select additional discovery goals and create models (Livingston, 2001). HAMB's heuristics that are most relevant to the subject of this paper are its heuristics which use domain knowledge to reduce the number of non-novel and uninteresting discoveries and its heuristics which use empirical support to eliminate statistically insignificant rules. HAMB reduces the number of non-novel discoveries by eliminating from the set of attributes that are used to form the induced rules (the feature set) those attributes that are known to be associated with the attribute HAMB is trying to learn rules for. HAMB reduces the number of uninteresting discoveries by filtering from the rules induced by RL those rules containing uninteresting attributes or values that the user has specified as uninteresting in the partial domain model. Finally, HAMB filters rules induced by RL by using statistical tests to remove statistically weak rules.

The Scienceomatic is another discovery program that uses different machine learning techniques (Phillips 2000). It may be run autonomously, or under user supervision. It uses a Minimum Message Length (or "MML", see Rissanen, 1978 or Cheeseman, 1995) inspired heuristic to schedule discovery tasks that could include adding new rules, consolidating the current set of rules, or transforming data into new tables and attributes. It also uses operator success counts to meta-learn how effective various operators have been in the past and uses this information to estimate how effective future operator applications will be. It made rediscoveries in geophysics when given tables of earthquake data and a weak domain theory. It also improved its discovery process given more meta-learned experience.

### 3.3.1 Empirical Support

A sample rule from RL's examination of data on the classification of chemicals as carcinogenic or non-carcinogenic in rodents, using the results of short-term assays as features, is shown below. The best rule proposed in the literature – that a chemical is carcinogenic in rodents if and only if the Salmonella genotoxicity assay is positive – results in 55.7% predictive accuracy. RL learned a rule that improved the predictive power of the accepted assay (Lee 1994).

IF: the Salmonella genotoxicity assay is positive
OR
the Salmonella genotoxicity assay is negative
AND the chromosomal aberrations assay is positive
OR
the Salmonella genotoxicity assay is negative
AND the sister chromatid exchange assay[1] is positive
THEN: the chemical is carcinogenic in rodents
Overall Accuracy = 70.5% (43/61)
Sensitivity = 81.4%
Specificity = 44.4%

### 3.3.2 Simplicity

The Scienceomatic used a Minimum Message Length (MML) inspired heuristic to make a principled decision about how complex a rule set should be given particular data (Phillips 2000). With this heuristic it decided to state some knowledge with very simple statements (*e.g.* "All aftershocks have an (average) magnitude of X."). Both its all encompassing scope ("All after-

[1] In Chinese hamster ovary cells.

shocks") and its straightforward prediction ("have a magnitude of X") show the extreme simplicity of this statement. (The statement correctly summarizes the data but is geophysically too simple. It is due to the occurrence of many more smaller earthquakes than larger ones, and to the fact that there is an sensitivity threshold for earthquake detection.) However, when the data supported it, the program found more complex relations. It found, for example, equations that predict the location of aftershocks on particular faults as a function of their magnitude and the magnitude and location of their original main earthquakes.

### 3.3.3 Novelty

HAMB has several heuristics to reduce the number of non-novel rules. One of them eliminates from the set of attributes those attributes already known to be associated with or discovered to be synonymous with the attribute for which RL will induce rules. When we applied HAMB to a database of cases of macromolecule crystallization experiments, this heuristic was responsible for avoiding the induction of 2,785 (out of 3,936) non-novel rules. This database has been problematic for machine learning techniques in the past because of the large amount of redundancy and known patterns, which made it difficult for a person examining the resulting discoveries because they contained a high proportion of non-novel discoveries. With this heuristic, HAMB was able to create a model that contained a greater proportion of novel discoveries and that was easier for a domain expert to examine.

### 3.3.4 Utility

RL has also been used to learn concept descriptions that are sensitive to the cost of making predictions. From a database of poisonous and edible mushrooms, RL can learn a set of rules that classifies mushrooms' edibility with high accuracy; alternatively, RL can learn a set of rules with very low risk. Training on 100 examples, a rule set achieved 96% classification accuracy, but less than 2% of its predictions were dangerous. An alternative rule set induced using knowledge about the cost of labeling a poisonous mushroom as non-poisonous achieved less than 0.1% dangerous predictions, but its accuracy was only 84%.

## 3.4 Future work on achieving good hypotheses

After having identified both the aspects of good hypotheses and the domain knowledge needed to obtain good hypotheses, we are challenged to put them together into one heuristic that will balance them appropriately. This is a difficult task. The experiments described above are the work of several programs, chiefly RL, HAMB and the Scienceomatic. Phillips (2000) has experimented with principled approaches to balancing several competing interest like Minimum Message Length (see Rissanen (1978) or Cheeseman (1995)) over sets of rules. Such concerns often are more applicable to models than individual rules. The creation of coherent models is our next topic.

# 4. GOOD MODELS

Although RL is able to learn good individual rules, and put them together into accurate classifiers, the rule sets often appear to be disconnected with no unifying principles. RL produces a set of rules that individually meet or exceed selection criteria specified by the user. While many of these rules may be interesting and useful, they cannot be said to constitute a model due their redundancy and lack of organization.

We have developed a model-building program, as part of the HAMB system that uses RL (Livingston, 2001), and we have experimented with preliminary criteria to guide the search and evaluate models. These criteria are essentially the same as the criteria for good hypotheses, but with somewhat different applicability:

1. *Empirical support* - criteria relating a set of hypotheses to the data, *i.e.*, overall how broad is the model's coverage and how accurate are its predictions.
2. *Simplicity* - (a) syntactic considerations such as a count of the number of rules or the total number of descriptive terms across all rules in the model, and (b) semantic considerations, which are sometimes called "coherence conditions" including how well rules and terms seem to "belong together".
3. *Novelty* - criteria encoding novelty are essentially the same as for individual hypotheses.
4. *Utility* - criteria encoding pragmatic considerations, which are essentially the same as for individual hypotheses.

## 4.1 Empirical Support

When several rules in a model apply to a new case for which predictions are to be made, a decision rule needs to be defined that wil! collect evidence for conflicting predictions and decide which of them, if any, is the prediction of the model. This is sometimes known as the evidence gathering problem (Buchanan, 1984). It is complicated by the fact that each rule carries a degree of evidential support based on its own past record of correct and incorrect predictions. Alternative methods of combining evidence have been suggested, such as using the prediction of the single best rule, of the majority vote of rules, of a weighted vote, and so on. More research on evidence gathering is needed, although most of the time weighted voting is reasonably accurate.

## 4.2 Simplicity

Syntactic simplicity of a model refers to countable items, such as the number of rules, the number of predicates across all the rules, and the number of distinct values of the predicates mentioned across all rules. Semantic simplicity is considerably more difficult.

We have introduced a criterion of systematicity as a first approximation to coherence of rules. Rules cover data systematically if (1) they make the same prediction from the same features, (2) all conditions except one are the same in the left-hand side, and (3) there is a systematic division of the data by systematic variations of values of the last feature. For example, rules R1, R2, and R3 systematically cover a subset of the data:

R1: IF (A=a1) & (B=b1) & (C=c1) THEN (D=d1)
R2: IF (A=a1) & (B=b1) & (C=c2) THEN (D=d1)
R3: IF (A=a1) & (B=b1) & (C=c3) THEN (D=d1).

Where values c1, c2, and c3 exhaust the possible values, leaving no data items out if they satisfy the first two conditions. The three conditions (AGE < 3), AGE 3-21) and (AGE > 21) use three such values. A set of one hundred rules following an enumeration of years would be equally systematic (but less simple by the criterion above). It would probably be more accurate as well, so a choice between these two models would rest on how much relative emphasis is put on each of the criteria.

We are experimenting with an operational definition of Goodman's concept of entrenchment of predicates (Goodman, 1955) as a way of preferring models that use predicates that are more familiar, more widely used in science, less dependent on particular space-time points, *etc.* We are also experimenting with tagging each predicate with a semantic type and reasoning about the homogeneity of the types used in a model. For example, a model mixing predicates from Western medicine and traditional Chinese medicine is less coherent than a model formed within one framework or the other.

## 4.3 Case study: some experiments with model building in structural biology (identifying calcium binding domains in proteins)

In collaboration with Dr. Russ Altman (Stanford University), we applied RL to training data on 116 sites in the 3-dimensional structures of proteins from the Protein Data Bank (PDB). Sixteen were known to be sites that bind calcium ions (Ca++); 100 were sites that do not bind Ca++. We asked RL to find rules for effective prediction of new sites and tested the rules on additional sites in new proteins.

RL was given a set of 63 features, which were organized in several different taxonomies. Some features are specializations of others (*e.g.*, "oxygen" is a specialization of "heteroatom"), and some are alternative characterizations of nearly the same underlying property (*e.g.*, one classification of polar residues includes the amino acids HIS and TRP while another one moves HIS to the class "Basic" and TRP to the class "Nonpolar"). We gave RL additional background knowledge about the semantic classes of features, classifying features as physical, chemical, or structural properties.

RL first found a collection of over 300 rules. Each satisfied our criteria of empirical support: (a) the ratio of true positive predictions to all predictions needed to be above 80% and (b) the number of true positive predictions needed to be above 12 (*i.e.*, each rule needed to cover more than 10% of the total number of training examples). And each satisfied our criterion of syntactic simplicity, viz., the number of conditions in any rule had to be three or fewer. But although each of these rules individually had some merit, together they lacked coherence.

The results of the two experiments presented below were evaluated using 16-fold cross-validation. Because there were only 16 positives, we randomly put exactly one positive example into each of the folds and then randomly assigned the negatives to the folds.

### 4.3.1 Experiment 1: prefer coherent models with strong rules

From an average base of 330 rules, we kept only rules that were strong and were coherent with the rest of the rules in the model. A rule was considered *strong* if it had a certainty factor greater than 0.90, covered more than 80% of the examples having the rule's RHS as their class (80% of POS), and incorrectly predicted no more than 20% of the negative examples. A rule was considered *coherent* with the rest of the rules in the model if every attribute used in the rule is in at least one other rule in the model. This last requirement captures the notion of using "entrenched" attributes.

Cross-validation tests of this rule pruning method revealed that the average model size dropped from an average size of 330 to an average of 124 rules for the pruned models and that predictive accuracy rose from an average of 90.5% on the base models to 94.9% on the pruned models.

### 4.3.2 Experiment 2: prefer models with fewer semantically redundant rules

For this experiment, semantic simplicity measured the amount of redundancy in the model, and redundancy was the amount of semantic overlap between rules. We defined a redundant rule as one that was intensionally generalized by another rule. A rule R1 is intensionally generalized by a rule R2 if and only if R1 and R2 predict the same class, and either (a) R1 subsumes R2 or (b) every feature in R1 has a unique corresponding feature in R2 that generalizes it.

We pruned the models using two different methods: (1) keeping the more general rule when a redundant pair of rules was identified and (2) keeping the more specific rule when a redundant pair of rules was identified. We evaluated these two methods using the models resulting from the first experiment. Cross-validation results was similar using either method, with the average model size dropping from 124 to ~60, and with the average accuracy not significantly changing.

From these experiments we conclude that we could define some of the semantic criteria needed to reduce a large, disconnected ruleset to a much smaller one, without loss of accuracy. Much remains to be done, but we are encouraged by the preliminary results.

## 5. CONCLUSION

Our work supports the idea that domain knowledge can be used by non domain-specific computer programs to aid in the formation of good hypotheses, and with the creation of coherent models. By making explicit the criteria for good hypotheses and good models, this framework allows changing the relative degrees to which each criterion contributes to an evaluation. For instance, in exploratory phases of science more attention is paid to finding hypotheses that account for a large amount of data than to eliminating small numbers of counterexamples. Also in an exploratory phase there is relatively less attention paid to simplicity of hypotheses than to their statistical significance. Finding any hypotheses that explain data better than the currently accepted explanations provides a starting place. Later phases of refinement and publishing emphasize utility, syntactic simplicity, and coherence. The point is not that we understand the progression of phases of scientific activity but that the computational framework proposed here provides a means for understanding them.

This research suggests how computer-based assistants for scientists might revolutionize scientific practice. Warehouses of data are becoming accessible to every investigator with a computer, both within and outside the narrow specialty areas for which the data were collected. Interdisciplinary questions will be easier to answer because sets of data from two or more databases may be joined. Data mining assistants are able to formulate millions of hypotheses and check them against data, which no human assistant would be able to do. We hope to help guide this process effectively

## REFERENCES

Brinkley, J. F., Altman, R. B., Duncan, R. B., Buchanan, R. B., and Jardetzky, O., 1988, The heuristic refinement method for the derivation of protein solution structures: Validation of cytochrome-b562, *J. Chem. Inf. Comp. Sci.* 28(4):194-210.

Buchanan, B. G., 1966, *Logics of Scientific Discovery*. Ph.D. Dissertation, Dept. of Philosophy, Michigan State University, E. Lansing, MI.

Buchanan, B. G., Smith, D. H., White, W. C., Gritter, R. J., Feigenbaum, E. A., Lederberg, J., and Djerassi, C., 1976, Application of artificial intelligence for chemical inference XXII: Automatic rule formation in mass spectrometry by means of the meta-DENDRAL program, *Journal of the American Chemical Society* 98:6168-72.

Buchanan, B. G. and Shortliffe, E. H., 1984, *Rule-Based Expert Systems: The MYCIN Experiments of the Stanford Heuristic Programming Project*. Addison-Wesley, Reading, MA.

Bunge, M., 1961, The weight of simplicity in the construction and assaying of scientific terms, *Philosophy of Science* 28(2):120-149.

Cheeseman, P, 1995, On Bayesian model selection, in: *The Mathematics of Generalization: Proceedings of the SFI/CNLS Workshop on Formal Approaches to Supervised Learning*, Wolpert, D. H., ed., Addison Wesley, Reading, MA, pp. 315-330.

Darden, L., 1991, *Theory Change in Science*, Oxford University Press, Oxford.

Goodman, N., 1949, The logical simplicity of predicates, *The Journal of Symbolic Logic* XIV:32-41.

Goodman, N., 1955, *Fact, Fiction and Forecast,* Harvard University Press, Cambridge, MA.

Hempel, C.G., 1945, Studies in the logic of confirmation, *Mind* 54:1-26, 97-120.

Hempel, C.G., 1985, Thoughts on the limitation of discovery by computer, in: *Logic of Discovery and Diagnosis in Medicine*, K.F. Schaffner, ed., University of California Press.

Hesse, M., 1972, Models and analogies in science, *The Encyclopedia of Philosophy*, MacMillan Publishing Co. & The Free Press, New York, pp. 354-359.

Kuhn, T., 1962, *The Structure of Scientific Revolutions*, University of Chicago Press, Chicago.

Lee, Y., Rosenkranz, H. S., Buchanan, B. G., and Mattison D. M., 1994, *Learning Rules to Predict Chemical Carcinogenesis in Rodents,* (Technical Report ISL-94-25), Intelligent Systems Laboratory, University of Pittsburgh.

Lindsay, R. K., Buchanan, B. G., Feigenbaum, E. A. and Lederberg, J., 1980, *Applications of Artificial Intelligence for Chemical Inference: The DENDRAL Project*, McGraw-Hill, New York.

Livingston, G. R., 2001, *A Framework for Autonomous Knowledge Discovery from Databases*. Ph.D. Dissertation. Department of Computer Science, University of Pittsburgh, Pittsburgh, PA.

Michalski, R. S. and Tecuci, G. (eds.), 1994, *Machine Learning: A Multistrategy Approach*, Vol. IV, Morgan Kaufmann, San Francisco.

Phillips, J. P., 2000, *A Representation Reducing Heuristics for Semi-Automated Scientific Discovery*. Ph.D. Dissertation. Department of Electrical Engineering and Computer Science, University of Michigan, Ann Arbor, MI.

Popper, K. R., 1968, *The Logic of Scientific Discovery*, Revised Edition, Hutchinson, London.

Provost, F. J., Buchanan, B. G., Clearwater, S. H., and Lee, Y., 1993, *Machine Learning in the Service of Exploratory Science and Engineering: A Case Study of the RL Induction Program* (Technical Report ISL-93-6), Intelligent Systems Laboratory, University of Pittsburgh.

Rissanen, J., 1978, Modeling by shortest data description, *Automata*, 14:445-471.

Ryle, G., 1949, *The Concept of the Mind*, Hutchinson Publishing, London.

Thagard, P., 1988, *Computational Philosophy of Science*, MIT Press, Cambridge, MA.

# Models as Parts of Distributed Cognitive Systems

Ronald N. Giere
*Department of Philosophy, Center for Philosophy of Science, University of Minnesota, Minneapolis, MN, USA, giere@umn.edu*

Abstract: Recent work on the role of models in science has revealed a great many kinds of models performing many different roles. In this paper I suggest that one can find much unity among all this diversity by thinking of many models as being components of *distributed cognitive systems*. I begin by distinguishing the relevant notion of a distributed cognitive system and then give examples of different kinds of models that can be thought of as functioning as components of such systems. These include both physical and abstract models. After considering several objections, I conclude by locating distributed cognition within larger movements in contemporary cognitive science.

## 1. INTRODUCTION

Recent work on the role of models in science has revealed a great many kinds of models performing different roles. In this paper I will suggest that one can find much unity among all this diversity by thinking of many models as being components of *distributed cognitive systems*. I will begin by distinguishing the relevant notion of a distributed cognitive system and then give examples of different kinds of models that can be thought of as functioning as components of such systems. After considering several objections, I will conclude by locating distributed cognition within larger movements in contemporary cognitive science.

*Model-Based Reasoning: Science, Technology, Values*, edited by
L. Magnani and N.J. Nersessian, Kluwer Academic/Plenum Publishers, New York, 2002

## 2. DISTRIBUTED COGNITION

Distributed cognition is much more than merely distributed processing, which has been around for a generation. There are a number of avenues by which one can reach distributed cognition. Here is one that leads directly to the role of models.

During the early 1980s, James McClelland, David Rumelhart and the PDP Research Group, based mainly in San Diego, began exploring the capabilities of networks of simple processors thought to be at least somewhat similar to neural structures in the human brain. They discovered that what such networks do best is recognize and complete *patterns* in input provided by the environment. The generalization to human brains is that humans recognize patterns through the activation of prototypes embodied as changes in the states of groups of neurons induced by sensory experience. But if something like this is correct, we face a big problem. How do humans do the kind of apparently *linear* symbol processing required for such fundamental cognitive activities as using language and doing mathematics?

Their suggestion, tucked away near the end of a long paper in Volume 2 of their massive 1986 publication, was that humans do the kind of cognitive processing required for these linear activities by creating and manipulating *external* representations. These latter tasks *can* be done well by a complex pattern matcher.

Consider the following simple example (McClelland et al., 1986, v. 2, 44-48). Try to multiply two three-digit numbers, say 456 × 789, in your head. Few people can perform even this very simple arithmetical task. Figure 1 shows how many of us learned to do it.

```
   456
   789
  ----
  4104
 3648
3192
------
359784
```

*Figure 1.* Multiplying two three-digit numbers.

This process involves an *external representation* consisting of written symbols. These symbols are manipulated, literally, by hand. The process involves eye-hand motor coordination and is not simply going on in the head of the person doing the multiplying. The person's contribution is (1) constructing the external representation, (2) doing the correct manipulations in the right order, and (3) supplying the products for any two integers, which educated humans can do easily from memory.

Notice that this example focuses on the *process* of multiplication; the task, not the product, and not knowledge of the answer. Of course, if the task is done correctly, one does come to know the right answer, but the focus is on the *process* rather than the *product*. The emphasis is on the *cognitive system* instantiating the process rather than cognition simpliciter.

Now, what is the cognitive system that performs this task? Their answer was that it is not merely the *mind-brain* of the person doing the multiplication, nor even the *whole person* doing the multiplication, but the *system* consisting of the person *plus* the external physical representation. It is this whole system that performs the cognitive task, that is, the multiplication. The cognitive process is *distributed* between a person and an external representation.

The external representation for this problem can be regarded as a *model* of a multiplicative process in several senses of the term. It is a model in the sense employed in logic and mathematics in that it *instantiates* various arithmetical operations. It is also a model in the sense that it *represents* these arithmetical operations. Furthermore, it is a model in the sense of a working model of a physical process. In the process of performing the multiplication, the person literally constructs the model on paper, using it to help perform the cognitive task at hand.

Of course, many people no longer perform multiplication using pencil and paper. They use an electronic calculator or a computer. In this case, the necessary arithmetical models are built into the machine, either as hardware or as software. The general situation, however, is the same a before. Calculators and computers are designed so as to be easily operated by a human with a pattern matching brain. In particular, a person *plus* a computer constitutes a powerful cognitive system that can carry out cognitive tasks far beyond the capabilities of an unaided human. On the other hand, a person alone is already a cognitive system with many capabilities, such as face recognition, that currently are little improved by adding a computer.

Here one is tempted to ask whether a computer *by itself* is a cognitive system. From the point of view of distributed cognition, I think we should *not* count solitary computers as cognitive systems. My reasons turn on general issues to which I will return later in this paper.

## 3. DIAGRAMMATIC REASONING

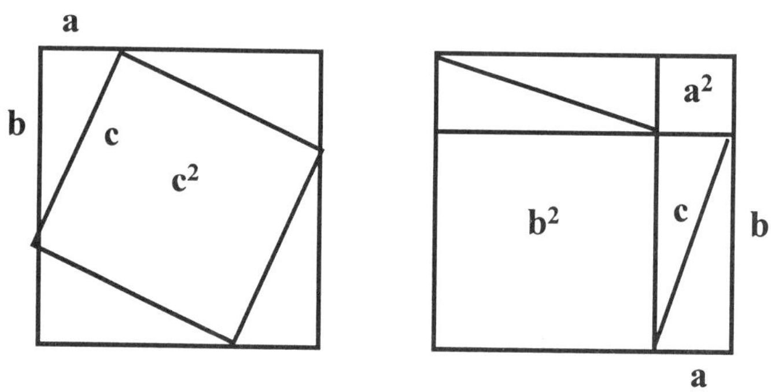


*Figure 2.* Diagrams used to prove the Pythagorean theorem.

Among those emphasizing the role of models in science, diagrams are typically regarded as models and diagrammatic reasoning a prominent kind of model-based reasoning. Figure 2 provides an example of mathematical reasoning that makes essential use of diagrams. The two diagrams in Figure 2 embody a famous proof of the Pythagorean theorem.

Note first that the areas of the two larger squares are identical, and, indeed, equal to $(a + b)^2$. The area of the single smaller square in the left diagram is equal to $c^2$ while the areas of the two smaller squares in the right diagram are $a^2$ and $b^2$ respectively. All of the triangles must be of equal area being right triangles with sides a and b respectively. In left top diagram the area of the smaller square, $c^2$, is equal to the area of the larger square minus the areas of the four triangles. But similarly, in the right diagram the combined area of the two smaller squares, $a^2$ plus $b^2$ equals the area of the larger square minus the areas of the same four triangles. So it follows that $c^2$ equals $a^2$ plus $b^2$.

In this reasoning, crucial steps are supplied by observing the relative areas of squares and triangles in the diagrams. Very little propositional knowledge, for example, that the area of a right triangle is proportional to the product of the lengths of its two sides, is assumed. The logicians Jon Barwise and John Etchemendy (1996) call this reasoning "heterogeneous inference" because it involves *both* propositional and visual representations. The issue is how further to characterize this reasoning. Here there are a number of views currently held.

A very conservative view is expressed in the following passage written by a contemporary logician. Referring to a geometrical proof, he writes:

> [The diagram] is only an heuristic to prompt certain trains of inference; [...] it is dispensable as a proof-theoretic device; indeed, [...] it has no proper place in the proof as such. For the proof is a syntactic object consisting only of sentences arranged in a finite and inspectable array (Tennant, 1986).

This position simply begs the question as to whether there can be valid diagrammatic proofs, or, indeed, even heterogeneous inference. Even granting that any diagrammatic proof is dispensable in the sense that it can be reconstructed as a syntactic object, it does not follow that diagrams can have "no proper place" in a proof. Only by assuming the notion of a "proper proof" to be essentially syntactic can one make this so.

Herbert Simon had a great interest in diagrams and took a more moderate position. He long ago (1978) claimed that two representations, for example, propositional and diagrammatic, might be "informationally equivalent" but not "computationally equivalent." By this he meant that there could be a method of computation that made the same information more accessible or more easily processed in one representational form rather than another. There remained, however, an important respect in which Simon's position was still relatively conservative, as illustrated in the following passage:

> Understanding diagrammatic thinking will be of special importance to those who design human-computer interfaces, where the diagrams presented on computer screens must find their way to the Mind's Eye, there to be processed and reasoned about (Simon, 1995, xiii).

Here Simon seems to be saying that all the cognition goes on in the head of the human. Others are more explicit.

A very common approach to diagrammatic reasoning in the AI community is to treat diagrams simply as perceptual input that is then processed in various ways (Chandrasekaran et al., 1995). Here the most conservative approach is first to translate the diagram into propositional form and then proceed with processing in standard ways (Wang, Lee and Hervat, 1995). Figure 3 shows diagrammatic reasoning as understood in standard artificial intelligence.

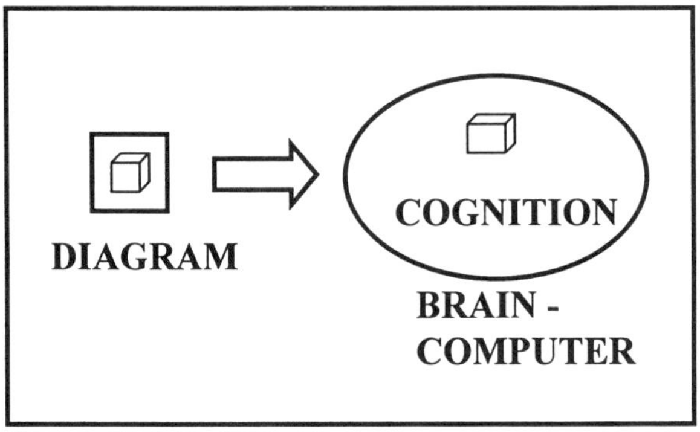


*Figure 3.* Diagrammatic reasoning in standard artificial intelligence.

From the standpoint of distributed cognition, treating a diagram merely as input to a standard logic-based computer misses most of what is important about diagrammatic reasoning. What is interesting about diagrammatic reasoning is the *interaction* between the diagram and a human with a fundamentally pattern-matching brain. Rather than locating all the cognition in the human brain, one locates it in the system consisting of a human together with a diagram. It is this system that performs the cognitive task, for example, proving the Pythagorean theorem. The system can fairly easily perform this task whereas the human alone might well not be able to do it at all. A large part of what makes the combined system more powerful is that the geometrical relationships are embodied in the diagrams themselves. Extracting the desired relationships from the diagrams is far easier (especially for a pattern-matcher) than attempting to represent them internally. Figure 4 shows diagrammatic reasoning as distributed cognition. As will become even clearer below, Figure 4 is oversimplified in that it pictures a brain in isolation from a body and without any means of physically interacting with the diagram.

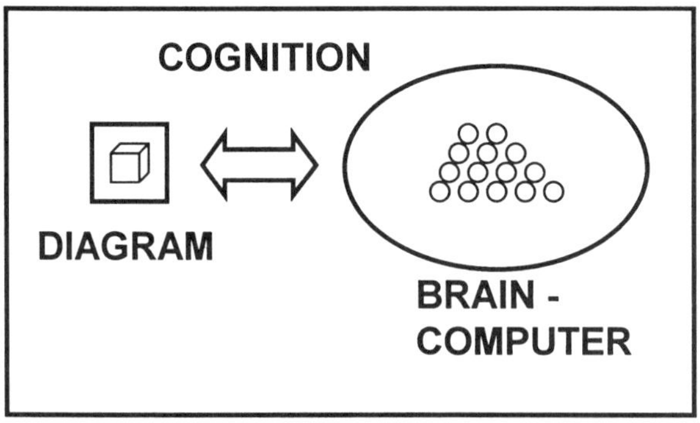


*Figure 4.* Diagrammatic reasoning as distributed cognition.

From this perspective, there is at least one clear way of studying diagrammatic reasoning within the framework of cognitive science. This is to investigate the characteristics of diagrams that most effectively interact with human cognitive capabilities. Stephen Kosslyn's *Elements of Graph Design* (1994) provides prototype for this kind of study. Unfortunately, he devotes only two pages explicitly to diagrams, but one of his examples shows the kind of thing one can do. The example concerns diagrams that show how pieces fit together in an artifact, as in Figure 5.

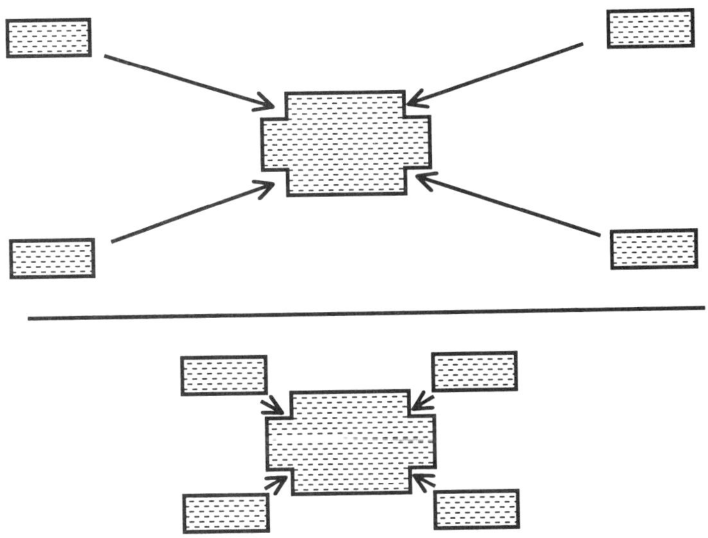

*Figure 5.* Good versus bad design of diagrams based on cognitive principles.

The relevant fact about human cognition is that humans process shape and spatial relationships using different neural systems, so the two types of information do not integrate very well. The lesson is that, in a diagram, the representations of the pieces should be near to each other as in the lower diagram. My guess is that, in addition to such general design principles, effective design of diagrams differs significantly for different sciences. Thus, features making for cognitively effective diagrams in molecular biology may be quite different from features effective in high-energy physics.

## 4. REASONING WITH PICTORIAL REPRESENTATIONS

One cannot draw a sharp distinction between diagrams and pictorial representations. At one extreme we find highly idealized line drawings such as the following picture of a diatomic molecule indicating three degrees of freedom, spinning, rotating, and vibrating. Such a drawing differs little from a diagram.

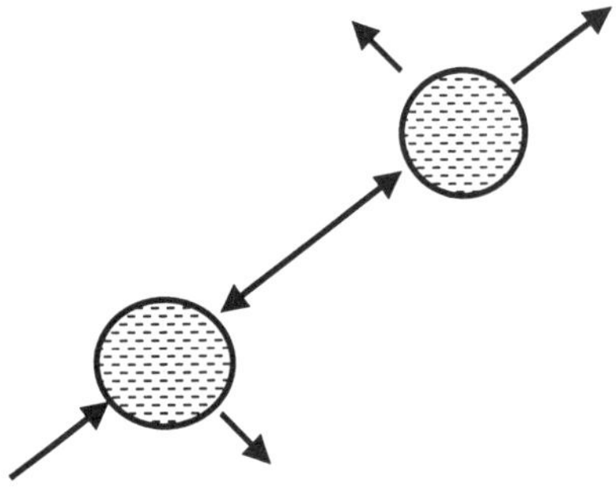

*Figure 6.* Idealized picture of a diatomic molecule.

At the other extreme are actual photographs, for example, of a volcano. In between we find a vast variety of such things as X-ray photographs of broken bones, telescopic pictures of distant nebulae, magnetic resonance images of brains, scanning electron microscope pictures of human tissues, magnetic profiles of the ocean floor, and so on.

Understanding pictorial reasoning as distributed cognition makes it possible to treat pictorial reasoning as more or less continuous with diagrammatic reasoning. The cognition is in the interaction between the viewer and the picture, as when geologists take magnetic profiles of the sea floor as indicators of sea-floor spreading (Giere, 1996). We need not imagine geologists as first forming a mental representation of the picture and reasoning with it. They can reason directly with the external representation of the phenomenon.

## 5. REASONING WITH PHYSICAL MODELS

From a distributed cognition perspective, there is again little fundamental difference between reasoning with diagrams and reasoning with physical models. A diagram is, after all, a kind of physical model, only two rather than three-dimensional.

With some physical models one can work with the relevant parts of the world itself, the world being its own best model. A simple example is the problem of fitting slightly different lids on a number of slightly different screw-top jars. Suppose a person cannot tell by simple inspection which lid fits which jar. There is, however, an obvious effective procedure for solving the problem using the same perceptual skills. Pick any lid and try in on the jars in any order until one finds the jar that fits. Then pick another lid and do likewise until all the jars are fitted with their appropriate lids. One need not construct a representation of the whole problem, whether internal or external. One simply interacts with the world itself in an appropriate manner and the world guides one's search for the correct solution.

In a more normal case, particularly in the sciences, a physical model is constructed to be an *external representation* of some aspect of the world. A well-known example is James Watson's model of DNA made from metal cut outs held together by a pole and clamps. By fiddling with the physical model so as to fit the base pairs inside a double-helical backbone, Watson came up with the right structure. Watson with his physical model turned out to be a more effective cognitive system than Roseland Franklin with her X-ray pictures and hand-drawn diagrams.

Physical models provide what is probably the best case for understanding model-based reasoning as an example of distributed cognition. Here it is very clear that one need not be performing logical operations on an internal representation. It is sufficient to perform and observe appropriate physical operations on an external physical representation. The interaction between a person and the model is physical as well as perceptual.

# 6. REASONING WITH ABSTRACT MODELS

Abstract models provide what is probably the most difficult case for understanding model-based reasoning as an example of distributed cognition. It is not clear in what sense an abstract model can be external. Nor is it clear how a person can interact with an abstract model. Yet many, if not most, models used in the sciences are abstract models. Think particularly of models in quantum physics or cosmology. So some account of abstract models is needed.

The only alternative to regarding abstract models as *external* is to regard them as *internal*. There are, however, many reasons for rejecting this alternative. One reason is that we could not really refer to *the* model of anything because is to be expected that every individual involved will have a somewhat different internal model. We would have to speak of A's model, B's model, etc. An even stronger reason is that it is highly implausible to suppose that all the details of the complex abstract models of contemporary science could be represented in the form of a mental model. Even experts, when asked to solve a difficult problem, typically proceed first to create an external representation of the problem in the form of diagrams or equations. The suggested account of this process is that the expert is using the external representations in order to *reconstruct* aspects of the abstract model relevant to the problem at hand. This no doubt requires some simpler mental models. It also requires acquired *skills* in constructing the sort of models in question.

Yet, even if we agree that abstract models are in some sense external, there remains a question of just what this means. This question threatens to lead us into the arid land of the philosophy of mathematics where one worries about what numbers might be. I think we would do well to avoid this detour and take a safer route. As in our understanding of time, we traffic in abstract models every day without worrying about what they are. Consider plans and planning, well-known topics in cognitive science. Here abstract models are simply taken for granted. Suppose three friends are planning a party. The planned party is, at least in part, really an abstract model of a party. It is assigned a date and time in the future and potential guests may even be designated in a written list. We know the party starts out as an abstract entity, a mere possibility, because it may in fact never materialize. Now I suppose almost everyone who as thought about the problem agrees that the ability to use language has a lot to do with the human ability make plans, which is to create abstract entities, which we may take to be abstract models. This is enough for present purposes.

Here we already have a plausible answer to the question how humans interact with abstract models. They do so by using language, both ordinary and highly technical. The three friends in my example build up their model of the party by talking about it. Moreover, they can reason about it as well, realizing, for example, that the number of potential guests has become too large for the intended space. It does not follow, however, that the possible party is itself in any way propositional, a mere linguistic entity. My three friends are not talking about what they are saying; they are talking about a possible party.

However important language might be in developing abstract models, it is not the only means for doing so. Diagrams, pictures and physical models may also be used to characterize aspects of an abstract model. Watson's physical model of DNA, for example, also served the purpose of specifying some features of an abstract model of DNA such as the pitch of the helix and the allowable base pairs. Other features of the physical model, such as being made partly of tin, have no counterpart in the abstract model.

## 7. OBJECTIONS

Before moving on to more general considerations I will pause to consider two objections to the whole idea of distributed cognition.

### 7.1 The individuation problem

Once we allow that cognition extends beyond the boundaries of individuals, where do we stop (Clark, 1997, chapter 10)? How do we individuate a cognitive system from its surroundings or, indeed, from other cognitive systems?

Imagine two developmental biologists standing in front of a chalkboard in an office talking and drawing pictures of a possible regulatory mechanism. From a distributed cognition point of view, the cognitive system developing models of the mechanism (both pictorial and abstract) includes at least the two scientists together with the sketches on the board. But does it include the chalkboard itself or, more questionably, the air in the office that sustains their lives and makes possible their verbal communication? As will be noted below, the idea of distributed cognition is typically associated with the thesis that cognition is *embodied*. In more standard terms, one cannot abstract the cognition away from its physical implementation. Adopting this point of view, we should say that the chalkboard and the air in the room are indeed part of this distributed cognitive system.

But now suppose that our two biologists work in different parts of the world and communicate electronically through their respective computers. Suppose part of the transmission is by means of satellite. One might easily agree that their computers are part of the cognitive system, but what about the transmission lines and the satellite? Yes, these too should count as part of the relevant cognitive system. But there are many things that clearly are not, for example, their automobiles and their homes, the supermarket down the street, and so on. It may sometimes be difficult to draw a sharp boundary between what is or is not part of a particular cognitive system, but that is true of many things.

It is, I think, more interesting to distinguish those parts of a cognitive system that *differentially* effect quantity or quality of the intended output from those parts that merely sustain the system, making it possible for it to do anything at all. Most likely the models considered by our two biologists would be no different if they were sitting together with a large sheet of paper rather standing in front of a chalkboard. I suspect it is this fact that fuels the initial feeling that the chalkboard is not part of the relevant cognitive system. But one must be careful not to discount too quickly the relevance of various components of a cognitive system. One might initially think that the satellite does not differentially effect the output of the system, but this is at least questionable. If our two biologists had to communicate by regular mail rather than by near instantaneous satellite transmission, this might well influence which model they eventually pursue in their subsequent work.

## 7.2 Minds

In countenancing extended cognition, are we not thereby committed to the existence of something like "extended minds" or even "group minds"? And are these not fairly disreputable ideas?

Now it is true that some advocates of distributed cognition are tempted by the notion of *distributed minds* (Clark, 1997, chapter 9). The presupposition behind this move seems to be that cognition is necessarily a property of minds. So, where there is cognition, there is also a mind. But I do not think that one is forced to take this route. I do not see why one cannot maintain an *ordinary* conception of *mind*, which restricts it to creatures with brains, that is, humans and perhaps some animals. Much cognition does, therefore, involve minds. Perhaps we could also agree that at least all of our currently existing cognitive systems have a human somewhere in the system. So every cognitive system has a mind somewhere. But as a concept in *cognitive science*, cognition need not be restricted exclusively to places where there are also minds. There is no compelling reason why we cannot extend the *scientific* concept of cognition to places where there are no minds.

This is question for which I do not think there is now a determinate answer. Rather, the question will be settled, if it is ever settled at all, in terms of considerations regarding how best further to develop a science of cognition.

# 8. CONNECTIONS

Although I have focused on just one source of recent interest in distributed cognition, the idea is much more widespread, as one can learn from Andy Clark's recent *Being There: Putting Brain, Body, and World Together Again* (1997). Lucy Suchman's *Plans and Situated Action* (1987), *The Embodied Mind* by Varela, Thompson, and Rosch (1993), and Edwin Hutchins' *Cognition in the* Wild (1995) are among earlier influential works.

A common theme in these works is that the human brain evolved primarily to coordinate movements of the *body*, thereby increasing effectiveness in activities such as hunting, mating, and rearing the young. Evolution favored cognition for effective action, not for contemplation. *Cognition is embodied.*

This point of view argues against there being a single central processor controlling all activities and for there being many more specialized processors. Central processing is just too cumbersome for effective action in the wild. Thus, an emphasis on distributed cognition goes along with a rejection of strongly computational approaches to cognition. In fact, some recent advocates of Dynamic Systems Theory have gone so far as to argue against there being any need at all for computation as the manipulation of internal representations (Thelan and Smith 1994).

Andy Clark, in particular, adopts the notion of "scaffolding" to describe the role of such things as diagrams and arithmetical schemas. They provide *support* for human capabilities. So the above examples involving multiplication and the Pythagorean theorem are instances of scaffolded cognition. Such external structures make it possible for a person with a pattern matching and pattern completing brain to perform cognitive tasks it could not otherwise accomplish.

The most ambitious claim made in behalf of distributed cognition is that *language* itself is an elaborate external scaffold supporting not only communication, but thinking as well (Clark, 1997, chapter 10). During childhood, the scaffolding is maintained by adults who provide instruction and oral examples. This is seen as analogous to parents supporting an infant as it learns to walk. Inner speech develops as the child learns to repeat instructions and examples to itself. Later, thinking and talking to oneself (often silently) make it seem as though language is fundamentally the *external* expression of *inner* thought, whereas, in origin, just the reverse is true. As argued by Vygotsky (1962) already in the 1920s, the capacity for *inner* thought expressed in language results from an internalization of the originally *external* forms of representation. There is no "language of thought". Rather, thinking in language is a manifestation of a pattern matching brain trained on external linguistic structures (See also Bechtel, 1996). This distributed view of language implies that cognition is not only embodied, but also *embedded* in a society and in a historically developed culture. The scaffolding that supports language is a cultural product (Clark 1997).

These views have recently been developed under the title of "cognitive linguistics" or "functional linguistics". Some advocates of cognitive linguistics, such as Tomasello, emphasize comparative studies of apes and humans (Tomasello, 1996). The most striking claim is that the major difference between apes and human children is *socialization*. Of course there are some differences in genotype and anatomy, but these are surprisingly small. The bonobo, Kanzi, raised somewhat like a human child by Duane Rumbaugh and Sue Savage-Rumbaugh (1993), is said to have reached the linguistic level of a two-year-old human child. Tomasello argues that what makes the difference between a natural and an enculturated chimpanzee is developing a sense of oneself and others as *intentional* agents. Natural chimpanzees do not achieve this, but enculturated ones can.

From the standpoint of computational linguistics, there has always been a question of how the neural machinery to support the necessary computations could possibly have evolved. From the standpoint of cognitive linguistics, this problem simply disappears. Language is not fundamentally computational at all, but the product of a pattern matching neural structure, which biologically *could* evolve, supported by an elaborate scaffolding of social interaction within an established culture.

Since I am myself sympathetic with these ways of thinking, it should now be clear why I earlier expressed reluctance to regard a solitary computer as a cognitive system. A computer may be embodied, but I am reluctant to call it a cognitive system because it fails to have the right kind of body for interacting with the world in appropriate ways. Syntactic inputs and outputs seem to me not to be enough. On the other hand, some present-day robots in some environments might be counted as cognitive systems roughly at the level of insects or other simple organic creatures (Brooks, 1994). Chimps surely should be counted as cognitive systems and even sometimes as parts of distributed cognition systems.

## 9. CONCLUSION

In conclusion, I would merely like to reiterate that the question as to exactly what is to be counted as a cognitive system or a distributed cognitive system seems not to have a determinate answer. Various groups in the cognitive sciences and related fields have strong interests in making the answer come out one way or another. It is not guaranteed that a general consensus will ever be achieved and it is possible that the question will eventually cease to be of any interest to anyone. In the mean time, individuals and groups remain free to promote whichever answer they think best.

## REFERENCES

Barwise, J. and Etchemendy, J., 1996, Heterogeneous logic, in: *Logical Reasoning with Diagrams*, G. Allwein and J. Barwise, eds., Oxford University Press, New York, pp. 179-200.

Bechtel, W., 1996, What knowledge must be in the head in order to acquire knowledge?, in: *Communicating Meaning: The Evolution and Development of Language*, B.M. Velichkovsky and D.M. Rumbaugh, eds., Lawrence Erlbaum, Nahwah, NJ, pp. 45-78.

Brooks, R., 1994, Coherent behavior from many adaptive processes, in: *From Animals to Animats 3,* D. Cliff et al., eds., MIT Press, Cambridge, MA.

Chandrasekaran, B., Glasgow, J., and Narayanan, N.H., eds., 1995, *Diagrammatic Reasoning: Cognitive and Computational Perspectives,* MIT Press, Cambridge, MA.

Clark, A., 1997, *Being There: Putting Brain, Body, and World Together Again*, MIT Press, Cambridge, MA.

Giere, R.N., 1996, Visual models and scientific judgment, in: *Picturing Knowledge: Historical and Philosophical Problems Concerning the Use of Art in Science*, B.S. Baigrie, ed., University of Toronto Press, Toronto, pp. 269-302.

Hutchins, E., 1995, *Cognition in the Wild*, MIT Press, Cambridge, MA.

Kosslyn, S.M., 1994, *Elements of Graph Design*, Freeman, New York.

McClelland, J.L., Rumelhart, D.E., and the PDP Research Group, eds., 1986, *Parallel Distributed Processing: Explorations in the Microstructure of Cognition*, MIT Press, Cambridge, MA.

Savage-Rumbaugh, S., Murphy, J., Sevcik, R.A., Brakke, K.E., Williams, S.L. Rumbaugh, D.M., 1993, Language comprehension in ape and child, *Monographs of the Society for Research in Child Development*, 58, Nos. 3-4.

Simon, H.A., 1978, On the forms of mental representation, in: *Perception and Cognition: Issues in the Foundations of Psychology*, C.W. Savage, ed., Minnesota Studies in the Philosophy of Science, vol. 9, University of Minnesota Press, Minneapolis, pp. 3-18.

Simon, H.A., 1995, Forward, in: B. Chandrasekaran, J. Glasgow, and N.H. Narayanan, eds., MIT Press, Cambridge, MA, pp. x-xiii.

Suchman, L., 1987, *Plans and Situated Actions*, Cambridge University Press, Cambridge, UK.

Tennant, N., 1986, The withering away of formal semantics, *Mind and Language*, 1:302-318.

Thelan, E. and Smith, L. 1994, *A Dynamic Systems Approach to the Development of Cognition and Action*, MIT Press, Cambridge, MA.

Tomasello, M., 1996, The cultural roots of language, in: *Communicating Meaning: The Evolution and Development of Language*, B.M. Velichkovsky and D.M. Rumbaugh, eds., Lawrence Erlbaum, Nahwah, NJ, pp. 275-307.

Varela, F.J., Thompson, E., and Rosch, E., 1993, *The Embodied Mind*, MIT Press, Cambridge, MA.

Vygotsky, L.S., 1962, *Thought and Language*, MIT Press, Cambridge, MA.

Wang, D., Lee, J., and Hervat., H. 1995, Reasoning with diagrammatic representations, in: *Diagrammatic Reasoning: Cognitive and Computational Perspectives*, B. Chandrasekaran, J. Glasgow, and N.H. Narayanan, eds., MIT Press, Cambridge, MA, pp. 501-534.

# Conceptual Models, Inquiry and the Problem of Deriving Normative Claims from a Naturalistic Base

Andrew Ward
*School of Public Policy, Georgia Institute of Technology, Atlanta, GA, USA, andrew.ward@pubpolicy.gatech.edu*

Abstract: John Dewey, in chapter ten of *The Quest for Certainty*, writes: "The problem of restoring integration and cooperation between man's beliefs about the world in which he lives and his beliefs about the values and purposes that should direct his conduct is the deepest problem of modern life". In cognitive science the problem plays itself out in the move from a description of what people such as scientists have done and how they reason, to claims about how science ought to be conducted and how scientists ought to reason. In this paper I will, adopting a broadly pragmatist account of cognitive inquiry, examine what sorts of processes go on in reasoning and scientific discovery, within a broadly connectionist framework. On such an account, the normative element of cognitive science results from understanding that inquiry is an activity of embodied, socially situated agents, and that model-based reasoning dynamically reflects this.

## 1. INTRODUCTION

In April and May of 1929, John Dewey delivered a series of Gifford Lectures at the University of Edinburgh that became the basis of his book *The Quest for Certainty*. In chapter 10 of that book, "The construction of good", Dewey writes:

*Model-Based Reasoning: Science, Technology, Values*, edited by
L. Magnani and N.J. Nersessian, Kluwer Academic/Plenum Publishers, New York, 2002

> The problem of restoring integration and cooperation between man's beliefs about the world in which he lives and his beliefs about the values and purposes that should direct his conduct is the deepest problem of modern life. It is the problem of any philosophy that is not isolated from that of life (Dewey, 1984).

It is in this spirit that Paul Churchland, in his essay "Moral facts and moral knowledge", explores the relevance of studies in cognitive science to issues in ethics. What emerges in the course of his explorations is a kind of ethical naturalism that, as Owen Flanagan notes, has two components. The first component is a descriptive-genealogical one whose purpose is to "specify certain basic capacities and propensities of *Homo Sapiens*, e.g., sympathy, empathy, and egoism, that are relevant to moral life" (Flanagan, 1996b). This component will, amongst other things, "explain how people come to feel, think, and act about moral matters in the way they do". The second component is a normative component whose purpose is to "explain why some norms (including norms governing choosing norms), values and virtues are good or better than others". As both Churchland and Flanagan recognize, the standard view is that even if the descriptive-genealogical component of ethics can be naturalized, the normative component of ethics cannot.

The reasons for doubting that the normative component of ethics can be naturalized are varied, but there are at least two interrelated problems. The first is the traditional view that descriptive is's and normative ought's "are on opposite sides of a great epistemic divide" (Lauden, 1990), and that naturalistic accounts fall on the descriptive, "is" side of that divide, while normative accounts fall on the other. For some this leads to the view that moral knowledge is not possible. Others believe that while moral knowledge does exist, the attempt to bridge the gap and provide a fully naturalistic account of the normative component of ethics commits the so-called "naturalistic fallacy". This fallacy, if indeed it is a fallacy, supposes that descriptive, naturalistic accounts of our beliefs, feelings and actions are subject to the same forms of adjudication as accounts of what we *should* believe and feel, and how we *should* act.

The second problem is that even if there is moral knowledge and we could somehow bridge the descriptive-normative gap, the resulting ethical account would be unacceptably relativistic. Briefly, the idea is that if the normative component of ethics is naturalized, its account will fall within the purview of the descriptive-genealogical component of ethics. In this case though, it seems that there is nothing more to moral epistemology and moral choice than successful social assimilation. However, if moral epistemology and moral choice are nothing more than successful social assimilation, then the normative component of naturalized ethics seems to become at best relativistic, and at worst nihilistic.

These are, indeed, difficult questions for those who would, like Dewey and Churchland, seek to integrate ethics into a naturalistic conception of the world. What is especially interesting in Churchland's case is his attempt to use the resources of cognitive science to affect such an integration. While, as Alvin Goldman writes, there is general acknowledgment that "[...] findings and theories in cognitive science have been increasingly important in many areas of philosophy, especially philosophy of mind, epistemology, and philosophy of language" (Goldman, 1993), the same cannot be said for the importance of such findings and theories to ethics. In many ways this is surprising. Although the expression "cognitive science" did not really emerge until the 1970s, the practice of what we now call "cognitive science" emerged as early as September 1948 with the Hixon Symposium on "Cerebral Mechanisms in Behavior". While it may sound as if such a conference would have little to say to people working in ethics, what emerged out of the conference was the beginning of a consensus about the importance of issues that bear directly on ethics. As Howard Gardner writes, what the conference laid the foundation for was a consensus that cognitive science asks "what it means to know something and to have accurate beliefs, or to be ignorant and mistaken [...] to understand what is known – the objects and subjects in the external world – and the person who knows [...] [to understand] the differences among individuals: who learns early or with difficulty; what can be known by the child, the inhabitant of a preliterate society, an individual who has suffered brain damage, or a mature scientist" (Gardner, 1987). So understood, it would be more surprising if the findings and theories of cognitive science did not have something substantive to contribute to ethics. Accordingly, in what follows I will argue that Churchland is right to believe that ideas from cognitive science can play an important role in addressing ethical issues. In addition, I will argue that by incorporating a pragmatist perspective along the lines suggested by Dewey makes it possible to bridge the descriptive-normative gap in a way not committed to an unpalatable relativism.

## 2. CHURCHLAND'S MODEL OF PSYCHOLOGICAL ARCHITECTURE

It is a central tenant of cognitive science, as well as a long tradition of philosophy, that a significant portion of human behavior is cognitive in character. Part of what this means is that an important and significant portion of human behavior is best understood as the product of representational activity on the part of human beings. Often bundled together with this is the view that the representational structures in question are sentential, and that the principal goal of knowledge acquisition is to maximally produce sentential representations (propositional attitudes) that appropriately "mirror" the world, and minimize those that don't. In contrast, basing his views on an acceptance of a parallel distributed processing (PDP) account of mind and cognitive development, Churchland writes:

> On these neurocomputational models, knowledge acquisition is primarily a process of learning *how*: how to recognize a variety of complex situations and how to respond to them appropriately. The quality of one's knowledge is measured not by any uniform correspondence between internal sentences and external facts, but by the quality of one's continuing performance (Churchland, 1993a).

Churchland goes on to explicate this account of knowledge acquisition by means of prototypes. For instance, within the PDP model of psychological architecture advocated by Churchland, children come, through their many experiences within the social world, "to recognize certain prototypical kinds of social situations, and they learn to produce or avoid the behaviors prototypically required and prohibited in each" (Churchland, 1993a). How the child subsequently reacts to novel learning situations will be a function of which of his or her many prototypes that situation activates. This, in turn, will be a "matter of the relative similarity of the new situation to the various prototypes on which she was trained" (Churchland, 1993a).

While there are many interesting and provocative features of Churchland's account of a PDP account of mind and cognitive development, a detailed discussion is beyond the scope of this paper. However, what is important in the present context is not Churchland's PDP account, but rather his use of prototypes. Although he quite consistently accounts for prototypes in PDP terms, Churchland does motivate his use of prototypes in characterizing the psychological architecture of human beings independent of his PDP account. For instance, in his *The Engine of Reason, The Seat of the Soul: A Philosophical Journey into the Brain*, Churchland ties his advocacy of prototypes into his acceptance of Kuhn's use of paradigms to account for the practice and progress of science. The way that Churchland understands Kuhn, what Kuhn was saying is that scientific understanding ought not be accounted for in terms of a person's grasp of a set of sentences constituting a scientific theory or part of a theory. Instead, Churchland claims that the moral to be drawn from the work of Kuhn is that a "student who is learning a scientific theory learns, first of all, the prototypical features of the central example, and then learns to extend that understanding, suitably modified, to further examples that radiate out from the central example already mastered" (Churchland, 1996b).

Thus, while Churchland does consistently go on to account for prototypes in PDP terms, the use of prototypes does not, by itself, demand that we do so. Instead, what is important in the use of prototypes is twofold. First, a prototype account moves us away from sententialism - the notion "that cognition is to be modeled largely on language and logical reasoning; having a thought is, functionally speaking, having a sentence in the head, and thinking is, functionally, doing logic, or at least running on procedures very like logic" (Churchland and Sejnowski, 1990). Although Churchland sometimes suggests the contrary, this need not be taken to exclude all sentential representational structures from the psychological architecture of human beings. Rather, it need only mean that many of the most significant of the representational structures constituting the psychological architecture of human beings ought not be understood as having the form of truth-functional sentences. Second, a prototype account suggests that learning - scientific and moral - is really the learning of a practice. Again, Churchland relates his views to Kuhn and his account of theory acceptance. In contrast to the orthodox view, Kuhn argued for a performance conception of theory evaluation. As Churchland interprets Kuhn, this means that a "theory is a vehicle whose virtue lies in its many uses: explanation, prediction, unification, and manipulation of the real world through the many technologies it makes possible" (Churchland, 1996b). What this means is that the traditional distinction between theory and practice has broken down. Learning theories is no longer a matter of grasping sentences with truth conditions; understanding a theory is no longer a matter of acquiring theoretical knowledge in the traditional sense. Learning a theory is acquiring a skill, and understanding a theory means to master a practice. In a somewhat provocative manner, Churchland expresses this point by writing that "the original distinction between *praxis* and *theoria* is both superficial and artificial" (Churchland, 1996a). Although there are several ways one might interpret this, I suggest the following. When we are talking about the knowledge of individuals, then the knowledge in question is practical - the mastery of a practice. However, when we are talking about the knowledge embodied in the practices of a culture or a society, then we are talking about codified practices. Here we may rightly talk about rules and principles that an individual may or may not follow. It is this "social knowledge" that may be rightly considered to be theoretical (See Clark, 2000a, and Clark, 2000b).

In his adoption of a prototype account, Churchland moves away from the classical theory about how concepts are represented in human beings. According to the classical theory, the origins of which, as Edward Smith and Douglas Medin note (Smith and Medin, 1981), go back to Aristotle in philosophy and to Hull's 1920 monograph on concept attainment in experimental psychology, concepts are represented by a collection of properties that the category members share. More importantly, the central idea of the classical view is that the collection of properties that the category members share is singly necessary and jointly sufficient to define the concept. Thus, according to the classical theory, no member of the category should be more representative than any other member of the category. In philosophy this classical theory was challenged by Wittgenstein and his claim that many categories are not structured by the presence of singly necessary and jointly sufficient properties, but rather by family resemblance. However, Wittgenstein's methodology leading him to this claim was not systematic, and, arguably, in many respects was *a priori*. It was in the 1970s that the psychologist Eleanor Rosch combined her own empirical research with that of earlier psychologists such as Brent Berlin, Paul Kay, and Roger Brown, to name only three, to come up with an alternative to the classical view. This alternative is called "the theory of prototypes and basic level concepts".

Although there are several different versions of "prototype theory", the one most directly relevant to Churchland's account is the probabilistic prototype theory. According to this theory an object is categorized as an instance or subset of a category if and only if the object possesses some critical sum of the weighted properties that constitute the category. This sum of weighted properties that constitute the category is represented by the prototype. Thus, rather than an object needing to possess a certain set of properties to be categorized as an instance or subset of a category, the process of categorization is a probabilistic function of the object's similitude to the prototype. As Medin and Heit write, "[...] membership is a category can thus be graded rather than all-or-none, where the better members have more characteristic properties than the poorer ones." (Medin and Heit, 1999) Moreover, although I have dwelt on Churchland's advocacy of prototypes, his points about sententialism (his rejection of sententialism) and a performance conception of theory evaluation carry over, *mutatis mutandis*, to other kinds of non-sentential representational structures. Thus, we can readily embrace Stephen Stich's remark that "[...] recent research strongly suggests that neither the prototype approach nor the exemplar approach can tell the whole story about conceptual representation, even for simple object concepts such as *fruit* and *bird* [...] The consensus seems to be that conceptual representation is a complex affair combining prototypes or exemplars with less observationally salient, more theoretical features" (Stich, 1993a).

What this shows is that Churchland is not alone in suggesting that an important part of the psychological architecture of human beings must incorporate prototypes and other non-discursive kinds of representational structures (e.g., exemplars). Most importantly, for those with an antipathy to PDP accounts of mind and cognitive development, the work of psychologists such as Barsalou, Medin, Rosch and others offer a variety of empirically based, non-PDP motivated reasons for believing that significant parts of human psychological architecture is non-sentential. Moreover, something to keep in mind is that in moving away from sententialism one also is moving away from the idea that prototype combinations must "consist in the linear addition of the properties of each contributing prototype" (Clark, 1993). I note this because one of the standard objections (Fodor, 1998) to accounting for the psychological architecture of human beings in terms of non-discursive structures like prototypes is that they cannot serve as the building blocks of complex contexts such as "United States capitals located west of the Mississippi river". However, the objection looses much of its force - though must still be addressed and does suggest as PDP account of non-discursive representational structures such as prototypes - once we give up sententialism and its requirement that concepts be combined in a linear fashion. Accordingly, in what follows I will, a bit tendentiously perhaps, assume that there are good reasons for accepting a wide range of concepts such as prototypes and exemplars in characterizing important parts of people's psychological architecture. Moreover, following Churchland, I will assume that an acceptance of prototypes - or other kinds of concepts - moves us away from sententialism, and towards a performance conception of theory evaluation.

## 3. PRAGMATISM

An acceptance of the claim that non-sentential representational structures play an important role in accounting for the psychological architecture of humans has important implications. For instance, truth and reference play an important role in evaluating propositional attitudes precisely because of their sentential structure. However, this does not seem to be the case for non-sentential representational structures. Because prototypes, exemplars and other such concepts do not have a sentential structure, their evaluation and role in the psychological architecture of human beings must be explicated in terms other than truth and reference. It is for this reason that Churchland proposes that we seek for some "more worthy and more penetrating evaluative and semantical notions" (Churchland, 1998) by which to assess the psychological architecture of human beings. In a discussion of Bas van Fraassen's critical empiricism, Churchland expands on this suggestion and writes that "if we are to reconsider truth as the aim or product of cognitive activity [...] we should move in the direction of *pragmatism*" (Churchland, 1993c).

It is just at this point that cognitive scientists and non pragmatically inclined philosophers differ in their reasons for supposing that the psychological architecture of human beings is representational. Whereas non-pragmatically inclined philosophers typically focus on issues of epistemology and how what appears to the senses may be different from the way the world really is, cognitive scientists typically focus on issues of dynamic adaptability. One of the distinguishing characteristics of intelligent systems is their ability to adapt to novel situations so as to achieve some particular goal. Such adaptability requires an ability to be sensitive to the changing conditions in which the system finds itself relative to the goal that is being sought by the system. A central thesis of cognitive science, then, is that any adequate explanation of such a dynamic relation between environmental (both inner and outer, perhaps) sensitivity and goal seeking behavior requires making reference to a representational structure possessed by the system. Thus, representational structures must be understood as distinguishing characteristics of our embodied agency; the loci of human beings' directions and actions within the world. It is only within a larger "ecological" context of success and failure that procedures and actions can be judged. Thus in assessing representational architecture we are led to an examination of how human beings evolved to have the psychological architectures they do. We must, in other words, seek to understand human beings as the products of a long process of natural selection within a dynamic environment. As Patricia Churchland writes, "[...] looked at from an evolutionary point of view, the principal function of the nervous system is to enable the organisms to move appropriately [...] a fancier style of representing is advantageous *so long as it is geared to the organism's chance for survival*. Truth, whatever that is, definitely takes the hindmost" (Churchland, 1987). What is important is not how human beings picture the world, but rather how they deal with it. This is the motivation for shifting the focus to how humans develop the principally non-sentential psychological architectures they have. It is these architectures that account for the successful performance of humans in their various environments - natural and social. We must, in other words, understand humans as organic creatures that, as Churchland expresses it, are "wonderfully designed by evolution to subserve the acquisition and deployment of a panoply of *skills* and *abilities*" within a dynamic environment (Churchland, 2000).

To better understand the connection of this to pragmatism, consider Dewey's account of inquiry. Like Peirce, Dewey claimed that inquiry is occasioned by some sort of irritation or trouble. For example, Dewey writes that inquiry starts "as an effort to get out of some trouble, actual or menacing" (Dewey, 1980). Accordingly, irritation, or what Dewey sometimes called states of "imbalance or disequilibration" (Dewey, 1941), is always contextually embedded - it is the instability specific to the interaction of an organism with its natural and/or social environment. It is this context of instability that Dewey refers to as the "indeterminate situation". When such an indeterminate situation occurs, there is an attempt by the organism to affect a transformation of the situation into one that is determinate and in which the instability is eliminated. Put differently, some "idea or mode of functioning adopted for its serviceability in resolving the problems of yesterday no longer fills the bill." (Ezorsky, 1958) In his *Logic: The Theory of Inquiry*, Dewey pulls these points together in his definition of "inquiry" as "the controlled or directed transformation of an indeterminate situation into one so determinate in its constituent distinctions and relations as to convert the elements of the original situation into a unified whole" (Dewey, 1991a). Thus, for Dewey the psychological architecture of a human evolves as it attempts to affect "an active organization of the given environment, a removal of some specific trouble or perplexity" (Dewey, 1991a). Inquiry that results in structures that systematically affect – both diachronically and synchronically – such active organizations are deemed as successful.

The goal of inquiry is not, for Dewey, a final, unshakable certainty. The goal of inquiry is the dynamic acquisition of successful orientations within the natural and social world. Actual problems and perplexities arise within specific circumstances, and inquiry is a problem-solving activity whose aim is to resolve these problems. As Dewey writes, "[W]e know that some methods of inquiry are better than others in just the same way in which we know that some methods of surgery, farming, road-making, navigating or what-not are better than others [...] They are the methods which experience up to the present time shows to be the best methods for achieving certain results" (Dewey, 1991a). In this respect our interactions with the natural and social worlds do not result in some sort of progressively sharper picture of the world but rather in the creation of cognitive maps (or models) of the world that govern our performances. Significantly, while maps and models may be more or less accurate, more or less detailed, and more or less useful in "getting us to where we want", neither has a simple binary relationship, a true or false relationship, to the world. Similarly, the concepts constituting important aspects of the psychological architecture of human beings are, in effect, dynamic models that permit the person to successfully navigate in the natural and social environments in which she or he is embedded. Stich puts the point quite nicely when he writes that cognitive processes, for pragmatists, "should be thought of as something akin to tools or technologies or practices that can be used more or less successfully in achieving a variety of goals" (Stich, 1993). This certainly accords with Dewey who thought of propositions as tools for resolving indeterminate situations, and whose meaning was a function of the role they played in such resolutions.

## 4. NATURALIZED ETHICS

We are now in a position to understand how the normative component of ethics is naturalized. As we have seen, for the pragmatist the psychological architecture of a human arises as a response to situated struggles and challenges within its natural and social environments. For Churchland as for the pragmatists, it does not matter for the development of the representational structures whether problematic situation is one involving the natural environment or the social environment. What matters is the resolution of the problematic situation and the development of the appropriate conceptual tools to navigate in both the natural and social environment. Recall though, that what created the descriptive-normative gap was the insistence that the adequacy of answers to descriptive, factual questions be gauged by a standard different than that of answers to normative questions. As the above suggests, this is precisely what the pragmatist and Churchland denies. For both, what we could call scientific descriptive inquiry on the one hand, and ethical inquiry on the other, is problem-solving activity structured as practical deliberation. Dewey, for example, writes that "[...] inquiry, discovery take the same place in morals that they have come to occupy in sciences of nature. Validation, demonstration become experimental, a matter of consequences" (Dewey, 1991b). Similarly, using the language of the PDP model, Churchland writes that all inquiry results in a "family of abilities encoded as a hierarchy of prototypes and prototypical sequences in the neuronal activation space" of the brain of the inquirer (Churchland, 1996b). The idea in both cases is that descriptive, naturalistic accounts of our beliefs, feelings and actions are subject to the same forms of adjudication as accounts of what we *should* believe and feel, and how we *should* act. In particular, both are gauged by how well they permit "successful navigation" within natural and social environments.

Given this account, we can now go on to ask for a bit more detail about what it is that people acquire when engaged in moral learning. Churchland's answer to this question is that "[...] they are learning how best to organize and administer their collective and individual affairs". (Churchland, 1993a). This is not intended to imply that ethics ignores the natural world. Quite the contrary; embodied agents always find themselves in both the natural and social worlds. However, often the relations with the social world are more unpredictably dynamic than those with the natural world and so require different kinds of skills. Thus, according to Churchland, in learning to make moral judgments one is learning "the *structure of social space* and *how best to navigate one's way through it*" (Churchland, 1993a). On such an account, the normative component of a naturalized ethics is precisely the pragmatic issue of inquiring into an optimal navigation through a specific social space. More technically, moral learning will be, according to Churchland, "a matter of slowly generating a hierarchy of moral prototypes, presumably from a substantial number of relevant *examples* of the moral kinds at issue. Hence the relevance of stories and fables, and above all, the ongoing relevance of the parental example of interpersonal behavior, and parental commentary on and consistent guidance of childhood behavior" (Churchland, 1996b). Naturally as the so-called social space changes, what counts as optimal navigation is likely to change as well. To use an example from Mark Johnson, a breach of promise is evaluated one way within a framework of lying to one's spouse about an extramarital affair, but has a very different significance in the context of failure to keep one's promise to an officer of a dictatorial police state (Johnson, 1993). Accordingly, our ability to navigate successfully in social space depends crucially on behavior patterns that are programmatically happy; behavior patterns that are dynamically sensitive to experiences and change as experiences fill in the gaps.

Now while it may seem that something non-natural has been smuggled into the account by referring to "optimal" navigation through social space, this criticism is misguided. Optimal navigation in social space is no less something natural than is flourishing within a biological context. Just as the flourishing of a plant or a species is something to be determined by its developmental history and its relation to other constituents of an ecosystem, an analogous claim can be made for optimal social navigation. Optimal social navigation is something to be decided by the success of navigation through social space as judged by the human interests and values which "evolve and get modified at the same time and in interaction with our evolving image of the world itself" (Putnam, 1990). As is true for flourishing, there is nothing here that does not readily lend itself to a naturalistic account. Of course this does not mean it will be a simple account. Just as the flourishing of a plant or species will depend crucially on the many features of the ecosystem in which it exists, so too will optimal navigation depend on the many features of the social (and natural) environment in which the navigation takes place. It is not coincidental, relative to Churchland's pragmatism, that while the language is different, a similar sentiment is found in Dewey. In his *Human Nature and Conduct*, Dewey writes that "[...] moral science is not something with a separate province. It is the physical, biological, and historic knowledge placed in a humane context where it will illuminate and guide the activities of man" (Dewey, 1988).

Finally then, this leads us to the question of whether the proposed naturalistic ethics is unacceptably relativistic. It is true that Churchland writes that "a human's moral virtues must always be turned to and measured against the abiding features of the local social economy in which he or she is unavoidably embedded" (Churchland, 1996a). Moreover, it seems right to say that the empirical facts of local social and natural ecology will sharply constrain what counts as successful modes of behavior. This reflects the facts that human beings are embedded within a rich natural-social environment in which their cognitive and social navigations take place, and that successful navigation must be judged against criteria provided by the specifics of the natural-social environment. It makes no sense to talk about any sort of navigation, optimal or otherwise, except by reference to the specifics of a natural and social ecology in which the navigation takes place. Put differently, natural world and social world navigation is always *local*.

While it follows that the proposed naturalized ethics is contextual. However, what does not follow is that a naturalized ethics must embrace some sort of "anything goes" ethical relativism. Ethics is not, as some relativists seem to claim, a matter of mere socialization. As Flanagan writes, "'Mere' socialization is socialization toward which no critical attitude is taken, for which there are no mechanisms for modification and refinement" (Flanagan, 1996). However, on the Churchland cum pragmatist account of naturalized ethics, there are always mechanisms for modification and refinement. The social space within which the individual is partially constituted and seeks for orientation is not some sort of homogenous, monolithic entity. Rather, the social space is made up of a combination of competing and collaborating narratives. As each narrative competes with myriad others, a critical reflexive attitude is created within the narrative whereby fallibilistic assessments are made as to the value of the social orientation offered by the narrative. All inquiry begin with the recognition that people find value in a variety of goods and practices, many of which conflict. This conflict, in order to be resolved, demands that we subject the goods and practices to critical scrutiny so as to assess their relative merits. This is the essence of pragmatism's experimentalism. Since all inquiry - all behavior - aims at the resolution of lived problems and perplexities, then the various specific practices of inquiry adopted will affect and be effected by each other. No particular practice can be "justified" as successful or unsuccessful in isolation from the myriad other practices seeking to resolve problems and perplexities. Just as problems and perplexities arise and are defined within a complex web of ways that inquirers find themselves situated in the world, so too it is for the practices of inquiry - scientific, ethical, economic, and otherwise. The likelihood of genuine success - long-lasting success that doesn't raise other problems and can be successfully integrated into all other aspects of the lives of people - depends upon an active critical examination of accepted practices. Thus, to the extent that the psychological architecture of a person is a function of these competing narratives there is always found a constitutive critical attitude conducive to continuing modification and refinement.

Moreover, it is crucial to remember that it is not only individuals who seek an optimal kind of navigation through the natural-social environment. In a larger sense it is the society created by the interweaving of various local narratives that itself seeks an optimal navigation. Viewed from this broader, social perspective, the legislative and judicial bodies of societies must themselves be understood as engaged in what Churchland calls a process of "continuously formulating and reformulating social policy of some sort or other - prohibiting certain kinds of behavior, regulating many others, and positively encouraging others still - in an ongoing response to the unfolding environment and to the observed social effects of policies already in place" (Churchland, 1996b). To the extent that a social narrative is not responsive to, and unwilling to engage the possibilities for social orientation offered by other narratives constituting the society, the narrative will (eventually) fail. Should large portions of the constituent narratives fail to be responsive and engaging, the entire social fabric will unravel. This is the moral of political theorists from Plato and Aristotle, to Machiavelli, to Marx and Sandel as to why it is that forms of political association arise, struggle, fail, and give birth to new forms of association. The larger systemic view of the dynamics of the social space in which individuals find themselves situated requires a social commitment to self-reflexive scrutiny built into the idea of justification, now construed quite broadly in the pragmatist spirit, as the goal of inquiry. As Margaret Walker writes, the success of societies requires "publicly recognized standards and forums which institutionalize and reward evaluation" of moral judgments and requires "examination of processes and relations of cognitive authority, which must not cloak cultural, political, or economic dominance or suppress relevant criticism from diverse viewpoints" (Walker, 1996). These requirements of self-reflexive scrutiny at the social level function as additional meta-level constraints on individuals' successful navigation through localized social spaces.

Thus, once the normative component of a naturalized ethics as a situated practice is understood, the charge of relativism is answered in two ways. First by recognizing that people exist and their psychological architecture formed, though not mechanistically determined, by what Lyotard calls a complex fabric of competing and agonistic social narratives (Lyotard, 1991). The psychological architecture of human beings is a complex, dynamic structure that "pragmatically represents" the agonistics of the natural and social worlds within which it evolved. Except in highly unusual or contrived situations, humans are not members of any single society or culture. Humans are members of multiple, competing cultures and societies, all of which contribute to the formation of the psychological architecture of individual humans. Relativism is answered in the second way by the presence of systemic social constraints governing societies' optimal navigation through the natural and larger social worlds. Successful resolutions to inquiries within social contexts require we work to integrate the interests and problems of the people constituting the society (and, of course, those of other societies with whom the members interact). Accordingly, from the pragmatist point of view the best way to answer challenges posed by people's interactions with one another is not by abstract theorizing about morality. Rather, we need to look in detailed ways at examples of social structures that seem to be successful, and to cull from them the elements that made them successful. In this respect, social organizations form, as Plato suggested in his *Republic*, macroscopic analogues of the cognitive maps/models constituting the psychological architecture of individual human beings. If all this is correct, then it follows that the worry that moral learning might turn out to be mere socialization, and ethics fall prey to some sort of Feyerabendian relativism, is misguided.

In conclusion, by incorporating insights from cognitive science and pragmatism, advocates of naturalized ethics are able to resolve at least two of the important problems with which such a view has traditionally been associated. The triangulation of cognitive science, pragmatism and naturalized ethics suggests, in the words of Dewey, "a living picture of the choice of thoughtful men about what they would have life to be, and to what ends they would have men shape their intelligent activities" (Dewey, 1991b).

## REFERENCES

Churchland, P., 1987, Epistemology in the age of neuroscience, *The Journal of Philosophy* 84: 544-553.

Churchland, P.M., 1993(a), Moral facts and moral knowledge, in: *A Neurocomputational Perspective: The Nature of Mind and the Structure of Science*, Bradford Books/MIT Press, Cambridge, MA, pp. 297-303.

Churchland, P.M., 1993(b), On the nature of explanation: a PDP approach, in: *A Neurocomputational Perspective: The Nature of Mind and the Structure of Science*, Bradford Books/MIT Press, Cambridge, MA, pp. 197-230.

Churchland, P.M., 1993(c), The ontological status of observables: In praise of the superempirical virtues, in: *A Neurocomputational Perspective: The Nature of Mind and the Structure of Science*, Bradford Books/MIT Press, Cambridge, MA, pp. 139-151.

Churchland, P.M., 1996(a), Flanagan on moral knowledge, in: *The Churchlands and Their Critics*, R.N. McCauley, ed., Blackwell Publishers, Inc., Cambridge, MA, pp. 302-306.

Churchland, P.M., 1996(b), *The Engine of Reason, the Seat of the Soul: A Philosophical Journey into the Brain*, Bradford Books/MIT, Cambridge, MA.

Churchland, P.M., 1996(c), The neural representation of the social world, in: *Mind and Morals: Essays on Ethics and Cognitive Science*, L. May, M. Friedman, and A. Clark, eds., Bradford Books/MIT Press, Camdidge, MA, pp. 91-108.

Churchland, P.M., 1998, Activation vectors v. propositional attitudes: How the *brain* represents reality," in: *On the Contrary: Critical Essays, 1987–1997*, P.M. Churchland and P.S. Churchland, eds., Bradford Books/MIT Press, Cambridge, MA, pp. 39-44.

Churchland, P.M., 2000, Rules, know-how and the future of moral cognition, *Canadian Journal of Philosophy* 26:291-306, *Supplementary Volume,* Bradford Books/MIT Press, Cambridge, MA, pp. 91-108.

Churchland, P. and Sejnowski, T.J., 1990, Neural representation and neural computation, *Philosophical Perspectives: Action Theory and Philosophy of Mind* 4:342-383.

Clark, A., 1993, *Associative Engines: Connectionism, Concepts, and Representational Change*, Bradford Books/MIT Press, Cambridge, MA.

Clark, A., 1996, Connectionism, moral cognition, and collaborative problem solving, in: *Mind and Morals: Essays on Ethics and Cognitive Science*, L. May, M. Friedman, and A. Clark, eds., Bradford Books/MIT Press, Cambridge, MA, pp. 109-127.

Clark, A., 1997, *Being There: Putting Brain, Body, and World Together Again*, Bradford Books/MIT Press, Cambridge, MA.

Clark, A., 2000a, Making moral space: A reply to churchland, *Canadian Journal of Philosophy* 26: pp. 307-312, *Supplementary Volume.*

Clark, A., 1991, *Microcognition: Philosophy, Cognitive Science, and Parallel Distributed Processing*, Bradford Books/MIT Press, Cambridge, MA.

Clark, A., 2001, *Mindware: An Introduction to the Philosophy of Cognitive Science*, Oxford University Press, Oxford.

Clark, A., 2000b, Word and action: Reconciling rules and know-how in moral cognition, *Canadian Journal of Philosophy* 26: 267-289, *Supplementary Volume.*

Dewey, J., 1941, Propositions, warranted assertibility, and truth, *The Journal of Philosophy* 38:169-186.

Dewey, J., 1980, Introduction to *Essays in Experimental Logic*, in *John Dewey - The Middle Works, 1899-1924, volume 10*, J.A. Boydston, ed., Southern Illinois Press, Carbondale, IL, pp. 320-365.

Dewey, J., 1984, *The Quest for Certainty: A Study of the Relation of Knowledge and Action - The Later Works of John Dewey, 1925-1953: volume 4: 1929*, J.A. Boydston, ed., Southern Illinois University Press, Carbondale, IL.

Dewey, J., 1988, *Human Nature and Conduct – 1922: The Middle Works of John Dewey, 1899-1924, volume 14*, J.A. Boydston, ed., Southern Illinois University Press, Carbondale, IL.

Dewey, J., 1991(a), *Logic: The Theory of Inquiry in John Dewey - The Later Works, 1925-1953: 1938, volume 12*, J.A. Boydston, ed., Southern Illinois University Press, Carbondale, IL.

Dewey, J., 1991(b), *Reconstruction in Philosophy: The Middle Works of John Dewey, 1899-1924: volume 12*, J.A. Boydston, ed., Southern Illinois University Press, Carbondale, IL.

Ezorsky, G., 1958, Inquiry as appraisal: The singularity of John Dewey's theory of valuation, *The Journal of Philosophy* 55:118-124.

Flanagan, O., 1996(a), Ethics naturalized: ethics as human ecology, in: *Self-Expressions: Mind, Morals, and the Meaning of Life*, Oxford University Press, New York, pp. 117-141.

Flanagan, O., 1996(b), The moral network, in: *The Churchlands and Their Critics*, R.N. McCauley, ed., Blackwell Publishers, Cambridge, MA, pp. 192-215.

Fodor, J., 1998, *Concepts: Where Cognitive Science Went Wrong*, Clarendon Press, Oxford.

Gardner, H., 1987, *The Mind's New Science: A History of the Cognitive Revolution*, Basic Books, New York.

Goldman, A., 1993, Ethics and cognitive science, *Ethics* 103:337-360.

Johnson, M., 1993, *Moral Imagination: Implications of Cognitive Science for Ethics*, University of Chicago Press, Chicago.

Johnson, M., 1996, How moral psychology changes moral theory, in: *Self-Expressions: Mind, Morals, and the Meaning of Life*, Oxford University Press, New York, pp. 45-68.

Lauden, L., 1990, Normative naturalism, *Philosophy of Science* 57:44-59.

Lyotard, J.-F., 1991, *The Postmodern Condition: A Report on Knowledge*, Geoff Bennington and Brian Massumi, translators, University of Minnesota Press, Minneapolis.

Medin, D.L. and Heit, E., 1999, Categorization, in: *Cognitive Science*, Benjamin Bly and David Rumelhart, eds., Academic Press, San Diego, pp. 99-143.

Putnam, H., 1990, *Realism with a Human Face*, Harvard University Press, Cambridge, MA.

Smith, E.E. and Medin, D.L., 1981, *Categories and Concepts*, Harvard University Press, Cambridge, MA.

Stich, S., 1993(a), Moral philosophy and mental representation, in: *The Origin of Values*, M. Hechter, L. Nadel, and R.E. Michod, eds., Aldine De Gruyter, New York, pp. 215-228.

Stich, S., 1993(b), *The Fragmentation of Reason*, Bradford Books/MIT Press, Cambridge, MA.

Walker, M.U., 1996, Feminist skepticism, authority, and transparency, in: *Moral Knowledge? New Readings in Moral Epistemology*, W. Sinnott-Armstrong and M. Timmons, eds., Oxford University Press, New York, pp. 267-292.

# Dynamic Imagery: A Computational Model of Motion and Visual Analogy

David Croft[*] and Paul Thagard[#]

[*]*Department of System Design Engineering, University of Waterloo, Waterloo, Ontario, Canada, dwcroft@uwaterloo.ca;* [#]*Department of Philosophy, University of Waterloo, Waterloo, Ontario, Canada, pthagard@watarts.uwaterloo.ca*

Abstract: This paper describes DIVA (Dynamic Imagery for Visual Analogy), a computational model of visual imagery based on the scene graph, a powerful representational structure widely used in computer graphics. Scene graphs enable the display of complex objects, include motion. Our model combines a semantic-network memory system with computational procedures based on scene graphs. The model can account for people's ability to imagine novel situations involving motion, and to use visual analogies that compare two systems of objects in motion.

## 1. INTRODUCTION

How is a gymnast's cartwheel like a windmill? Answering this question cannot rely just on the verbal information that we have about gymnasts and windmills, or just on the static visual images we have of them. Rather, the comparison becomes interesting when we mentally construct dynamic images of a person whose limbs are in motion and a windmill whose blades are turning.

Mental imagery is often used to visualize a static scene, but it can also be dynamic, when people imagine objects in motion. For example, Pedone, Hummel, and Holyoak (2001) found that the use of animated diagrams sig-

*Model-Based Reasoning: Science, Technology, Values*, edited by
L. Magnani and N.J. Nersessian, Kluwer Academic/Plenum Publishers, New York, 2002

nificantly improved analogical problem solving. How do people construct and use dynamic mental images? It has become increasingly accepted that problem solving sometimes uses mental imagery that employs pictorial representations in addition to the verbal representations that commonly support problem solving (Larkin and Simon, 1987). But little research has been done on how people represent motion and use it in analogical problem solving and other kinds of thinking.

This paper describes a computational model of dynamic imagery based on a powerful representational structure widely used in computer graphics, the scene graph. We hypothesize that human thinking uses dynamic representations that have some of the same computational powers. Scene graphs employ a hierarchical organization that makes possible the visual display of complex objects, including the motions of individual objects. Organizing visual information in a hierarchy allows motions to be represented as discrete behaviors that can be applied to specific objects in a scene. We adapt this technology to explain how the mind can apply motions to objects that it has not previously seen in action. For example, people can imagine a computer monitor falling off a desk even if they have never previously perceived such an event.

Our computational model of human dynamic imagery combines a semantic-network memory system with computational procedures based on scene graphs. We have exploited the computational power of a recently developed programming library, Java 3D, to model how people are able to construct and transform complex visual scenes. Our model simulates visual operations such as object combination, rotation, and imagined motion, as well as mapping of scenes in dynamic visual analogies. The scene graph representation makes possible a computationally and psychologically richer account of mental imagery than has previously been achieved in cognitive science and artificial intelligence (see Chella, Frixione, and Gaglo, 2000; Davies and Goel, 2000; Forbus, 1995; Funt, 1989; Glasgow, Narayanan, and Chandrasekaran, 1995; Glasgow and Papadias, 1992; Tabachneck-Schuf, Leonardo, and Simon, 1997).

## 2. ARCHITECTURE AND REPRESENTATION

Dynamic imagery requires a cognitive architecture that uses working memory to integrate and operate on representations derived from perception and long term memory. After describing our theoretical assumptions about the interactions of perception and memory, we will describe how scene graphs can be used for the representation of visual information including motion.

## 2.1 Perception and memory

Our model of dynamic imagery assumes that mental imagery is tied in with three major processing systems: perception, working memory, and long-term memory. Perception provides the input, and long-term memory serves as a high-level store that integrates visual memories with semantic information. Working memory serves as the arena for combining and processing visual information. Later, when we present our computational model of dynamic imagery, we will say much more about the processes involved in perception and memory.

Our model emphasizes the organizational role of long-term memory in mental imagery. For example, given a visual concept such as *dog*, long-term memory stores images of particular dogs derived from perception, and it also provides non-visual information about how *dog* relates to other concepts (eg. *dog* isa *animal*, Spot isa *dog*). Working memory serves to extract and compare visual concepts, as well as to update our visual database when a novel concept is imagined. For example, you can imagine a futuristic all-terrain vehicle with no wheels but rather large spider-like legs for moving it about. Imagining involves retrieving from long-term memory visual representations of concepts such as *vehicle* and *spider,* combining the two together in working memory, and finally storing the assembled spider-vehicle back into long-term memory. Similarly, processing the cartwheel-windmill analogy requires retrieving from long-term memory representations of gymnasts and windmills derived from previous perceptions, then using working memory to analyze motion similarities between the two concepts. Working memory is able to infer information that was not previously stored in long-term memory, as when we answer the question: How many dialing keys are there on a standard phone? To answer this question, our working memory constructs a mental image of the telephone and we can count the 12 keys.

## 2.2 Scene graphs as mental representations

What is the representational format in which visual information is stored in long-term memory and manipulated in working memory? Skeptics about mental imagery such as Pylyshyn (1984) maintain that no special pictorial representations need to be postulated, but there is much evidence that human thinking employs visual as well as verbal representations (e.g. Larkin and Simon, 1989; Kosslyn, 1994). But the exact nature of these visual representations remains a major problem in cognitive science. It is not helpful to think of them as collections of pixels like the display on a computer screen, because a pixel representation does not support operations such as object combination and motion analysis.

Imagine a picture corresponding to the verbal representation: "There is a rectangle with a circle on the left half, and a star on the right half. A pyramid of three triangles is centered inside the circle". With the verbal representation, it is awkward to make inferences that are automatic from a visual representation, such as that the top triangle is to the left of the star. A pixel representation would not support interesting visual operations such as imagining the star moving into the circle and that the triangles moving to the original location of the star.

The human mind has to organize visual information in a way that enables complex visual inputs to be broken apart, stored and recombined to form new wholes. Computer graphics researchers have converged upon the idea of a *scene graph* to represent visual information. A scene graph is a hierarchical representation that provides a powerful means of representing the world. It consists of an inverted tree in which each branching point or leaf of the tree is used to store a piece of information about a visual scene. This piece of information could be about the size of an object, the position of a group of objects, or the color of an object. Any branching point or leaf of the tree is referred to as a node. The graphical hierarchy in the scene graph helps to capture the structure of the picture of the rectangle and star. The rectangle contains all the other objects in the image and is therefore featured at the top of the hierarchy. The next two prominent features, the circle and the star, are represented at the next level within the hierarchy. Finally, the three triangles are stored below the circle to capture the fact that they are contained within that part of the image.

Procedures such as translation or rotation are controlled by adding nodes to the graph structure. For example, a node could be added to rotate an object contained in the graph. If a rotation node was placed above the section of the graph containing the information for the arch, the entire arch would be rotated while maintaining the spatial relationship between the pieces of the arch.

## 2.3 Representing motion

Unlike verbal and pixel representations, scene graphs provide a powerful means to store and process motion. Imagine a car driving down the road, losing control, and swerving off the road. Scene graphs provide a way of representing the motion of the car by inserting nodes into the graph that describe the motion of an object. These motion nodes, called *behaviors*, contain information about how a property of an object changes over a given period of time. Any property of an object can change over time, whether it be position, shape or color. Multiple behavior nodes can be added into a visual hierarchy to represent complex changing scenes. For example, one behavior node might specify the pathway of the car swerving off the road and hitting the tree. Another behavior node could specify that after the collision, the tree falls down across the hood of the car.

Note the advantage of associating motion with particular objects in the form of behaviors rather than with the entire scene. Dynamic images are not like movies retrieved from memory and run as a whole. Rather, individual visual concepts can be combined with behaviors to create any number of novel scenarios. We can imagine the car starting up and driving around all the trees, or even the trees starting to dance away from the car. In scene graphs, complex motions can be decomposed into a set of interpolations that model the behavior of an object. For example, a simple interpolator behavior could specify the path followed by the car moving from one tree to another. To model the revolution of one object around another, a behavior node can be added to the graph that specifies the rotation of an object about a central point over a given period of time. For example, we have constructed a scene graph for a simple model of an atom in which the electrons orbit the nucleus.

Behavior nodes can be combined together to create very complex motions. For example, starting with a 3-dimensional representation of a stick-man, an interpolator behavior can be added to each arm that changes its position over time. Done properly, these interpolators can simulate the swinging motion we observe watching the arms of someone walking down the street. Similarly, an interpolator can be added to each leg that simulates the stepping motion.

The use of behaviors within a visual hierarchy permits a dynamic scene to be represented in a very compact fashion, not as a complete movie composed of 2-dimensional images where each image is slightly different from the others. If people stored dynamic scenes in movie-like format, we would not be able to create new movies, let alone solve problems that involve reasoning about the motions involved. Our ability to compose mental movies relies upon the creative ability to combine behaviors with static visual concepts stored in our long-term memory. For example, our minds are able to visualize a cat typing at the computer even though we have probably never perceived this situation. To visualize the typing cat, we transfer the behaviors responsible for the human typing motion to our representation of the cat.

Evidence for a recombinant set of motions in human thinking is provided by the dream research of Hobson (1988). He carefully analyzed the dream journal of "Engine Man" who copiously recorded his dreams over a 3-month period in 1939. Of the 110 diagrams contained in Engine Man's dream entries, 25 contain explicit representation of "dream trajectories" that use a dashed line to represent the path and an arrowhead to label the direction. These trajectories describe a set of motions that could be represented as behaviors in scene graphs.

In sum, scene graphs have many advantages as a form of representation for visual information and imagery. First, their hierarchical form enables them to capture a broad range of information about the properties of objects and the relations between them. Second, the inclusion of behaviors in scene graphs makes possible the representation and simulation of complex motions and allows novel scenes to be created from a finite set of objects and motions. Third, scene graphs are computationally feasible - in fact, this representation was developed by the computer graphics community for applications such as virtual reality and digital animation.

# 3. COMPUTATIONAL IMPLEMENTATION

We have developed a computer program which produces animated visual displays and which can be used for simulating visual analogies involving motion. The program is called DIVA: Dynamic Imagery for Visual Analogy. DIVA is written in the language Java, which includes a library named Java 3D for representing and animating 3-dimensional information. Java 3D uses scene graphs in which different types of nodes store different pieces of information. Some nodes are used to group objects together, whereas another type of node might specify a translation or rotation. Behaviors are represented as another type of node that includes information about changing properties of an object, whether it be position, shape or color.[1] DIVA combines visual representations based on scene graphs with a network that represents semantic components of long-term memory. Simulated perceptual input to the program is provided by VRML, a tool developed to exchange and display 3-dimensional models on the Web.

## 3.1 Perception via VRML

The Virtual Reality Modeling Language (VRML) was developed to serve as a standard means of representing 3-dimensional information on the Internet (see http://www.vrml.org). The second proposed standard for VRML was called Moving Worlds, reflecting the fact that it permits motion information to be included within a model. VRML is a text-based descriptive file

[1] Technical details about Java 3D can be found at the Web site: http://java.sun.com/products/java-media/3D/.

format, not a full-fledged programming library, and consequently does not provide any tools for manipulating 3-dimensional scenes.

We use VRML for input into DIVA because there is an abundance of VRML models available on the Internet. VRML models of common objects, from chairs to castles to avatars (human shapes), are easily located on the Internet and parsed into the program. This allows us to simulate perception without having to create our own visual representations.

## 3.2 Working memory via Java 3D

Java 3D provides the representation and some of the processing for DIVA's working memory component. Java 3D provides functionality to parse the text-based VRML files and load them into memory as a scene graph. This 3-dimensional structure can then be rendered to the screen of the computer from any viewing perspective. Information is extracted from the scene graph by traversing up, down or across the graph and querying information stored at each node. Some examples of the types of node used within the Java 3D scene graph are:

Transform: specifies a translation, rotation or scale for all children below this node.

*Group*: groups one or more children nodes together.

*Shape*: specifies a shape using geometric primitives such as squares and spheres or using complex groupings of triangles to represent curved or irregular surfaces.

*Material*: specifies a color or texture that should be applied to the 3 dimensional surface of a shape node.

*Behavior*: specifies a changing condition such as movement for some section of the scene graph.

Java 3D assembles the information from all the nodes in the scene graph into a 3-dimensional model that can be viewed on the computer screen.

Java 3D provides a rich set of behaviors that can be added to a scene graph. These include rotation interpolators (such as the motion of planets about the sun), complex path interpolators (such as walking through a winding wilderness trail) and scaling interpolators (something that grows or shrinks over a period of time). All of the interpolators operate using $x$, $y$ and z co-ordinates to describe positions, and matrices to describe rotations about one or more of the $x$, $y$ or $z$ axes.

Java 3D includes functionality for specifying the location where the viewer is looking at the current visual scene. For example, if we load in a VRML model of a car, we can view the car from an external perspective watching it drive by the street, or from the perspective of the driver sitting behind the steering wheel. Given a certain perspective or "viewpoint", the Java 3D library is able to convert the 3-dimensional model into a 2-dimensional image that can be observed on the computer monitor.

The operation of working memory in DIVA employs built-in Java 3D functions for simulating basic operations such as translation, rotation and scaling. These operations involve finding and modifying a node within the scene graph.

## 3.3 Long-term memory via a semantic network and scene graphs

Long-term memory can be thought of as a database containing memories of sensory input and relationships between these memories. It requires storage of visual scenes, but also includes semantic information about objects or abstract concepts that we have learned. Our program models long-term memory using a structure like the semantic networks in the ACT model of Anderson (1983). It can store simple propositions such as "stick-man is drawing." or "stick-man represents human." A concept such as stick-man is represented by a node within the network, and is defined by its associations with other nodes.

It is important to distinguish between a node within the semantic network, which represents a complete concept, and a node within a scene graph, which is simply part of a hierarchy representing a single visual object. A node within a semantic network can have a chunk of encoded data attached to it that stores a visual representation. For example, a stick-man node in a semantic network might have a scene graph structure attached to it that could be decoded in working memory and viewed.

A basic set of relationships is established in long-term memory when a VRML model is input into the system. These include part-whole relationships that we naturally process when looking at an object. For example, when imagining the model of the stick-man, we also recognize the stick-man's head, arms, legs, and body. When we imagine the head of the stick-man, we cannot help but bring to mind the whole of the stick-man. So the visual information for the stick-man's head should be related to the whole of the stick-man.

DIVA uses a simple mechanism for exchanging information between long-term memory and working memory. Concepts can be retrieved from memory using their names or by association with concepts that might already be active in working memory. For example, a working memory process could retrieve all the concepts that are associated with *arch*. These might include *column*, *crosspiece*, *support*, or *Romans,* depending on the semantic associations that have been input into the model. The long-term memory system is able to extract associated concepts using spreading-activation algorithms similar to those in ACT.

# 4. MODELING VISUAL ANALOGIES

Dynamic imagery can contribute to many important kinds of thinking, including:

*Problem solving*, when we imagine objects (possibly ourselves) moving physically from a starting position to a goal;
*Diagnosis*, when we visualize the operation of a physical device or biological system to determine why it is faulty;
*Thought experiments*, when scientists imagine unusual motions to draw insights about the principles governing the motions of objects.

Our modeling efforts to date have largely concentrated on simulating visual analogies which involve comparisons of visual representations, including ones in which the source and target analogs both involve objects in motion.

## 4.1 Algorithm for analogical mapping of scene graphs

If visual images are represented by scene graphs, then the problem of finding analogies between images is at least as hard as the computational problem of subgraph isomorphism, which requires deciding if there is a subgraph of one graph which is isomorphic to another graph. This problem is known to be NP-complete, that is, it belongs to a large class of problems that are believed to be computationally intractable. However, visual analogical mapping can be performed by an algorithm adapted from the constraint-satisfaction algorithm used for mapping verbal representations in the program ACME (Holyoak and Thagard, 1989). DIVA performs analogical mapping by creating a constraint network representing possible mappings between a source and target scene graph and then uses a connectionist algorithm to settle the network and select the best mappings between the two graphs.

Creating the constraint network requires generating mapping hypotheses concerning the possible correspondences between the nodes in the two scene graphs, establishing positive constraints between mapping hypotheses that fit together, and establishing negative constraints between incompatible mapping hypotheses. Here is an outline of the algorithm that generates mapping hypotheses and positive constraints, and then selects the best mappings. The visual similarity comparison made in the first step of the algorithm is discussed in the next section.

1. For each node *Si* in the source scene graph, search through the nodes in the target graph, identifying each *Tj* that is visually similar to *Si*.
2. Create a mapping unit *Si* = *Tj* that hypothesizes that *Tj* corresponds to *Si*.
3. Create a positive constraint (symmetric excitatory link) between a SPECIAL unit that is always active and each mapping unit concerning two nodes that are visually similar. The weight on these links is proportional to the degree of visual similarity between the two nodes.
4. For the parent node *SiP* of *Si* and the parent node *TjP* of *Tj*, create a mapping unit *SiP* = *TjP*, unless (as is often the case) it has been already been created. Establish a positive constraint between *Si* = *Tj* and *SiP* = *TjP*, i.e. make a symmetric excitatory link between the two units.
5. For each pair of incompatible mapping units, e.g. *Si* = *Tj* and *Si* = *Tk*, create a negative constraint using inhibitory links between the units, so that the two hypotheses militate against each other.
6. Spread activation through the network until it settles. For each node Si in the source scene graph, determine which mapping unit *Si* = *Tj* has the highest activation, which indicates that *Tj* is the best mapping for *Si*.

This algorithm is similar to the one used in ACME in that it employs the constraints of similarity and structure that are part of the multiconstraint theory of analogy (Holyoak and Thagard, 1995), but it implements the constraints based upon the scene graph representation. In ACME, similarity is semantic similarity, whereas DIVA uses visual similarity as defined below. Unlike ACME, DIVA uses similarity to help limit the number of mapping hypotheses; in this respect it is like the DRAMA model of analogical mapping between distributed representations (Eliasmith and Thagard, 2001). As in ACME, one-one mappings are encouraged by the inhibitory links established by step 5 of the algorithm, but they are not strictly required as they are in the SME model (Gentner, 1983; Falkenhainer, Forbus, and Gentner, 1989).

Step 4 of the algorithm that establishes constraints between parent-child mappings (*Si* = *Tj* and *SiP* = *TjP*) is an essential aspect of the algorithm since it conserves the structural organization of the source and target scene graphs. The parent-child constraints help to conserve relational consistency or parallel connectivity: *R*(*a*,*b*) tends to map to *S*(*c*,*d*) only if *R* tends to map to *S*, *a* tends to map to *c*, and *b* tends to map to *d*. In the scene graphs there is no verbal representation of relations and objects, so relational consistency is established by connecting mappings between nodes with mappings between their parents as specified in step 4.

## 4.2 Visual similarity in scene graphs

The algorithm for mapping between scene graphs requires a way of computing the visual similarity of two nodes. This is accomplished by considering the type of each node, and then querying the properties corresponding to that node type. Only nodes of the same type are compared, as a node that represents a transform will never map with a node representing a shape. A very important aspect of the visual similarity comparisons is that wherever possible, a ratio is established that represents the degree of visual similarity. For example, if we are comparing two group nodes, *Si* and *Tj*, if *Si* has 2 children and *Tj* has 3 children, then a similarity ratio of 2/3 = 0.667 is used as the initial activation for the mapping unit between these nodes. This reduces the number of hard-coded variables in the model, and makes flexible comparisons between nodes of the same type. Unlike ACME, which implements similarity constraints by means of differential weights from a special, always active units to units representing mapping hypotheses, DIVA implements similarity constraints by means of differential initial activations of units representing mapping hypotheses. Another important aspect of the similarity comparisons is that wherever possible, comparisons are made relative to the overall dimensions of the scenes. For example, if a 50m-wide field contains a pyramid with a 10m a base, this would be considered visually similar to a pyramid-shaped object with a base of 10cm sitting on a 50cm-wide tabletop. The two pyramid objects are the same size relative to the scenes that contain them. The specific similarity comparisons that establish the initial activation for a mapping unit between a source node *Si* and a target node *Tj* are as follows:

If the nodes are both group types, which group subordinate nodes together, then the initial activation for this mapping unit will be a ratio of the number of children for each group node. Group nodes with only one child are ignored.

If the nodes are both transform nodes, which specify three-dimensional translation, rotation, or a scaling factors, then establish a similarity ratio based upon the $4 \times 4$ matrices containing the *x*, *y*, *z* translation, rotation and scaling information. This is a detailed comparison involving considerable matrix algebra. Note that the ratio is made relative to the size of the scenes containing the transforms.

Some nodes include group and transform information (items 1 and 2 above), and if this is the case, the similarity ratio is composed of the group similarity in addition to the transform similarity. The consequence of this is that some nodes can have similarity ratios greater than 1.0 (but still always less than 2.0).

If the nodes represent a primitive shape (e.g. box, cylinder, sphere, cone), then establish a similarity ratio only if the nodes represent the same type of shape (e.g. both nodes represent a sphere), and make the initial ratio based upon the dimensions of the objects. Again, the ratio is made relative to the size of the scene containing each object.

If the nodes represent an irregular geometric shape, then compare the arrays containing the vertices for this shape. After comparing the co-ordinates for each vertex, establish a ratio that is the average for all the vertices. Note that vertices are compared relative to the size of each shape.

If the nodes represent a shape (items 4 and 5 above), it is possible they also contain appearance information. If the appearance for both nodes is a color value specified using three decimals representing red-green-blue colour components, then establish a ratio based upon these decimals. For example, the color value red (1, 0, 0) would compare with another color (1, 0.5, 0) with a ratio of (1.0 + 0.5 + 1.0)/3 = 0.833. The ratio that is established for the similarity of the appearance is then added to the ratio generated in steps 3 and 4. Note that this will potentially generate a similarity ratio greater than 1.0 (but still less than 2.0). By enabling nodes to have an activation greater than 1.0, nodes that contain two types of properties (group and transform information, or shape and color information) are able to stand out amongst other nodes that specify only a single piece of information.

If the nodes are both behavior nodes that contribute to a specific motion pattern (eg. convergence), then a comparison is made based upon the type of motion pattern. For example, if both the scenes contain a convergence pattern, then behavior nodes are compared based upon the angle and pathway they follow when converging towards the center of the scene.

## 4.3 Dynamic visual similarity

Mapping dynamic visual analogies requires comparing the motions of the objects in the source and target scenes. Pedone, Hummel and Holyoak (2001) investigated the use of animated diagrams in analogical problem solving. As a target, they used the much-investigated radiation problem, in which a way must be found to destroy a tumor without injuring the flesh that surrounds it. One solution to the problem is to use a number of radiation beams from all sides of the patient that converge on the tumor but are not intense enough individually to harm the surrounding tissue. Pedone et al. found that spontaneous retrieval and noticing the relevance of a problem analogous to the radiation problem were increased markedly by animating displays representing converging forces. Subjects were more successful using the convergence animations as a source analogy to solve the tumor problems than were subjects given static diagrams or animated divergence diagrams.

To test the applicability of our dynamic imagery program to complex visual analogies, we produced VRML representations of the tumor problem and the fortress problem, in which a general manages to destroy a fortress by dispersing his forces and having them converge on the fortress. Holyoak and Thagard (1995) review the many experiments that have been done using these problems. Unlike Pedone, Hummel and Holyoak, who used an abstract diagram to represent a dynamic convergence solution, we implemented dynamic versions of both the tumor and fortress problems using animated Java 3D representations. Animation occurs on the screen when the soldiers converge on the castle and the rays converge on the head.

## 5. FUTURE DIRECTIONS

Our model of dynamic imagery can generate moving objects and simulate visual analogies, but it captures only a small part of human capabilities for imagining objects in motion. A richer model will need to add some knowledge of how objects in the world interact physically with each other, and add more inferential mechanisms to enable it to simulate problem solving, diagnosis, and thought experiments.

The current program lacks a physics model that would enable the behaviors of different objects to connect with each other. For example, although the stick-man in our simulation mentioned earlier appears to be walking, there is no simulated frictional force between the stick-man's feet and the ground, so that the stick-man would appear to remain in the same position relative to the ground, always remaining in the same spot. Similarly, when we imagine a ball travelling though the air, it rarely just hovers around at a constant vertical level but is subject to our understanding of gravity. Hobson (1988) gives the example of a dream in which a golf ball travels through the air and rebounds from an object that it collides with. By itself, the scene graph used to represent visual information is not subject to any natural laws. Hence a more powerful dynamic imagery program would use physical laws to constrain motions produced through the behaviors nodes present in a scene graph.

The addition of a physics model would improve a dynamic imagery program's ability to solve problems. It would be interesting, for example, to model how children use visual information to build new structures out of LEGO blocks. A problem could be posed using an image of a structure to be built, and a solution to the problem would require a series of visual inferences about how the blocks could be combined to produce the desired structure. It would also be desirable to simulate the role of imagery in solving problems that require the generation of explanations. Shelley (1996) describes numerous examples of abductive inference, the formation of explanatory hypotheses, based on visual imagery. For example, anthropologists have explained the structure of stone tools by visualizing them as being produced by controlled flaking using a stone hammer. It would be natural to extend our model of dynamic imagery to simulate hypothesis generation by visual reasoning. In addition, it should be possible to apply a theory of dynamic imagery to the use of thought experiments and visual analogies in scientific discoveries, as discussed by Brown (1991) and Nersessian (1992).

Finally, it would be desirable to provide a more neurologically realistic representation of visual structure than is provided by the Java 3D scene graph. We conjecture that the techniques that have been used to convert complex verbal representations into distributed neural representations could also be used to produce artificial-neural-network versions of our imagery program. Scene graphs could be converted to complex vectors that preserve structural information (Smolensky, 1990; Eliasmith and Thagard, 2001); or they could be translated into neuronal groups that coordinate information by means of neural synchrony (Shastri and Ajjanagadde, 1993; Hummel and Holyoak, 1997). Either way, we would have steps toward a more neurologically interesting model of dynamic imagery. A major challenge for both these techniques, however, would be to find a means of representing motion.

## 6. CONCLUSION

In this paper, we have borrowed from computer graphics the idea of a scene graph to suggest new structures and algorithms for modeling how people generate and use dynamic visual images. Scene graphs are more flexible and inferentially powerful than representations such as bitmaps and arrays that have been previously use to model mental imagery. Most importantly, they provide a natural way to model how people can generate images of objects in motion. Representation of physical objects involves not only their static properties such as shape, color, and position, but also their behaviors. Dynamic imagery is not like running a pre-compiled movie in the mind, but instead requires a constructive process of combining the behaviors of different objects into novel and flexible animations. Scene graph representations can support analogical mappings between visual images, including ones that involve complex motions such as convergence. We hope future work on dynamic imagery will extend our model to apply to a wide range of cognitive processes, including problem solving, explanation, and thought experiments.

## NOTES

Because of space limitations, we have not been able to include figures that show the structure of scene graphs. For a longer version of this paper with many illustrations, see: http://cogsci.uwaterloo.ca/Articles/Pages/Coherence.html.

## ACKNOWLEDGEMENTS

We are grateful to Rob Krajcarski and Micah Zarnke for very helpful ideas and programming assistance. Thanks to Cameron Shelley and Keith Holyoak for comments on an earlier draft. This research has been supported by the Natural Sciences and Engineering Research Council of Canada.

## REFERENCES

Anderson, J.R., 1983, *The Architecture of Cognition*, Harvard University Press, Cambridge, MA.

Brown, J.R., 1991, *TheLlaboratory of the Mind*, Routledge, London.

Chella, A., Frixione, M., and Gaglo, S., 2000, Understanding dynamic scenes, *Artificial Intelligence* 123:89-132.

Conklin, D. and Glasgow, J.I., 1992, Spatial analogy and subsumption, in: *Machine learning: Proceedings of the Ninth International Conference*, D. Sleeman and P. Edwards, eds., Morgan Kaufmann, San Mateo.

Davies, J.R. and Goel, A.K., 2000, A computational theory of visual analogical transfer, Technical report GIT-COGSCI-2000/3, Georgia Institute of Technology.

Eliasmith, C. and Thagard, P., 2001, Integrating structure and meaning: A distributed model of analogical mapping, *Cognitive Science* 25:245-286.

Falkenhainer, B., Forbus, K.D., and Gentner, D., 1989, The structure-mapping engine: Algorithms and examples, *Artificial Intelligence* 41:1-63.

Forbus, K., 1995, Qualitative spatial reasoning: framework and frontiers, in: B. Chandrasekaran, J. Glasgow, and N.H. Narayanan, eds., AAAI Press/MIT Press, Cambridge, MA.

French, R., 1995, *Tabletop: An Emergent, Stochastic, Computer Model of Analogy-Making*, MIT Press, Cambridge, MA.

Funt, B., 1980, Problem solving with diagrammatic representations, *Artificial Intelligence* 13:201-230.

Gentner, D., 1983, Structure mapping: A theoretical framework for analogy, *Cognitive Science* 7:155-170.

Glasgow, J., Narayanan, N.H., and Chandrasekaran, B., eds., 1995, *Diagrammatic reasoning: Cognitive and computational perspectives*. AAAI Press/MIT Press, Cambridge, MA.

Glasgow, J.I. and Papadias, D., 1992, Computational imagery, *Cognitive Science* 16:355-394.

Hobson, J.A., 1988, *The dreaming brain*, Basic Books, New York.

Holyoak, K.J. and Thagard, P., 1989, Analogical mapping by constraint satisfaction. *Cognitive Science* 13:295-355.

Holyoak, K.J. and Thagard, P., 1995, *Mental Leaps: Analogy in Creative Thought*, MIT Press/Bradford Books, Cambridge, MA.

Hummel, J.E. and Holyoak, K.J., 1998, Distributed representations of structure: a theory of analogical access and mapping, *Psychological Review* 104:427-466.

Kosslyn, S. and Koenig, O., 1992, *Wet Mind: The New Cognitive Neuroscience*, Free Press, New York.

Kosslyn, S.M., 1994, *Image and Brain: The Resolution of the Imagery Debate*, MIT Press, Cambridge, MA.

Larkin, J.H. and Simon, H.A., 1987, Why a diagram is (sometimes) worth ten thousand words, *Cognitive Science* 11:65-100.

Nersessian, N., 1992, How do scientists think? Capturing the dynamics of conceptual change in science, *Cognitive Models of Science*, R. Giere, ed., University of Minnesota Press, Minneapolis, pp. 3-44.

Pedone, R., Hummel, J.E., and Holyoak, K.J., 2001, The use of diagrams in analogical problem solving, *Memory and Cognition* 29:214-221.

Pylyshyn, Z., 1984, *Computation and Cognition: Toward a Foundation for Cognitive Science*, MIT Press, Cambridge, MA.

Shastri, L. and Ajjanagadde, V., 1993, From simple associations to systematic reasoning: a connectionist representation of rules, variables, and dynamic bindings, *Behavioral and Brain Sciences* 16:417-494.

Shelley, C.P., 1996, Visual abductive reasoning in archaeology, *Philosophy of Science* 63:278-301.

Smolensky, P., 1990, Tensor product variable binding and the representation of symbolic structures in connectionist systems, *Artificial Intelligence* 46:159-217.

Tabachneck-Schuf, H.J.M., Leonardo, A.M., and Simon, H.A., 1997, CaMeRa: A computational model of multiple representations, *Cognitive Science* 21:305-350.

Thagard, P., Gochfeld, D., and Hardy, S., 1992, Visual analogical mapping, in: *Proceedings of the Fourteenth Annual Conference of the Cognitive Science Society*, Erlbaum, Hillsdale, NJ, pp. 522-527.

# Model-Based Reasoning and Similarity in the World

Qiming Yu
*Department of Philosophy, Central University for Nationalities, Beijing, China,*
*yuqiming@ihw.com.cn*

Abstract: Scientific reasoning is commonly regarded as the core problem in science studies. This paper suggests that model-based reasoning is an important part of scientific creativity. I approach this problem from the standpoint of the origin of human thinking and genetic epistemology to understand representative practices of scientists. I also argue that objective similarity in the world is what makes model-based reasoning a powerful tool.

## 1. THE OCCURRENCE OF THE PROBLEM

Scientific reasoning has been commonly considered to be the central problem in studies of scientific methodology. Scientific reasoning includes aspects of both discovering and proving. This domain was studied two thousand years ago, beginning with Aristotle. This continued with modern science through four hundred years since Bacon, but it still has yet not been given an answer that is entirely satisfactory. In the XX century the rise of logical positivism excluded scientific discovery from considerations of scientific method. Since the seventies and eighties of the past century the circles of philosophy of science inside and outside China have attached importance to studies of the methods of scientific discovery about, providing analyses of typical cases and suggesting some discovery patterns from these cases.

Conventional viewpoints of philosophy about reasoning in science are limited and do not accord with scientists' practices. These viewpoints limit reasoning to induction and deduction, applied to linguistic or formulaic representations. This is not in accord with scientific practices exhibited in the

*Model-Based Reasoning: Science, Technology, Values*, edited by
L. Magnani and N.J. Nersessian, Kluwer Academic/Plenum Publishers, New York, 2002

historical records. Thus, many interrelated practices and methods of scientists were excluded from accounts of scientific reasoning. Practices provide the criteria for testing the truth of philosophers' hypotheses. The positivists were putting the cart before the horse. The study of many typical cases of scientific reasoning has demonstrated that there are multiple methods of performing fruitful and creative reasoning that cannot be described with the traditional viewpoint. We need to extend our idea of scientific reasoning to include complex and creative reasoning to completely and thoroughly understand it. The study of these reasoning methods is situated at the intersection of philosophy, artificial intelligence, cognitive psychology, and logic.

What is logic? What is reasoning? It is the time to survey them again. The word "logic" has its roots in the Greek "logos" and can be understood as the laws of thinking. However in philosophy these have often been understood as the formal structure of reasoning, so the contents are reduced in importance. It seems that those laws only involve the syntax of arguments. In fact, "law (theorem)" means the essential relations and necessity in the developing process of things. Laws can be decisive, like Newton's laws of mechanics; can also be statistical, like Maxwell's law of velocity distribution; or can be probabilistic, like Heisenberg uncertainty principle.

Moreover, reasoning is a form of thinking by which an unknown conclusion is inferred from one or several premises. How does one infer the conclusion from the premises? Traditional notions claimed that this was done with formal rules and guaranteeing the truth of the conclusion by the truth of the premises is necessary for a piece of "reasoning" to be worthy of the name. Everything else is excluded from scientific reasoning. In this case reasoning is restricted to the strait scope of deductive reasoning and symbolic logic, so this has tied the hands of those who study scientific methodology as employed in scientific practices.

## 2. THE LONG HISTORY OF MODEL-BASED REASONING

What is model-based reasoning? I understand it to be a form of thinking which infers a conclusion from premises with aid of a mental model. What is a mental model? The word "model" has roots in the Latin "modulus"; its original meaning was copy, standard and yardstick. Now its connotation has been greatly opened up and it has commonly two kinds of usage.

First, it means the duplicate or copy of a certain object. It is the kind of model that is gained through eliminating or idealizing according to some characteristics of typical properties, configurations, and functions. Some models are used as an analog to the original, like the model ship, the building model etc. The notion is also applied, in experimental studies as a substitute for the original, like the airplane model in the wind tunnel laboratory, and the riverbed model in the liquid laboratory. In addition, manufactured objects, which are created to simulate biological properties, configurations, functions or laws of the original are sometimes called models, such as the "sonar" localizer designed in accordance with the bat's hearing of ultrasonic frequencies and the "robot" simulating partial actions, skills, or intelligence of humans.

Second, it can also mean a mental analog model of properties, configurations, functions or laws of the referent: this is the mental model we are discussing. These are created by employing perceptual mechanisms in cognition. Also, the mental model reflects only some characteristics of the original: those related to particular cognitive aims. Others are eliminated.

Examples include the "Lorentz model" in chaos, Rutherfold's model of the planetary structure of the atom, and the twin-helix structure model of DNA. Mental models can be either formal mental models or visual mental models, depending on whether the representation is propositional, mathematical, or a structural or functional analog. People can create idealized models, which do not exist in the real world, like the particle model in mechanics, and the elevator model in the theory of relativity. Therefore, model-based reasoning can be divided into formal model-based reasoning and visual model-based reasoning in accordance with the type of model employed.

Studies have indicated that the appearance of model-based reasoning came earlier than traditional logical reasoning in the evolution of human cognition. In fact, the development of human thinking itself has evolved. It is hypothesized that our ancient ancestors, like animals, had no logical thinking. Later they developed a simple language, due to the necessity of communicating with one another in labor. This provided the conditions for the development of logical thinking. In the lengthy pre-historical period, intuitive, non-logical factors held primary status in the thinking of hominids. Simple visual model-based reasoning was employed, while logical thinking was in the budding state. Logical thinking only began to make a breakthrough when humans entered the slavery society. Early in 1910 the French sociologist Lucien Lévy-Bruhl published the book *Les fonctions mentales dans les sociétés inférieures*. In this book he introduced the hypothesis of the "generic law of primitive thinking" for the first time. He considered that primordial thinking, as performed through speaking, did not use concepts, but had some "pre-conceptual" relations with perceptual phenomena and that judgments were made by perceptual cognition. He pointed out that, in thinking, language had almost a material property for primitive man. He quoted Steensby: "The method of reasoning of the Eskimo gives us the impression of being very superficial, because they are not accustomed to retain what we call a definite line of reasoning or a single, isolated subject for any length of time; their thoughts, namely, do not rise to abstractions or logical formulas, but keep to picture of observation or situations which change according to laws we find it difficult to follow" (Steensby, 1910, pp. 374-5). 1) He concluded that: "According to the ideas of most primitives, the vault of heaven rest like a dome upon the flat surface of the earth or of the ocean. Thus, the world ends on the circle of the horizon. In it space is weighted with qualities, and each of its regions, as we have already seen, participates in all that is usually found there. Frequently a future event, if considered certain to happen, and if provocative of great emotion, is felt to be already present". 2) Thus it can be seen that here what he called "judgements made by perceptual thinking" are just a form of what we are calling primitive visual model-based reasoning. Of course we need to account for the differences between it and our modern model-based reasoning, especially because the nature of the causal inferences allowed in the modern world are different. For example, in the light of primitive thinking, "any circumstance, however slightly unusual it may be, is at once regarded as the manifestation of one or another of occult powers". 3) However these are largely eliminated as causal mechanisms in the modern world. So "we find it difficult to follow". But this does not rule out the possibility of our concluding that model-based reasoning has been of long standing in the history of human cognition.

## 3. ENLIGHTENMENT BY GENETIC EPISTEMOLOGY

Piaget's genetic epistemology has important status in the field of studying the origins and development of human thinking. His epistemology was the fruit of interdisciplinary study. He called his view "genetic epistemology" and stated: "Genetic epistemology, then, aims to study the origins of the various kind of knowledge, starting with their most elementary forms, and to follow their development to later level up to and including scientific thought [...]. The passage from an inadequate, relatively poorer kind of knowledge to one richer both in intension and extension" (Piaget, 1972, pp. 15-16). In Piaget's view, the evolutionary origins and growth of human cognition are parallel to and reflected in individual psychogenesis, i.e. the development of thinking in the child. This is the basic premise of Piaget's doctrine. Given that evidence of human thinking before history is absent, Piaget inaugurated other way of gaining knowledge about primitive cognitive development. He claimed: "The only course open to us, therefore, is to follow the example of biologists who supplement their scanty stock of phylogenetic knowledge by turning to embryogenesis in the case of psychology. This means studying the mental ontogenesis of the child at every age". In a similar way, Piaget wanted to proceed with the ontogenetics of child psychology to study the origins and development of human thinking.

After studying intensively for a few decades, Piaget brought forward the theory of the origin, phases, and structure of the child's mind. He considered it originated neither in the object nor in the subject, but from the reciprocity between subject and object. The newborn infant is in a state of chaos, without differentiations between subject and object. The construction of the concept of subject and object in the child's thinking is through the action (also called "movement") of the child herself. Only though the child's acting on a certain object repeatedly does she have the possibility to realize that the object's existence is independent from herself. This kind of movement involves first the child's exterior operations on the object, and then step by step it transforms into interior mental operation. Piaget held that cognizance of the object by the subject depends on her having the requisite cognitive structure, which develops from low-grade to high-grade in the process of differentiating subject from object. This "structure" is the organizing structure of mental action that is employed in what is called a "scheme". According to the Piaget's partitioning, the child who is in the first phase of *apperceptive movement* (0-2 years) or in the second phase of *pre-operational thought* (2-7 years) cannot yet relate to the emergence of operational structure. This develops in the third phase: the *concrete operational phase* (7-12 years). The fourth phase, the *formal operation phase* (12 year-adult) is the last phase of development of the child's mind. At this phase the ability of the child to perform logical operations attains a new level. There may be intersections between phases and the age of emergence of every phase could be moved forward or delayed in an individual, but the order of the four phases is invariable.

In Piaget's view, the cognitive experience of the child (of physical objects, numbers, or logical operations) is rooted in child's operating on objects. For example, the cognition of the child about size, color, shape, degree, texture and capability is obtained through perception by repeating operations on that object. For example, the child does not abstract the understanding of the number 8 simply by looking at 8 toys, but from performing repeated actions on the toys such a lining them up in some order. As Piaget said, the concepts of symbolic logic contain a series of operations, they are not abstracted immediately from the sensed thing, but from the actions performed on the thing. These are widely different things.

The object the child acts on can be the original and can also be an indirect model. In early activity, what the child operates on is the object itself, for example, a green plot of grass or the sand of the beach. In later activity, what the child operates on is not the actual object, but a substitute, for example, various toys, teaching aids, learning aids, and drawings. These models simulate some obvious or visualized aspects and correlations; the child can implement the operations of contrast, removal, composition, change, and measurement with them. The virtue of acting on models is that not only is intuition of the specificity of movement easy to comprehend, but it can also be used to ascertain spatio-temporal localization. Many of the objects on which a child operates are just models, so the series of operations leading to the establishment of symbolic logic relate to reasoning by means of models. The basic developmental process of the child's exterior operating on objects "translating step by step into the kind of interior psychological operation" is, thus, "model-based". In sum, much of the origin of the child's cognitive development is associated with models and model-based reasoning.

## 4. MODEL-BASED REASONING AND ANALOGY

What is the model-based reasoning? Why does model-based reasoning hold very important status in the origin and development of human thinking?

If we start from Piaget's genetic epistemology, the thinking of early hominids was on the pre-concept period and not aware of logical rules. The process of comprehending the instance of a thing (pre-judgment) occurs only by simple visual model-based reasoning; it might as well be called "primitive model-based reasoning". This is intuitive and very useful practically, but it was not self-conscious. Later, human language came into being and developed. Logical thinking evolved constantly, so it seems that model-based reasoning should possibly have retreated from the field of human thinking and have been replaced completely. However, this is not what happened. On the one hand, logic made great progress, from Aristotle's syllogistic logic to Hilbert's symbolic logic, from Bacon and Mill's inductive logic to Reichenbach's probabilistic logic. On the other hand, cognition using models and model-based reasoning also became more clear-headed and thorough. It was employed self-consciously in science from Galileo and Kepler to Einstein and onwards. Philosophers of science from Duhem and Hesse to Harre led the way to making "model-based reasoning in scientific discovery" a part of the standard contents of study of philosophy of science. Of course, this development was not direct, but has been reached circuitously.

The evolution of human thinking, from primitive model-based reasoning to logical reasoning is marvelous advancement. However, logical reasoning is insufficient for scientific thought. Logical thinking deepens human cognition and improves model-based reasoning. However, the discovery processes of scientific theories are essentially rooted in model-based reasoning.

Scientific theory is the lawful explanation of experience and offers a more profound and more accurate understanding, allowing one to have an understanding of not only how something is, but also to know why it is. Therefore, "theory is the emperor's headpiece". To reach this goal, theory explains the phenomena as the behavior of entities and processes that are not directly observable. For instance, the particle interpretation of photon theory explains the refraction of light in this way: it is the result of a denser medium having stronger attraction to photons; the wave theory interpretation attributes the refraction of light to the fact that light waves travel more slowly in denser medium. How, then, were the particle and wave theories of photons put forward and discovered? Since theories deal with the behaviors of theoretical entities that are invisible and imperceptible, they cannot be directly discovered through observation and experiment; theory cannot directly posit laws through the generalization and systematization of behaviors of observable objects. Therefore theory is only the result of guessing and assuming.

How do scientists guess at the causes and effects working in a mechanism? The reasoning is as follows:

1. Correlative experience law *L*;
2. If *H*, the theoretical hypothesis is true, then *L* can be explained;
3. Therefore, we have reason to recognize *H* as true.

This reasoning route is often called "abduction," i.e., ascertaining the cause from the effect. The knowledge that Einstein brought could not arise only from experience, but through both intellectual invention and observed fact. Here "the knowledge," is "the theory"; here "invention" means the "guessing" of the theory.

How can scientists guess at a theory? The methods of discovery relate not only to their individual knowledge and experience and that of the scientific community to which they belong, but also to how they guess as individuals. How are these guesses – or hypotheses - made? Because the theoretical entities and their movements are invisible and imperceptible, scientists are little aware of the cause and effect mechanisms involved. Therefore they can only guess at what they might be, starting off from an analogue with which they are already well-informed, and transferring knowledge to describe the properties, configurations, and functions for the new law. The facsimile or blueprint, which is created in this way, is the mental model. Although the mental model is created by referring to an analogue, it represents the cause and effect mechanism involved with the phenomena under investigation. Thus, there is the "model" of the analogue and the "model" of the studied object. The relation between the three of them goes from the more familiar analogue to the mental model to the proposed cause and effect of studied object.

So, in the discovery process of scientific theorizing scientists make hypotheses about the entities and their movements by constructing mental models. This is the method of model-based reasoning. This reasoning method leads to contingent, possible, fair and reasonable conclusions and not to the direct, necessary conclusions of deductive reasoning methods. Considering models as the guesses made by an investigator, models are only facsimiles. The true aim of the investigator is to cognize the causes and effects of theoretical entities. Mental modeling is only a tool that the investigator brings to bear. Mental models occupy a special status in the discovery process because of their dual nature. By means of the model the hypothesis finds the main direction for development far from the source analogy. Or it can lead nowhere. The cause and effect mechanism that explains correlative experience laws, is found by means of the modification and perfection of model. Understanding about cause and effect mechanism will be constantly refined, deepened and extended.

*Model-based reasoning* can be partitioned into the following types:

1. Model-based reasoning of *qualitative similarity* characterizes the qualitative similarity between objects under study and known object. An example is the creation of the atom-planet model, which British physical scientist Rutherford put forward in 1911.

2. Model-based reasoning of *functional similarity* is based on the similarity between function and behavior, and does not require similarity of qualitative and structural aspects between the object under study and the known object. For instance, the British physiologist Harvey discovered the circulation of the blood in 1628. In that time a kind of water bellows, different from the modern force pump, had been created. It was a kind of soft structure, using a piece of skin as the valve in the interior, lest water in the pipe flow out. Hydraulic pressure could be enhanced while water passes through the two valves. Harvey guessed that blood circulation, the center of which is the heart, worked analogously (using a known cause and effect mechanism). The heart is not a respiratory organ (as Galen had described), but is an organ of transporting blood. The constriction and stretching of the heart causes blood circulation. This guess provided a new understanding about the function of the heart and was key in putting forward a theory of blood circulation. The theory was arrived at through analogy with the function of water bellows and belongs to the class of model-based reasoning of functional similitude. Nowadays the study of cognition itself also uses this type of reasoning in using the functional analogy to the computer.

3. Model-based reasoning of *formal similarity* characterizes the formal similarity between objects under study and known objects. Formal similarity indicates cases where the mathematical formulas of each are sufficiently similar that they lead to the inference that possibly the cause and effect mechanisms are similar. For example, French physicist L. de Broglie used such reasoning while he found the theory of matter wave in 1924. In geometry optics the movement of light wave obeys the Feynman principle, namely $\int \frac{1}{u} ds = \min$, where $u$ is the wave velocity, and $s$ is the distance. Corresponding with this, in the displacement of matter, particles obey the Maupertuis principle, namely, $\int \frac{1}{v} ds = \min$, where $v$ is the corpuscular velocity. Because of the comparison of the two integral formula L. de Broglie thought that if photons have wave properties, then might electrons as well? And then, from the wave-particle duality he deduced that practically all particles have also the wave-particle duality.

From a logical point of view, model-based reasoning can be also said to be a kind of *analogical reasoning*. Analogy is the comparison of two objects that are not the same but have some similar properties and relations. From the similarity of known properties and functional or causal relations, similar properties in the new object can be hypothesized. For instance, L. de Broglie demonstrated such reasoning: photons and practically all particles have wave-particle duality; photons satisfy also Plank relation

$E = h\nu$ and Einstein relation $P = \frac{h}{\lambda}$; therefore practically all particles should also satisfy these two relations. Therefore E is the energy of photons or practically all particles, P is the momentum, $\nu$ is the frequency, $\lambda$ is the wavelength, $h$ is Plank constant.

Analogy is different from inductive and deductive reasoning. Deduction infers the result from the hypothesis, from generic to individual; induction infers the hypothesis from support of evidence, from individual to the generic. Analogy obtains the hypothesis first, then infers from specific object or domain to other specific objects or domains. Although the above three kinds of reasoning methods all infer from known premises, there are differences with respect to the degree that the premises restrict the conclusion. This restriction is strongest with deduction; induction takes second place; and analogy is the weakest. This makes analogy more helpful to scientific discovery. In the fields where induction and the deduction are helpless, analogy can generate ideas, widen thinking, and enable inferences about other cases from one instance.

Analogy is different from induction and deduction, which differ also with the inevitability of the conclusions drawn. The conclusion of a deduction is necessary - if the premises are true the conclusion is also true; the degree of inevitability of the conclusion for an induction takes second place; but the inevitability for an analogy is worst. When stressing that static view, this is the disadvantage of the analogy; but if reasoning is viewed from the dynamic processes of scientific discovery, this is not such a shortcoming. For resolving new problems, setting one's foot in an unknown field, constructing new mental models, or putting forward new hypotheses there are no the reliable methods scientists can use to lead to correct conclusions. Making mistakes is hard to avoid. However, since analogy places the least restrictions for the conclusions that can be reached. The process is fit for moving about freely and quickly in the wide universe of scientific exploration and is likely to lead to innovative discoveries.

## 5. SIMILARITY IN THE WORLD

Model-based reasoning holds a very important status in the origin and development of human thinking. Leibniz said that nature is similar. Scientific study has made it clear that there are a lot of similar things in the nature, great as the unlimited vast universe and little as the electron and quark. Taking plants, for example, pines and cypresses reaching the sky in a forest share photosynthesis with the dark green muskeg on the side of a creek. Likewise, all plants are similar in this way, containing chlorophyll. From humans, the most intelligent animals, to the low mollusk, there are similar haemachrome in their cells. Do animals and plants contain some of the same things? Yes, there are similar contents: chlorophyll and haemachrome are a porphyrin compound, chlorophyll is the compound of porphyrin and magnesium, haemachrome is the compound of porphyrin and ferric. From the levels of molecular biology, the codes of heredity, DNA molecules are in all living things. The most forceful evidence of this is that a virus gene can supervise the compound of RNA and protein in the cell. Analyzing them with respect to chemical elements, there are also similar contents between the living and non-living: the bioplasm of any living thing has many elements (carbon, hydrogen, oxygen, nitrogen etc.), biological substances are made from non-biological substances. Quantum field theory in contemporary physics already makes it clear that the basic modality of living and non-living includes the field of all particles. So it looks as though there are similarities between everything in universe, and this should be the accepted fact. Why should this be so? Big bang theory holds that the universe is rooted in a "primitive fireball" in a state of super-high heat and super-high density at first. How could things in the universe not be similar each another?

I contend that this shows that the fundamental role similarity plays in nature and in our understanding nature provides a foundation for model-based reasoning. Things in nature are not absolutely the same, neither are they absolutely different. Similarity is just the unification of congruence and difference. Unless there are similarities in nature, there would not be any experience or laws. Much of human reasoning is based on similarity and cannot function without the search for similar information.

Scientific theories require similarity. On the principle of simplicity it is quite likely that the similarities that exist in nature are captured in the human focus on similarities. For scientific research, if we do not investigate the similarities in nature and do not understand how to discover congruencies despite differences, we cannot form any theory and the emergence of science would itself have been impossible. Model-based reasoning is a means of discovering these similarities and thus discovering new theories. Such reasoning relates similar things to each another, enables selections from among a lot of similar information and comparisons of series of qualitative, functional, and formal similar relations.

## ACKNOWLEDGMENTS

I would like to thank Jim Davies and Nancy Nersessian for editorial assistance.

## REFERENCES

Steensby, H.P., 1910, *Contributions to the Ethnology and Anthropology of the Polar Eskimos*, B. Luno, Copenhagen. (Reprinted by Meddelser om Grönland, vol. XXXIV).

Lévy-Bruhl, L., 1910, *Les fonctions mentales dans les sociétés inférieures*, F. Alcan, Paris (translation by C. Scott Littleton, *How Natives Think*, Knopf, New York, 1926, reprint, Princeton University Press, Princeton, 1985).

Lévy-Bruhl, L., 1922, *La mentalité primitive*, Presses Universitaires de France, Paris (translation by L.A. Clare, *Primitive Mentality*, Beacon Press, Boston, 1966).

Piaget, J. 1972, *The Principles of Genetic Epistemology*, English translation by W. Mays, Basic Books, New York.

# Epistemic Artifacts: Michael Faraday's Search for the Optical Effects of Gold

Ryan D. Tweney
*Department of Psychology, Bowling Green State University, Bowling Green, OH, USA, tweney@bgnet.bgsu.edu*

Abstract: In 1856, Michael Faraday (1791-1867) carried out an extensive program of research to explore the properties of thin films of metallic gold. Such films had long been of interest to him because they possess the peculiar property of appearing gold in color by reflected light, but green by transmitted light. Faraday hoped this behavior would model the general interaction of light and matter, perhaps extending his earlier finding that magnetic fields could affect a beam of polarized light passing through a highly refractive substance. In the course of this work, Faraday prepared over 600 microscope slides with thin metallic films adhering to them. Most of these *epistemic artifacts* have survived, as have his extensive diary records. Examination of this material permits reconstruction of his successive attempts to create a kind of dialogue with nature and provides evidence for his sophisticated ability to interactively create a microverse that could expand his mental microverse of field lines of force. While he failed to achieve his larger goals, the surviving specimens constitute a series of increasingly articulated representations of the structure of metallic matter. The case is thus one in which the "model" converges with the very thing intended to be modelled.

## 1. INTRODUCTION

Michael Faraday's (1791-1867) research has served as the source for a number of interesting case studies that reveal the extent to which his research relied upon model-based reasoning. There is now little doubt that understanding his strategies of experimentation and theorizing requires appre-

*Model-Based Reasoning: Science, Technology, Values*, edited by
L. Magnani and N.J. Nersessian, Kluwer Academic/Plenum Publishers, New York, 2002

ciation for his construction, evaluation, and use of models. Thus, Gooding (1990) explored Faraday's 1821 experiments on the rotation of current-carrying wires in a magnetic field and replicated the constructive processes, the "makings," that went into Faraday's discovery of the phenomenon. In this work, Faraday gradually moved from the vaguest of construals to a fully-formed mental model of the tangential character of electric and magnetic forces. Similarly, Nersessian (1984; 1985) showed how Faraday's emerging notion of field was developed over time and later extended by Maxwell using a model-based approach to elaborate the laws of electromagnetism.

In the present paper, I explore the model-based reasoning of Faraday along dimensions that have previously received little attention. In particular, I give an account of his constructive making of physical specimens in a microscopic domain. In this way I bridge the gap between accounts of mental models, as such, and in which the mental models are intertwined with physical models. In effect, I open inquiry into the nature of *epistemic artifacts*, a term suggestive of a merging of recent work on "cognitive artifacts" (e.g., Hutchins, 1995; Zhang and Norman, 1995) and "epistemic things" (Rheinberger, 1997).

Nersessian (1999) noted that traditional accounts of science have been excessively "propositional" and hence have been restricted to the linguistic expression of results. As one consequence, many analysts have felt that any account of discovery processes was "merely" a psychological question. Yet, as Magnani (2001) and others have argued, discovery processes can be analyzed under the general heading of "abduction," that is, as processes which emphasize the creation of hypothetical explanations, rather than just their justification once formulated. Magnani's notion of "manipulative abduction" is especially relevant in the present case.

To be sure, Faraday's gold research, like most of his other research, reflects cognitive processes that are now often used to describe *in vivo* scientific thinking. Thus, uses of analogy are present, imagery is extensive, much of the record could be interpreted as search through a problem space, and Faraday can be seen as struggling to develop new classifications (see, e.g., Andersen, 2001; Dunbar, 2001; Gentner et al., 2001; Gorman, 1997; Langley and Jones, 1988; Nersessian, 1999; Tweney, 2001). Yet none of these processes taken singly can fully capture the way in which Faraday *interacts* with the materials and objects of his laboratory to shape his model construction activity. In the present paper, I hope to show how such interactions can be studied, and their implications for understanding scientific activity in general.

## 1.1 Light, matter, and the field in Faraday's time

For Faraday, the 1840s and 1850s represented a period of consolidation of his field theory (Nersessian, 1985). Faraday himself had played an important role in the transition from earlier simply "materialistic" conceptions of the ether to the characteristic British accounts that followed the mid-century, accounts that emphasized the imponderable character of the ether (Cantor and Hodges, 1981). These later accounts emphasized the continuous character of the ether, and its special status - if the ether was "matter," it was matter of a very peculiar sort. Faraday had not taken direct part in earlier debates about the undulatory theory of light, perhaps because many of the relevant papers were mathematical in character, and, as is well known, he did not use (or perhaps even understand) the formal mathematical methods characteristic of a George Gabriel Stokes or William Thomson (see Gooding, 1982; Williams, 1965).

As James (1985) has shown, Faraday was committed in many ways to an "Optical mode of investigation," using optical criteria, for example, in his development of the distinction between diamagnetic and paramagnetic substances, and manifesting a long-standing concern with whether or not there was an ether. Chen (2000), in examining the role of instruments in optics during the 1800s, distinguished between a "visual tradition" in optics, in which the eye played an intrinsic role (and the wave vs. particle dispute was secondary), and a "geometric tradition," in which instruments were used to both generate and detect optical effects, and in which the wave vs. particle issue was paramount.

From an early date, Faraday felt that the optical behavior of metals was an important domain in which to learn about the nature of light and matter. His 1845 discovery of magneto-optical rotation, the rotation of the plane of polarization of light when a magnetic field is applied to a highly refracting medium, suggested that a unified account of light, electricity, magnetism, and matter might be possible. After 1845 he emphasized how important it was to understand the interaction of field forces with the material substrates within which they were manifested (Fisher, 2001). His research on diamagnetism (part of which was also optical) served to solidify Faraday's account of the lines of electric and magnetic force as *physical* lines of force (Faraday, 1852). Thus, by 1856, convinced of the reality of physical lines of force in the electric and magnetic domains and committed to a hunch that light might be similar, Faraday reopened his "ancient query" of gold's interaction with light (Faraday, *Diary,* 14722, 21 April, 1856, in Martin, 1936, p. 108).

## 1.2 Situating Faraday: the problem of gold

The research carried out in 1856, the focus of the present paper, occurred toward the end of Faraday's productive career.[1] In this year, Faraday actively explored the action of matter on light. From the outset of the program, partly conducted using microscopical techniques, he sought to examine the optical effects of thin films of metal (primarily gold leaf) on light. Faraday began by asking a deceptively simple question: Why does gold leaf appear to be yellow by reflected light but *green* by transmitted light? This question had appeared in Faraday's 1822 "Notes" (Tweney and Gooding, 1991). His renewed interest in the question in the 1850s may have been partly stimulated by the discovery of fluorescence by George Gabriel Stokes (1852). Stokes argued for a "resonance" account to explain the emission of visible light when certain substances were illuminated by invisible ultraviolet light. His explanation was related to Faraday's conjecture that light must involve "transverse" vibrations (i.e., across the line of propagation) as well as the longitudinal vibrations characterized by the known wavelength and frequency of light (Faraday, 1846).

Faraday began his 1856 program (which lasted almost a year) by using a powerful achromatic microscope at Warren De la Rue's[2] laboratory to examine commercial gold leaf. This merely confirmed the limits of microscopical vision; one can see holes in gold leaf at all magnifications, but the resolution of even the finest microscope was simply too limited to reveal whether the structure of the leaf was continuous or whether the holes continued "all the way down." If the holes did continue downward, then the transmitted light colors could conceivably be a diffraction effect.

Using procedures first developed by De la Rue, Faraday explored ways of making films that would be thinner than commercial gold leaf (which is made by mechanical means, by hammering gold to thin it). In the process of making thin films by chemical means (first with De La Rue and then in his own laboratory), Faraday noted a "ruby colored fluid" that seemed to be generated when gold salts were reduced by phosphorus to make gold films. These fluids possessed rather remarkable properties, as Faraday soon discovered. In particular, he soon characterized what later became known as the "Tyndall Effect" (or, more properly, the "Faraday-Tyndall Effect"), namely,

[1] It has been suggested (Williams, 1965) that Faraday was working with "diminished capacity" at this stage of his life, implying perhaps that the research on gold should be taken as uncharacteristic of his full ability to understand nature. At the present stage of my investigation, however, I can see no evidence that any of his 1856 research was affected in such a way. This issue will be addressed at a later date.

[2] De la Rue (1815-1889), an accomplished microscopist, is best known for his photographic expertise, having made stereo photos of the moon and sun in 1861. He had long corresponded with Faraday (James, 1999; Williams, 1971).

that the fluids scattered light in a fashion similar to that of suspensions of particulate matter. Unlike other suspensions, however, the fluids did not settle with time. Today, Faraday's work is most often remembered for this; he is acknowledged as the discoverer of *metallic colloids*.[3]

During the subsequent months of research Faraday prepared hundreds of thin films, dozens of colloids, and a variety of other specimens as well. Although the program was largely unsuccessful in terms of its major theoretical goal, to integrate the interaction of light and matter with Faraday's developing ideas about electrical and magnetic fields, nevertheless, the discovery of gold colloids, the Faraday-Tyndall Effect, and Faraday's explorations of the molecular-level interactions of particles and light, were important outcomes. The research culminated in what was to be Faraday's last major publication in the *Philosophical Transactions* (Faraday, 1857).[4] Sadly, later investigators (particularly Zsigmondy, 1909) mischaracterized what he did achieve in this work, and it has continued to be regarded (except by colloid chemists!) as "minor."

## 1.3 The "discovery" of Faraday's gold

Recently I was privileged to "discover" the surviving specimens used by Faraday in his 1856 research on gold. In fact, it has long been known that some of his colloids survive in the collections of the Royal Institution of Great Britain, where Faraday worked and lived during his entire career; these colloids are frequently mentioned as proof of the longevity of colloidal suspensions. Yet it had not been noted that over 600 other specimens also survive, mostly in the form of thin films of metal deposited on ordinary 1" x 3" microscope slides (Tweney, 2000). The surviving films are mostly gold, but also include silver, platinum, palladium, copper, and other metals. Each slide was numbered (by Faraday) and keyed to reference lists in his Diary entries covering the period of his research on gold (Martin, 1936; the gold research is recorded in entries 14243 to 15403, dated February 2, 1856 to December 20, 1856, pp. 11-254). Because of his careful indexing, we know that nearly all of the specimens are preserved. The combined Diary and specimen collection is one of the most complete records in existence of the work of a great XIX century scientist.

Together with the Diary, the surviving specimens allow us to see deeply into the processes of Faraday's constructive scientific thought, even given his characteristically thorough records. Metallic films (especially gold films) are inherently stable and non-reactive, so we can be sure the specimens re-

[3] Thomas Graham (1805-1869) first coined the term "colloid" (Graham, 1861).

[4] Later the same year he published a two-page "Note on the irregular fusibility of ice", (Faraday, 1858,) in the same journal, as a comment on a longer work by John Tyndall.

main much as they were in Faraday's time. Furthermore, it is possible to reconstruct our own versions of many of the specimens, permitting insight into the dynamics of the physical and chemical makings used by Faraday, and allowing us also to redo many of his manipulations (destructive and otherwise) of the specimens. One could not ask for a more complete "thought record!"

The present paper previews our work with this material. Currently, we are photographically documenting many of the surviving specimens with an eye toward creating an "enhanced" diary, one in which the reader can access not only Faraday's textual comments about his work, but the images associated with it as well. In addition, we are beginning to recreate some of the chemistry used by Faraday, and we are working to prepare slides and colloids that resemble the originals. In the present paper, I give a brief account of some of our preliminary results and suggest ways in which these results may contribute to further understanding of the role of epistemic artifacts in the development of mental models.

## 2. FARADAY'S STRATEGY OF RESEARCH

There is now little dispute over the nature of the complex optical properties of gold and other metals; these can be explained as the outcome of the interaction of light and the free conducting electrons that constitute the defining characteristic of metals (Hecht and Zajac, 1974; Tilley, 2000). In Faraday's time, however, metals constituted a serious puzzle for scientific accounts of matter.[5] One of Faraday's concerns was whether metals were continuous material substances or particulate in nature. Since Faraday did not accept the view of matter as composed of hard material particles ("Daltonian atoms"), he was disposed toward any evidence that gold was continuous and not particulate, and he actively sought such evidence in the 1856 research.

Yet how can the continuity of a film be established? The principal strategy used by Faraday was to explore the optical properties of thin transparent metallic films, particularly of gold. At the outset, he knew that commercial gold leaf was far thinner than the wavelength of light, and he was able to prepare films that were even thinner. Yet, if the films are so thin and still affect light, an explanation is needed for how that could happen. Thus, the colorful transmitted light appearance of gold could be due to diffraction effects caused by extremely small holes in the film. In ordinary gold leaf, such

[5] An account of these disputes is beyond the scope of the present paper; see Roscoe and Schorlemmer, 1886, for a relevant historical overview that still reflects uncertainty about the nature of metals.

holes are visible at all magnifications, even at ones approaching the limit of resolution of a light microscope. Could even smaller holes be causing the colors? This hypothesis is present in the text of 1822 "Notes" and is probably not unique to Faraday. Yet, as noted, he began his 1856 research by using De la Rue's state-of-the-art achromatic microscope to examine gold leaf - without success.

## 2.1 Faraday's chemical and optical approach

Following his inconclusive microscopical examination of gold leaf, Faraday pursued other strategies for attacking the issues. It is clear that he spent a fair amount of time planning his program of research, since the first Diary entries (dated 2 February, 1856) lay out a series of queries to be answered, along with notes of his observations at De la Rue's a few days earlier. In these initial notes, Faraday ponders both the theoretical issues and the specific procedures that might be used to obtain thinner and more regular films of gold. He also laid out a plan to examine precipitates of gold and gold salts, and to compare these to films produced electrolytically and chemically. In the end, electrochemical deposition never worked for Faraday as a means of preparing such films, and so chemical means were used in most of the research.

Preparation aside, how are such films to be studied? Both physical and chemical manipulation played a large role here. Faraday subjected his slides to extensive treatment by acids, bases, heat, pressure, and even "boiling in oil" to determine their properties. Of special note are the many optical explorations he carried out. Does gold leaf polarize light? Can it refract light and hence act as a lens? What is the relation between the reflectivity of such films and their transmitted light color? Over time an inquiry that began with films alone interrelated both films and fluids, raising fundamental questions about the possible states of matter.

## 2.2 The 1856 program[6]

Almost at the beginning of his work with the gold films, in the course of preparing films by using phosphorous to reduce gold chloride, Faraday noticed the "ruby red fluid" produced as a byproduct of "the mere washing" of the equipment (Diary, section 14321, 6 February, 1856, in Martin, 1936, p. 22). This serendipitous remark initiated a new focus in his research, to investigate these ruby-colored fluids. Many others had noticed such fluids in connection with the reactions of gold (e.g., they were described by Samuel Parkes, 1822), but Faraday was the first to describe their connection to the appearances of metallic films, and he was probably the first to closely examine them. As soon as he began to make the fluids deliberately, he found that some were blue or purple (rather than red) and that blue and purple fluids were more subject to rapid settling. Using an optical technique (described in Faraday, 1827) to examine whether these fluids were suspensions, he found that even the ruby ones, which did not settle and appeared clear, would disperse light when a thin beam of sunlight was passed through them. The fluids were indeed a puzzle; like solutions, they appeared clear and did not settle over time, but, like suspensions, which *do* settle, they dispersed light.

Two kinds of specimens thus formed the focus of his program, thin films and the very interesting "fluids." The similarities and the differences in the color of such specimens intrigued him; both films and fluids could be red, blue, or purple by transmitted light, but only films ever appeared green.

By the beginning of March, 1856, Faraday had convinced himself that the films and the fluids were both gold in an uncombined state. He then turned to a more intensive examination of the properties of each, looking in part for ways to quantify the thickness of his films. This problem was never solved directly (other than to establish that the phosphorous films were much thinner than commercial gold leaf). Thickness of the films was an important variable at this stage; could the varying color of deposited films of gold be due to varying thickness? He tested this option by depositing multiple films on the same slide, arranging them in a "staircase" fashion, thus a gradually increasing change in thickness could be seen in one view. The results were clear; while the amount of light transmitted became less with increasing

[6] The characterization of Faraday's program given here is necessarily very sketchy. A complete account should also discuss his inquiry into the reflective and transmissive properties of other metals, his work on gels, his work on the color of organic substances stained by gold, and - of most relevance in the present context - Faraday's reports of his visits to craftspersons, e.g., goldsmiths, gold beaters, and scientific instrument makers. In addition, his own skill as a bookbinder meant that Faraday had a long acquaintance with gold leaf and gilding processes. All of these aspects must be ignored here, but will be dealt with in future publications.

thickness, there was no change in the hue for a given kind of film. Thickness alone did not account for the color.

Using an agate burnisher to subject the films to high mechanical pressure, Faraday found that he could turn a blue or purple film into a green film, suggesting that the mechanical disposition of the particles of the film had something to do with their color. By the end of March he knew that the color of most films could be destroyed by heat (made into a gray, generally) but that a green color could be restored by burnishing. He then turned to a more intensive examination of the color of the ruby fluids.

The fluid colors were also mutable; the red fluids seemed most resistant to settling, but they could be changed to blue or violet fluids by certain chemical agents (which did not react chemically with the gold), as well as by heat. Was heat therefore changing the nature of the gold particles? Their size? Their shape? And if the fluids indeed manifested the optical effects of particulate gold, are the similar colors produced by the films also the result of particles of gold? Or are the films really continuous (non-particulate) in character?

In April, Faraday attacked the question of particles in a more direct fashion, first by showing that thin gold films on glass do conduct electricity (implying their continuity), and then by "deflagration," that is, by exploding gold wires using sudden currents of electricity (Wilkinson, 1804, had described this technique, as part of a larger work on "galvano-electricity"). When wires are exploded in proximity to a glass slide, a colorful deposit of gold can be left on the slide; such specimens are arguably particulate in character (most do not conduct electricity, for example), and yet they show a similar range of color as the thin films, and they exhibit clear metallic reflections, just like the films. By the 21$^{st}$ of April, he was determined to find whether the ruby fluids were in fact red, blue, or purple because of the influence of multiple particles on the transverse vibrations of light. Accordingly, he carefully constructed a prism apparatus that allowed him to separate a narrow band of the spectrum of sunlight into a thin beam of nearly monochromatic light. Using this, he examined "Very many of the films" (14722), as well as the ruby fluids. The results merely confirmed what the unaided eye had already told him; that the films and fluids absorb some colors and transmit others, but there was no evidence that the specimens were directly changing the color of light, as seemed to occur in fluorescence when "Stokes's Ray" (ultraviolet light) was changed into visible light.

In May and June, Faraday concentrated upon careful observations of the color of slides and films (and, to a lesser extent, of dried precipitates). He also evaporated some of the colloids, noting that these also show metallic reflection. The inquiry now appeared broader; the problem of color was not restricted to the transmitted color, but also involved reflected metallic color.

Remarkably, evaporated colloids and exploded wire deposits both showed metallic reflections - and he was increasingly sure that both were particulate. Thus, even the metallic color of a metal may be due to an interaction among particles and light!

Relatively few entries occur for July of 1856, but they are important ones, since Faraday reached the conclusion that the green appearance of films is due to the presence of *elongated* particles. And in August and the first part of September, concentrating his attention on the ruby fluids, he carefully explored the chemical properties of the fluids, ruling out the possibility that the colors were a product of dissolved salts or any substance other than metallic gold. At the same time, he sought to understand why some of the fluids are blue and some red. Once again, was it size of particle? Or shape? The results suggested to him that *size* was the important factor here; blue fluids have larger particles than the red.

In October, Faraday carefully sought to characterize the optical properties of films and fluids in polarized light. At the beginning of the program (in February), Faraday had established that films of gold did not affect a polarized ray of light sent through them. Warren De la Rue later informed Faraday, however, that he had found an effect *if the film was inclined at 45$^o$ to the beam of polarized light.* Glass will act the same way, however, so is the effect due just to the glass? Or are the films alone sufficient? Needless to say, it is difficult to mount so thin a film across an open space, but Faraday (and De la Rue) finally succeeded in doing so. In addition, Faraday worked out a way to optically remove the optical effect of the glass by immersing a glass-mounted gold film in carbon disulfide, which has the same refractive index as glass. Glass "disappears" in such a fluid and no longer rotates a polarized beam, even when tilted. But the gold films, now mounted on "invisible" glass, still rotated a polarized ray, as did the particulate deposits from exploded gold wires. There were now at least two strong grounds (rotation of a polarized beam and the metallic reflection of dried colloids and exploded wire deposits) for asserting that the interaction of light with separate particles is responsible for optical activity, even in films. And, most importantly, Faraday's model of the interaction of light with gold now manifested a true field effect; color is produced when a light beam traverses an *array* of particles; that is, color is something that happens *between* particles!

By the end of October, Faraday still believed that green is the true color of "continuous" gold, and that red and blue are the true colors of "divided" gold. Alas, however, he could not, in the end, prove this, since all of the evidence was indirect, and he could never "see" the particles directly, nor could he rule out other factors in all of the cases. By November he knew that the chemistry is very complex, that the reaction kinetics (to use a modern term) are similarly complex and diverse, and that mechanical effects can occur at a very small scale.[7] Thus, since green happens only when some mechanical force operates to elongate otherwise symmetrical particles, the obvious conclusion would seem to be that *all* color effects in gold are due to field effects among particles. Still, Faraday was cautious in phrasing the conclusion; "[This] would seem to imply that the particles are so small and so near, that two or more can act at once upon the individual atoms of the ether. Their association is such as to present as it were an optical continuity" (Faraday, 1857, p. 439).

Faraday's research on gold ended abruptly in December, 1856, and there are no further entries dealing with gold after this date. In February, Faraday read a paper to the Royal Society (published as Faraday, 1857). He characterized the results almost apologetically; "I do not pretend that [the results] are of very great value in their present state, but they are very suggestive" (p. 393). In March, 1857, Faraday took up the completely different problem of "time in magnetism."

## 3. REPLICATING FARADAY

The value of replications as a means to historical understanding has been established by a number of studies.[8] Our own attempts to replicate Faraday's work have so far been "opportunistic" rather than systematic, insofar as we have had to learn skills that, for Faraday, could be taken for granted. Among other things, we have had to work under the constraints of modern safety regulations and procedures. Thus, Faraday's most important method for producing thin films of gold, the reduction of gold chloride by the phosphorous vapor found over a carbon disulfide solution of phosphorous, has not yet been attempted (both carbon disulfide and phosphorous are extremely hazardous materials!)

We have succeeded in producing gold colloids that closely resemble Faraday's by using a modern method (reducing a weak solution of gold chloride with sodium citrate solution). These produce an excellent dispersion effect and constitute a replication of the discovery of the Faraday-Tyndall Effect. When evaporated, the colloids produce slides and watch glasses having the same metallic reflective colors that struck Faraday as noteworthy.

[7] For example, he found that the lateral pressure of a slowly dissolving phosphorous particle floating on a fluid was sufficient to create at least some green areas in the films that form around the phosphorous particle.

[8] For example, the papers in Blondel and Dörries (1994), especially that by Heering (1994), suggest that Coulomb's experimental determination of the law of electrostatic attraction and repulsion could not have been carried out as Coulomb described it.

In replicating some of his chemical manipulations we have used aqua regia to thin a glass-mounted gold leaf, and, in the process, verified Faraday's observation that such chemical action does not in itself change the green color of such films. We have also noted some of the unusual morphology that can emerge from such treatment, in particular, a strange array of connected block-like "cells" that Faraday referred to as the "Chinese Wall."

Our friend and associate, Harry Brown, of Jackson Mississippi, has also produced a thin gold film by reducing dissolved dental gold using dextrose. This exceptionally thin film displays a blue color by transmitted light and an apparent continuity under the microscope that closely resembles some of Faraday's slides. The film does not adhere well, however, and we have not been able to repeat Faraday's burnishing of such slides since the film simply rubs off. Producing such films is also highly unreliable using "soft" reductants like dextrose; we now know why Faraday relied upon the phosphorous method!

At present, as we accommodate to the hazardous materials that we will need to use eventually, we are beginning experiments to "deflagrate" metallic wire with sudden surges of electricity. This also has been a trickier project than we had first imagined, and makes clear why Faraday went to the expense and bother of using a "Grove's Cell" for these experiments, an acid battery using zinc and platinum electrodes, characterized by very low internal resistance (Grove, 1839). When we tried a variety of lead acid batteries (commercial automobile storage batteries), we found that their internal resistance was so high that the resulting slow discharge across a thin wire merely melted the wire instead of exploding it. Since then, and given the huge expense of recreating a Grove's Cell, we have set up a large capacitor bank driven by a standard laboratory power supply. When the capacitors are fully charged they discharge rapidly across the thin wire and produce a satisfying "explosion." Copper wires deflagrated in this manner resemble the kinds of deposits seen by Faraday, and we are beginning the same experiments with pure gold wire, obtaining slides which also resemble Faraday's specimens.

Our replication work is still ongoing, but it has already served to "open" many aspects of the Diary to our view. Photography of the slides is alone helpful in this regard, since much of the Diary text relies upon verbal descriptions of what Faraday saw. Good though his Diary descriptions are, they are necessarily incomplete as records of this overwhelmingly visual domain; accordingly, much time has been devoted to photographing some of the surviving specimens, and we are beginning to create a "digitized" version of his Diary that will be enhanced with appropriate images.[9]

[9] Some of the images can be seen at my web site;, which will also include links to the digitized Diary, when this is complete; http://personal.bgsu.edu/~tweney.

Clearly, however, the replications can carry us even further; if photographs open the Diary text by adding the visual elements, replications open the processes by which the visual objects of the text were constructed. These successive openings have revealed just how much the original Diary record is telegraphed and incomplete.

## 4. CONCLUSION - EPISTEMIC ARTIFACTS

The larger goals of the present project are centered on the process of uncovering the interactive strategies used by Faraday in his research on gold, thus situating the research more carefully in the context of its time and as an important episode in Faraday's own career. In addition, there are implications that are relevant to understanding the general nature of model-based reasoning.

Our "hands on" work replicating the procedures used by Faraday to produce his specimens has the promise of uncovering his tacit skills and knowledge.[10] As we proceed with the work, more and more of the Diary text becomes transparent to us. Until we had begun the replication work, much of the Diary was puzzling; "Why did he do this?" "Why didn't he do that?" The value of our replications is thus similar to the value of thought experiments in science; like thought experiments, replications assist in the reconstruction of an argument (Nersessian, 1999), noting, of course, that the argument is one we are making. By contrast, only some of Faraday's constructive practices were intended to aid argument construction; in fact, most of his work was not finished enough to permit a final argument to be constructed. Instead, Faraday was most often using constructive practices and the resulting objects to either pose a question or ask a question. Each slide was an element in a dialogue between Faraday and the world of gold.

In casting the specimens as elements in a dialogue, I mean to call attention to two related points. First, as Fisher (1992) noted, Faraday's "voice," even in his published writings, was highly dialectical in the sense that questions were posed, tentative answers given, new questions raised, and so forth. As the gold project makes clear, this is especially true when Faraday is uncertain of the outcome of his work; even in the end, the specimens were as much questions as they were answers. Faraday's goal, unfulfilled here but present in all of his research, was to make nature *legible* (to use Fisher's, 2001, term), that is, to allow something like a direct reading of the nature of nature from his specimen preparations. In this case, he succeeded only partially.

[10] For similar views on the value of replication, see Cavicchi, 1997; Kurz & Hertwig, 2001; and Gooding, 1990.

The specimens are more than just discursive elements; they are *epistemic artifacts*. They are "epistemic" insofar as they participate in the discourse; things can come to be known by means of the artifacts. And they are "artifacts" because they are made by Faraday as parts of that discourse; he is an *agent* in their construction (Kurz and Tweney, 1998). Much of science reflects a similar use of "preparations," of course, but note that several important distinctions need to considered. First, *cognitive artifacts* (Hutchins, 1995; Zhang and Norman, 1995) are computational artifacts which can lead to new results but do not lead to new knowledge (except in the restricted sense that the outcome of a computation is new information). Second, Rheinberger's (1997) *epistemic things*, which do lead to new knowledge, differ from Faraday's preparations in that they were constructed to serve as a consensually agreed means used by a community of scientists; they are finished products. By contrast, Faraday's specimens are mostly private objects; only a very small fraction ever became publicly known, and only a few, even for Faraday, represented anything like a finished, unambiguously legible, object. Instead, most of his slides were rather like the construals noted by Gooding (1990) or the "inceptions" discussed by Ippolito and Tweney (1995; see also Tweney, 1992) in that they remain vague and unformed as carriers of meaning. Legibility emerged only after long series of artifacts, arguments, and experiments had been deployed.

It has long been noted that Faraday's universe was "force-centered," in that force was given priority over matter as a fundamental physical construct (Cantor, 1991; Williams, 1965). In my view, even Faraday's "ways of knowing" (Cavicchi, 1997) are force centered; for him, one came to know the world partly by making the proper epistemic artifacts. In similar fashion we seek to make Faraday's model-based world of gold films and ruby fluids legible to us.

## ACKNOWLEDGEMENTS

Initial work on this project was supported by a grant from the British Academy, (sponsored by the University of Bath), and by NSF #SES-0100112, awarded through Bowling Green State University. I am deeply grateful to the staff of the Royal Institution Center for History, Science, and Technology, London, and especially to Frank A.J.L. James, Reader in History and Philosophy of Science, and Keeper of the Collections, at the RI. Without his enthusiastic support, this project could not have been conducted. My laboratory work on the replication experiments was assisted by Ryan

Mears and Andy Wickiser, and I have had the benefit of chemical and photographic advice from Harry Brown and Tom Grill.

## REFERENCES


Andersen, H., 2001, The development of family resemblance concepts, in this volume, pp.95-111.

Blondel, C. and Dörries, M., eds., 1994, *Restaging Coulomb: Usages, Controverses et Répli-cations Autour de la Balance de Torsion,* Leo S. Olschki, Firenze.

Cantor, G., 1991, *Michael Faraday: Sandemanian and Scientist,* St. Martin's Press, New York.

Cantor, G. and Hodges, M.J.S., eds., 1981, *Conceptions of Ether: Studies in the History of Ether Theories,* Cambridge University Press, Cambridge.

Cavicchi, E., 1997, Experimenting with magnetism: Ways of learning of Joann and Faraday, *American Journal of Physics* 65:867-882.

Chen, X., 2000, *Instrumental Traditions and Theories of Light: The Uses of Instruments in the Optical Revolution*, Kluwer/Plenum, Dordrecht.

Dunbar, K., 2001, What scientific thinking reveals about the nature of cognition, in: *Designing for Science: Implications from Everyday, Classroom, and Professional Settings*, K. Crowley, C.D. Schunn, and T. Okada, eds., Lawrence Erlbaum Associates, Mahwah, NJ, pp. 115-140.

Faraday, M., 1827, *Chemical Manipulation; Being Instructions to Students in Chemistry*, W. Phillips, London.

Faraday, M., 1846, Thoughts on ray vibrations, *Philosophical Magazine* 38:345-350.

Faraday, M., 1852, On the physical character of the lines of magnetic force, *Philosophical Magazine* 3:216-220. Reprinted in M. Faraday, ed., 1855, *Experimental Researches in Electricity,* Vol. 3, Richard Taylor & William Francis, London, pp. 407-437.

Faraday, M., 1857, Experimental relations of gold (and other metals) to light, *Philosophical Transactions*, 145-181 (read Feb. 5, 1857). Reprinted in M. Faraday, ed., 1859, *Experimental Researches in Chemistry and Physics,* Richard Taylor & William Francis, London, pp. 391-442.

Faraday, M., 1858, Note on the irregular fusibility of ice, *Philosophical Transactions*, 228-229 (read Dec. 17, 1857). Reprinted in M. Faraday, ed., 1859, *Experimental Researches in Chemistry and Physics,* Richard Taylor & William Francis, London, pp. 374-377.

Fisher, H.J., 1992, Faraday's two voices, *Physis: Rivista Internazionale di Storia della Scienza* 29:165-186.

Fisher, H.J., 2001, *Faraday's Experimental Researches in Electricity: Guide to a First Reading,* Green Lion Press, Santa Fe, NM.

Gentner, D., Brem, S., Ferguson, R.W., Markman, A.B., Levidow, B.B., Wolff, P., and Forbus, K.D., 1997, Analogical reasoning and conceptual change: A case study of Johannes Kepler, *Journal of the Learning Sciences* 6:3-40.

Gooding, D., 1982, A convergence of opinion on the divergence of lines: Faraday and Thomson's discussion of diamagnetism, *Notes and Records of the Royal Society of London* 36:243-259.

Gooding, D., 1990, *Experiment and the Making of Meaning*, Kluwer, Dordrecht.

Gorman, M.E., 1997, Mind in the world: Cognition and practice in the invention of the telephone, *Social Studies of Science* 27:583-624.

Graham, T., 1861, Liquid diffusion applied to analysis, *Philosophical Transactions*, 151:183-224.

Grove, W., 1839, On a small voltaic battery of great energy, *Philosophical Magazine* 15:287-293.

Hecht, E. and Zajac, A., 1974, *Optics,* Addison-Wesley, Reading, MA.

Heering, P., 1994, The replication of the torsion balance experiment: the inverse square law and its refutation by early XIX century German physicists, in: *Restaging Coulomb: Usages, Controverses et Réplications Autour de la Balance de Torsion*, C. Blondel and M. Dörries, eds., Leo S. Olschki, Firenze, pp. 47-66.

Hutchins, E., 1995, How a cockpit remembers its speeds, *Cognitive Science* 19:265-288.

Ippolito, M.F. and Tweney, R.D., 1995, The inception of insight, in: *The Nature of Insight*, R.J. Sternberg and J.E. Davidson, eds., The MIT Press, Cambridge, MA, pp. 433-462.

James, F.A.J.L., 1985, "The optical mode of investigation": light and matter in Faraday's natural philosophy, in: *Faraday Rediscovered: Essays on the Life and Work of Michael Faraday, 1791-1867*, D. Gooding and F.A.J.L. James, eds., Macmillan, Basingstoke, UK, pp. 137-162.

James, F.A.J.L., ed., 1999, *The Correspondence of Michael Faraday;* Volume 4, 1849-1855, Institution of Electrical Engineers, London.

Kurz, E.M. and Hertwig, R., 2001 (in press), To know an experimenter ..., in: *The Essential Brunswik: Beginnings, Explications, Applications*, K.R. Hammond and T.R. Stewart, eds., Oxford University Press, New York.

Kurz, E.M. and Tweney, R.D., 1998, The practice of mathematics and science: From calculus to the clothesline problem, in: *Rational Models of Cognition*, M. Oaksford and N. Chater, eds., Oxford University Press, Oxford, pp. 415-438.

Langley, P., and Jones, R., 1988, A computational model of scientific insight, in: *The Nature of Creativity*, R. Sternberg, ed.,Cambridge University Press, Cambridge, pp.177-201.

Magnani, L., 2001, *Abduction, Reason, and Science: Processes of Discovery and Explanation,* Kluwer Academic/Plenum Publishers, New York.

Martin, T., ed., 1936, *Faraday's Diary: Being the Various Philosophical Notes of Experimental Investigation Made by Michael Faraday During the Years 1820-1862,* vol. 7, G. Bell, London.

Nersessian, N., 1984, *Faraday to Einstein: Constructing Meaning in Scientific Theories*, Kluwer, Dordrecht.

Nersessian, N., 1985, Faraday's field concept, in: *Faraday Rediscovered: Essays on the Life and Work of Michael Faraday, 1791-1867*, D. Gooding and F.A.J.L. James, eds., Macmillan, Basingstoke, UK, pp. 175-188.

Nersessian, N., 1999, Model-based reasoning in conceptual change, in: *Model-Based Reasoning in Scientific Discovery*, L. Magnani, N.J. Nersessian and P. Thagard, eds., Kluwer Academic/Plenum Publishers, New York, pp. 5-22.

Parkes, S., 1822, *The Chemical Catechism, With Tables, Notes, Illustrations, and Experiments,* Tenth edition, Baldwin, Cradock, and Joy, London.

Rheinberger, H.J., 1997, *Toward a History of Epistemic Things: Synthesizing Proteins in the Test Tube,* Stanford University Press, Stanford, CA.

Roscoe, H.E. and Schorlemmer, C., 1886, *A Treatise on Chemistry, Volume II(1): Metals*, D. Appleton and Co., New York.

Stokes, G.G., 1852, On the change of refrangibility of light, *Philosophical Transactions.* Reprinted in G.G. Stokes, *Mathematical and Physical Papers*, Vol. 3., Johnson Reprint Corp., New York, 1966, pp. 277-413.

Tilley, R., 2000, *Colour and the Optical Properties of Materials,* John Wiley & Sons, Chichester, UK.

Tweney, R.D., 1992, Stopping time: Faraday and the scientific creation of perceptual order, *Physis: Rivista Internazionale di Storia della Scienza* 29:149-164.

Tweney, R.D., 2000, *Faraday's Microscope Slides; A Catalog of the Royal Institution Collection,* Unpublished manuscript, The Royal Institution of Great Britain, London.

Tweney, R.D., 2001, Scientific thinking: A cognitive-historical approach, in: *Designing for Science: Implications from Everyday, Classroom, and Professional Settings,* K. Crowley, C.D. Schunn and T. Okada, eds., Lawrence Erlbaum Associates, Mahwah, NJ, pp. 141-176.

Tweney, R.D. and Gooding, D., eds., 1991, *Faraday's 1822 "Chemical Notes, Hints, Suggestions, and Objects of Pursuit",* The Science Museum and Peter Peregrinus, Ltd., London.

Wilkinson, C.H., 1804, *Elements of Galvanism in Theory and Practice*, John Murray, London.

Williams, L.P., 1965, *Michael Faraday: A Biography*, Basic Books, New York.

Williams, L.P., ed., 1971, *The Selected Correspondence of Michael Faraday,* Cambridge University Press, Cambridge.

Zhang, J. and Norman, D.A., 1995, A representational analysis of numeration systems, *Cognition* 57:271-295.

Zsigmondy, R., 1909, *Colloids and the Ultramicroscope: A Manual of Colloid Chemistry and Ultramicroscopy*, translated by J. Alexander, John Wiley & Sons, New York (originally published 1905).

# Epistemic Mediators and Model-Based Discovery in Science

Lorenzo Magnani
*Department of Philosophy and Computational Philosophy Laboratory, University of Pavia, Pavia, Italy, and Georgia Institute of Technology, Atlanta, USA, lmagnani@cc.gatech.edu*

Abstract: What I call *theoretical abduction* (sentential and model-based) certainly illustrates much of what is important in abductive reasoning, especially the objective of selecting and creating a set of hypotheses that are able to dispense good (preferred) explanations of data, but fails to account for many cases of explanations occurring in science or in everyday reasoning when the exploitation of the environment is crucial. The concept of *manipulative abduction* is devoted to capture the role of action and of external representations in many interesting situations: action provides otherwise unavailable information that enables the agent to solve problems by starting and performing a suitable abductive process of generation or selection of hypotheses. I will present some aspects of this kind of reasoning derived from the "cognitive" history of the discovery of the non-Euclidean geometries. Geometrical diagrams are considered external representations which play both a *mirror* role (to externalize rough mental models), and an *unveiling* role (as gateways to imaginary entities). I describe them as *epistemic mediators* able to perform various abductive tasks (discovery of new properties or new propositions/hypotheses, provision of suitable sequences of models as able to convincingly verifying theorems, etc.).

## 1. MANIPULATIVE ABDUCTION AND EPISTEMIC MEDIATORS

An interesting and neglected point of contention about human reasoning is whether or not concrete manipulations of external objects influence the generation of hypotheses. I am focusing on the first features of what I call *manipulative abduction* showing how we can find in scientific and everyday

*Model-Based Reasoning: Science, Technology, Values*, edited by
L. Magnani and N.J. Nersessian, Kluwer Academic/Plenum Publishers, New York, 2002

reasoning methods of constructivity based on external models and action-based reasoning. I am analyzing the problem in the light of the so-called historical-cognitive method (Nersessian, 1998). While it tries to integrate findings from research on cognition and findings from historical-epistemological research into models of actual scientific practices, assessments of the fit between cognitive findings and historical-epistemological practices aid in elaborating richer and more realistic models of cognition. The relevance of the concept of abduction can contribute to a central issue in cognitive science: hypothesis formation both in science and in everyday reasoning.

More than a hundred years ago, the great American philosopher Charles Sanders Peirce coined the term "abduction" to refer to inference that involves the generation and evaluation of explanatory hypotheses. Peirce says that mathematical and geometrical reasoning "consists in constructing a diagram according to a general precept,[1] in observing certain relations between parts of that diagram not explicitly required by the precept, showing that these relations will hold for all such diagrams, and in formulating this conclusion in general terms. All valid necessary reasoning is in fact thus diagrammatic" (*CP*, 1.54).[2] I contend that geometrical construction is a kind of abductive reasoning, and more precisely, what I call manipulative abduction.

What is abduction? Many reasoning conclusions that do not proceed in a deductive manner are *abductive*. For instance, if we see a broken horizontal glass on the floor[3] we might explain this fact by postulating the effect of wind shortly before: this is certainly not a deductive consequence of the glass being broken (a cat may well have been responsible for it). Hence, what I call *theoretical* abduction, sentential and model-based (cf. Figure 1), is the process of *inferring* certain facts and/or laws and hypotheses that render some sentences plausible, that *explain* or *discover* some (eventually new) phenomenon or observation; it is the process of reasoning in which explanatory hypotheses are formed and evaluated.[4]

The type of inference called abduction was studied by Aristotelian syllogistics, as a form of ἀπαγωγή, and later on by mediaeval reworkers of syllogism. A hundred years ago, Peirce interpreted abduction essentially as an

[1] That is a kind of definition that prescribes "what you are to *do* in order to gain perceptual acquaintance with the object of the world" (*CP*, 2.330).

[2] This kind of reasoning is also called by Peirce "theorematic" and it is a kind of "deduction" necessary to derive significant theorems: "is one which, having represented the conditions of the conclusion in a diagram, performs an ingenious experiment upon the diagram, and by observation of the diagram, so modified, ascertains the truth of the conclusion". The "corollarial" reasoning, mechanical and not creative, "is one which represents the condition of the conclusion in a diagram and finds from the observation of this diagram, as it is, the truth of the conclusion" (Peirce, *CP*, 2.267, cf. also Hoffmann, 1999 and forthcoming).

[3] This event constitutes in its turn an *anomaly* that needs to be solved/explained.

[4] A general theory of abduction is given in Magnani, 2001a.

"inferential" *creative process* of generating a new hypothesis. Abduction and induction, viewed together as processes of production and generation of new hypotheses, are sometimes called reduction, that is ἀπαγωγή.

There are two main epistemological meanings of the word abduction: 1) abduction that only generates "plausible" hypotheses (*selective* or *creative*) and 2) abduction considered as *inference to the best explanation*, which also evaluates hypotheses.

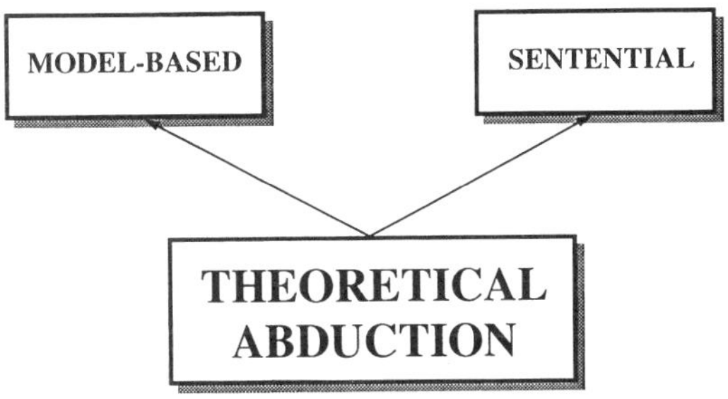


*Figure 1.* Theoretical abduction.

To illustrate from the field of medical knowledge, the discovery of a new disease and the manifestations it causes can be considered as the result of a creative abductive inference. Therefore, creative abduction deals with the whole field of the growth of scientific knowledge. This is irrelevant in medical diagnosis where instead the task is to *select* from an encyclopedia of pre-stored diagnostic entities (Ramoni et al., 1992). We can call both inferences ampliative, selective and creative, because in both cases the reasoning involved amplifies, or goes beyond, the information incorporated in the premises.

All we can expect of our "selective" abduction, is that it tends to produce hypotheses for further examination that have some chance of turning out to be the best explanation. Selective abduction will always produce hypotheses that give at least a partial explanation and therefore have a small amount of initial plausibility. In the syllogistic view (see below) concerning abduction as inference to the best explanation advocated by Peirce one might require that the final chosen explanation be the most "plausible".

Many attempts have been made to model abduction by developing some formal tools in order to illustrate its computational properties and the relationships with the different forms of deductive reasoning. Some of the formal models of abductive reasoning are based on the theory of the *epistemic state* of an agent (Boutilier and Becher, 1995), where the epistemic state of an individual is modeled as a consistent set of beliefs that can change by expansion and contraction (*belief revision framework*). This kind of *sen-*

*tential frameworks* seems to exclusively deal with selective abduction[5] (diagnostic reasoning) and relates to the idea of preserving *consistency*. If we want to provide a suitable framework for analyzing the most interesting cases of conceptual changes in science we do not have to limit ourselves to the *sentential* view of theoretical abduction but we have to consider a broader *inferential* one which encompasses both sentential and what I call *model-based* sides of creative abduction.

Hence, if we want to deal with the nomological and most interesting creative aspects of abduction we are first of all compelled to consider the whole field of the growth of scientific knowledge. Related to the high-level types of scientific conceptual change (Thagard, 1992) are different varieties of *model-based abductions* (see, for examples, Magnani, 1999b).

Following Nersessian (1995a and b, 1999), I use the term "model-based reasoning" to indicate the construction and manipulation of various kinds of representations, not mainly sentential and/or formal, but mental and/or related to external mediators. Obvious examples of model-based reasoning are constructing and manipulating visual representations, thought experiment, analogical reasoning, occurring when models are built at the intersection of some operational interpretation domain - with its interpretation capabilities - and a new ill-known domain.

What exactly is model-based abduction from a philosophical point of view? Peirce stated that all thinking is in signs, and signs can be icons, indices, or symbols. Moreover, all *inference* is a form of sign activity, where the word sign includes "feeling, image, conception, and other representation" (*CP* 5.283), and, in Kantian words, all synthetic forms of cognition. That is, a significant part of the thinking activity is *model-based*. Geometrical construction is clearly a kind of model-based reasoning. Of course model-based reasoning acquires its peculiar creative relevance when embedded in abductive processes, such that we can individuate a *model-based abduction* (Magnani, 1999a and b).

Peirce gives an interesting example of model-based abduction related to sense activity: "A man can distinguish different textures of cloth by feeling: but not immediately, for he requires to move fingers over the cloth, which shows that he is obliged to compare sensations of one instant with those of another" (*CP* 5.221). This idea surely suggests that abductive movements also have interesting extra-theoretical characteristics and that there is a role in abductive reasoning for various kinds of manipulations of external objects. When manipulative aspects of external models prevail, like in the case

[5]As illustrated above, we have to distinguish between selective and creative abduction. Abduction that merely *selects* from an encyclopedia of pre-stored hypotheses is called selective. Abduction that generates *new* hypotheses (Magnani, 1992) is called creative.

of manipulating diagrams in the blackboard, we face what I call *manipulative* abduction (or action-based abduction, Magnani, 2000).

*Manipulative* abduction (cf. Figure 2) happens when we are thinking *through* doing and not only, in a pragmatic sense, about doing. For instance, when we are creating geometry constructing and manipulating a triangle, like in the case given by Kant in the "Transcendental Doctrine of Method" (Kant, 1929).[6] In the case of natural sciences the idea of manipulative abduction goes beyond the well-known role of experiments as capable of forming new scientific laws by means of the results (the nature's answers to the investigator's question) they present, or of merely playing a predictive role (in confirmation and in falsification). Manipulative abduction refers to an extra-theoretical behavior that seeks to create communicable accounts of new experiences to integrate them into previously existing systems of experimental and linguistic (theoretical) practices.

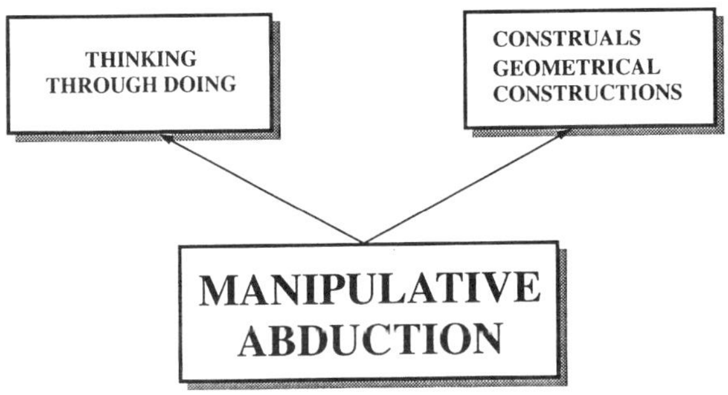


*Figure 2.* Manipulative abduction.

Some common features (in some cases present in geometrical construction) of the tacit templates of manipulative abduction, that enable us to manipulate things and experiments in science are related to: 1. sensibility to the aspects of the phenomenon which can be regarded as *curious* or *anomalous*; manipulations have to be able to introduce potential inconsistencies in the received knowledge (Oersted's report of his well-known experiment about electromagnetism is devoted to describe some anomalous aspects that did not depend on any particular theory of the nature of electricity and magnetism; 2. preliminary sensibility to the *dynamical* character of the phenomenon, and not to entities and their properties, common aim of manipulations is to practically reorder the dynamic sequence of events into a static spatial one that should promote a subsequent bird's-eye view (narrative or visual-diagrammatic); 3. referral to experimental manipulations that exploit *artificial apparatus* to free new possible stable and repeatable sources of infor-

[6] On the philosophical, computational, and cognitive aspects of the relationships between geometry and space, also from a historical point of view, cf. Magnani, 2001b.

mation about hidden knowledge and constraints (Davy well-known set-up in term of an artifactual tower of needles showed that magnetization was related to orientation and does not require physical contact); 4. various contingent ways of epistemic acting: *looking* from different perspectives, *checking* the different information available, *comparing* subsequent events, *choosing*, *discarding*, *imaging* further manipulations, *re-ordering* and *changing relationships* in the world by implicitly *evaluating* the usefulness of a new order (for instance, to help memory).

It is clear that in geometrical construction all these requirements are fulfilled. Geometrical constructions present situations that are curious and "at the limit". These are constitutively dynamic, artificial, and offer various contingent ways of epistemic acting, like looking from different perspectives, comparing subsequent appearances, discarding, choosing, re-ordering, and evaluating. Moreover, they present some of the features indicated below, typical of all the so-called abductive epistemic mediators, not just of the ones which play a scientific role in manipulative reasoning: simplification of the task and the capacity to get visual information otherwise unavailable.

The whole activity of manipulation is devoted to building various external *epistemic mediators*[7] that function as an enormous new source of information and knowledge. Therefore, manipulative abduction represents a kind of redistribution of the epistemic and cognitive effort to manage objects and information that cannot be immediately represented or found internally (for example exploiting the resources of visual imagery).[8]

From the point of view of everyday situations manipulative abductive reasoning and epistemic mediators exhibit very interesting features (we can find the first three in geometrical constructions): 1. action elaborates a *simplification* of the reasoning task and a redistribution of effort across time (Hutchins, 1995), when we need to manipulate concrete things in order to understand structures which are otherwise too abstract (Piaget, 1974), or when we are in presence of *redundant* and unmanageable information; 2. action can be useful in presence of *incomplete* or *inconsistent* information - not only from the "perceptual" point of view - or of a diminished capacity to act upon the world: it is used to get more data to restore coherence and to improve deficient knowledge; 3. action enables us to build *external artifactual models* of task mechanisms instead of the corresponding internal ones, that are adequate to adapt the environment to agent's needs. 4. action as a

[7] I derive this expression from the cognitive anthropologist Hutchins (1995), who coined the expression "mediating structure" to refer to various external tools that can be built to cognitively help the activity of navigating in modern but also in "primitive" settings. Any written procedure is a simple example of a cognitive "mediating structure" with possible cognitive aims, so mathematical symbols and diagrams.

[8] It is difficult to preserve precise spatial and geometrical relationships using mental imagery, in many situations, especially when one set of them has to be moved relative to another.

*control of sense data* illustrates how we can change the position of our body (and/or of the external objects) and how to exploit various kinds of prostheses (Galileo's telescope, technological instruments and interfaces) to get various new kinds of stimulation: action provides some tactile and visual information (e. g, in surgery), otherwise unavailable. Also natural phenomena can play the role of external artifactual models: under Micronesians' manipulations of their images, the stars acquire a structure that "becomes one of the most important structured representational media of the Micronesian system" (Hutchins, 1995, p. 172). The external artifactual models are endowed with functional properties as components of a memory system crossing the boundary between person and environment (for example they are able to transform the tasks involved in allowing simple manipulations that promote further visual inferences at the level of model-based abduction). The cognitive process is *distributed* between a person (or a group of people) and external representation(s), and so obviously *embedded* and *situated* in a society and in a historical culture: the human cognitive agents are *intentional* (Giere, 2002).[9]

## 2. GEOMETRICAL CONSTRUCTION IS A KIND OF MANIPULATIVE ABDUCTION

Let's quote again Peirce's passage about constructions. Peirce says that mathematical and geometrical reasoning "consists in constructing a diagram according to a general precept, in observing certain relations between parts of that diagram not explicitly required by the precept, showing that these relations will hold for all such diagrams, and in formulating this conclusion in general terms. All valid necessary reasoning is in fact thus diagrammatic" (*CP*, 1.54). Not dissimilarly Kant says that in geometrical construction "[...] I must not restrict my attention to what I am actually thinking in my concept of a triangle (this is nothing more than the mere definition); I must pass beyond it to properties which are not contained in this concept, but yet belong to it" (Kant, 1929, A718-B746, p. 580).

We have seen that manipulative abduction is a kind of abduction, usually model-based, that exploits external models endowed with delegated (and often implicit) cognitive roles and attributes.

[9] Nonverbal organisms like monkeys and apes present many (spatial) cognitive abilities (static and dynamic) and sensorimotor intelligence related to the exploitation of external objects (for example: location of food in a manipulative space - by using the presence of landmarks - and recovery of hidden objects, ability to travel efficiently by using external cues rather than mental records, tool use and tool manufacturing) (Call, 2000).

1. The model (diagram) is *external* and the strategy that organizes the manipulations is unknown *a priori*.

2. The result achieved is *new* (if we, for instance, refer to the constructions of the first creators of geometry), and adds properties not contained before in the concept (the Kantian to "pass beyond" or "advance beyond" the given concept, Kant, 1929, A154-B193/194, p. 192).[10]

## 3. EXTERNAL REPRESENTATIONS

Certainly a big portion of the complex environment of a thinking agent is internal, and consists of the proper software composed of the knowledge base and of the inferential expertise of that individual. Nevertheless, any cognitive system is composed by a "distributed cognition" among people and "external" technical artifacts (Hutchins, 1995; Norman 1993).

In the case of the construction and examination of diagrams in geometrical reasoning, specific experiments serve as states and the implied operators are the manipulations and observations that transform one state into another. The geometrical outcome is dependent upon practices and specific sensory-motor activities performed on a non symbolic object, which acts as a dedicated external representational medium supporting the various operators at work. There is a kind of an epistemic negotiation between the sensory framework of the geometer and the external reality of the diagram. This process involves an external representation consisting of written symbols and figures that are manipulated "by hand". The cognitive system is not merely the mind-brain of the person performing the geometrical task, but the system consisting of the whole body (cognition is *embodied*) of the person plus the external physical representation. In geometrical discovery the whole activity of cognition is located in the system consisting of a human together with diagrams.

An external representation can modify the kind of computation that a human agent uses to reason about a problem: the Roman numeration system eliminates, by means of the external signs, some of the hardest parts of the addition, whereas the Arabic system does the same in the case of the difficult computations in multiplication (Zhang and Norman, 1994; Zhang, 1997). The capacity for inner reasoning and thought results from the internalization of the originally external forms of representation. In the case of the external representations we can have various objectified knowledge and structure (like physical symbols - e.g. written symbols, and objects - e.g. three-

[10] Of course in the case we are using diagrams to demonstrate already known theorems (for instance in didactic settings), the strategy of manipulations is not necessarily unknown and the result is not new.

dimensional models, shapes and dimensions), but also external rules, relations, and constraints incorporated in physical situations (spatial relations of written digits, physical constraints in geometrical diagrams and abacuses) (Zhang, 1997). The external representations are contrasted to the internal representations that consist of the knowledge and the structure in memory, as propositions, productions, schemas, neural networks, models, prototypes, images.

The external representations are not merely memory aids: they can give people access to knowledge and skills that are unavailable to internal representations, help researchers to easily identify aspects and to make further inferences, they constrain the range of possible cognitive outcomes in a way that some actions are allowed and other forbidden. The mind is limited because of the restricted range of information processing, the limited power of working memory and attention, the limited speed of some learning and reasoning operations; on the other hand the environment is intricate, because of the huge amount of data, real time requirement, uncertainty factors. Consequently, we have to consider the whole system, consisting of both internal and external representations, and their role in optimizing the whole cognitive performance of the distribution of the various subtasks. It is well-known that in the history of geometry many researchers used internal mental imagery and mental representations of diagrams, but also self-generated diagrams (external) to help their thinking.

In the construction of mathematical concepts many external representations are exploited, both in terms of diagrams and of symbols. I am interested in my research in the diagrams which play an *optical* role - microscopes (that look at the infinitesimally small details), telescopes (that look at infinity), windows (that look at a particular situation), a *mirror* role (to externalize rough mental models), and an *unveiling* role (to help to create new and interesting mathematical concepts, theories, and structures).[11] I describe them as the *epistemic mediators* (cf. the previous section) able to perform various abductive tasks (discovery of new properties or new propositions/hypotheses, provision of suitable sequences of models able to convincingly verifying theorems, etc.).

[11] The epistemic and cognitive role of optical diagrams in "perceiving the infinite and the infinitesimal world" in the calculus is illustrated in Magnani and Dossena (2002).

# 4. UNVEILING DIAGRAMS: GATEWAYS TO IMAGINARY ENTITIES

## 4.1 Internal and external representations

For over two thousand years some of the greatest mathematicians tried to prove Euclid's fifth postulate. Geometers were not content merely to manipulate proofs in order to discover new theorems and thereby to resolve the anomaly without trying to answer questions about what the symbols of the principles underlying Euclidean geometry represent. Lobachevsky's strategy for resolving the anomaly of the fifth postulate was to manipulate the symbols, rebuild the principles, and then to derive new proofs and provide a new mathematical apparatus. Needless to say, Lobachevsky was working in a specific cultural and scientific environment and his analysis depended on the previous mathematical attempts to demonstrate the fifth postulate. The failure of the demonstrations of his predecessors induced Lobachevsky to believe that the difficulties that had to be overcome were due to causes other than those which had until then been focused on).

We simply can assume that many of the internal visualizations of the working geometers of the past were spatial and imaginary because those mathematicians were precisely operating with diagrams and visualizations. By using internal representations Lobachevsky has to create new external visualizations and to adjust them tweaking and manipulating (Trafton, Trickett, and Mintz, forthcoming) the previous ones in some particular ways to generate appropriate spatial transformations (so-called *geometrical constructions*).[12] In cognitive science many kinds of spatial transformations have been studied, like mental rotation and any other actions to improve and facilitate the understanding and simplification of the problem. It can be said that when a spatial transformation is performed on external visualizations, it is still generating or exploiting an internal representation.

Spatial transformations are used to create new internal/mental images and subsequently new related internal/mental images. Lobachevsky mainly takes advantage of the transformation of external diagrams to create and modify the subsequent internal image. So mentally manipulating both external diagrams and internal representations is extremely important for the geometer that uses both the drawn geometrical figure and his own mental representation. An active role of these external representations as *epistemic mediators* of discoveries can be curiously seen also beyond Euclid, at the beginning of

[12] We maintain that in general spatial transformations are represented by a visual component and a spatial component (Glasgow and Papadias, 1992, Magnani, 2001a, chapter 5).

modern mathematics, when new abstract, imaginary, and counterintuitive non-Euclidean notions are invented and developed.

In general, during the comparison the scientists are looking for discrepancies and anomalies, but also equivalences and coherent shapes (like in the case of geometers, as we will see below). The comparison between the transformations acted at the level of external representations and their "internal" counterpart forces the geometer to merge or to compare the two sides (some aspects of the diagrams correspond to information already represented internally at a symbolic propositional level). Usually scientists try to determine *identity*, when they make a comparison to determine the identity of one of the objects; *alignment*, when they are trying to determine an estimation of fit of one representation to another (e.g. visually inspecting the fit of a rough mental triangular shape to an external constructed triangle); and *feature comparison*, when they compare two things in terms of their relative features and measures (size, shape, color, etc.) (Trafton, Trickett, and Mintz, forthcoming).

External geometrical diagrams activate perceptual operations, such as searching for objects that have a common shape and inspecting whether three objects lie on a straight line. They have permanent and invariant geometrical information that can be immediately perceived and kept in memory without the mediation of deliberate inferences or computations, such as whether some configurations are spatially symmetrical to each other and whether one group of entities has the same number of entities as another one. Internal operations prompt other cognitive operations, such as making calculations to get or to envision a result. Moreover, internal representations may have information that can be directly retrieved, such as the relative magnitude of angles or areas.

## 4.2 Mirror diagrams and the infinite

As previously illustrated the failure of the demonstrations of his predecessors induced Lobachevsky to believe that the difficulties that had to be overcome were due to causes other than those which had until then been focused on.

Lobachevsky was obliged first of all to rebuild the basic principles: to this end, it was necessary to consider geometrical principles in a new way, as neither ideal nor *a priori*. New interrelations were created between two areas of knowledge: Euclidean geometry and the philosophical tradition of empiricism/sensualism. From this perspective the abductive explanation of the basic concepts of any science is in terms of senses: the basic concepts are always acquired through our *sense impressions*. Geometry is built upon the concepts of body and bodily contact, the latter being the only "property" common to all bodies that we ought to call geometrical. The well-known concepts of depthless surface, widthless line and dimensionless point were constructed considering different possible kinds of bodily contact and avoiding, *per abstractionem*, everything but contact itself: these concepts "exist only in our representation; whereas we actually measure surfaces and lines by means of bodies" for "in nature there are neither straight lines nor curved lines, neither plane nor curved surfaces; we find in it only bodies, so that all the rest is created by our imagination and exists just in the realm of theory" (Lobachevsky, 1897, p. 16). The only thing that we can know in nature is movement "without which sense impressions are impossible. Consequently all other concepts, e.g. geometrical concepts, are generated artificially by our understanding, which derives them from the properties of movement; this is why space in itself and by itself does not exist for us" (Lobachevsky, 1897, p. 9).

On the basis of these foundations Lobachevsky develops the so-called absolute geometry, which is independent of the fifth postulate:

> Instead of commencing geometry with the plane and the straight line as we do ordinarily, I have preferred to commence it with the sphere and the circle, whose definitions are not subject to the reproach of being incomplete, since they contain the generation of the magnitudes which they define (Lobachevsky, 1929, p. 361).

The basic unit is the manipulation of diagrams. Before the birth of the axiomatic method the geometers still and strongly have to exploit external diagrams, to enhance their thoughts. It is impossible to mental imaging and evaluating the alternative sequences of symbolic calculations being only helped by the "analytic" tools, such as various written equations and symbols and marks: it is impossible to do a complete anticipation of the possible outcomes, due to the limited power of working memory and attention. Hence, because of the complexity of the geometrical problems space and the limited power of working memory, complete mental search is impossible or difficult. Geometer may use perceptual external biases to make decisions.

Moreover, in those cognitive settings, lacking in axiomatic theoretical awareness, certainly perceptual operations were epistemic mediators which needed less attentional and working memory resources than internal operations:

> The directly perceived information from external representations and the directly retrieved information from internal representation may elicit perceptual and cognitive biases, respectively, on the selections of actions. If the biases are inconsistent with the task, however, they can also misguide actions away from the goal. Learning effect can occur if a task is per-

formed more than once. Thus, the decision on actions can also be affected by learned knowledge (Zhang, 1997, p. 186).

The external diagram (for example the diagram of the drawn parallel lines - cf. Figure 5) is a kind of *analogous* both of the mental image we depict in the mental visual buffer and of the symbolic-propositional level of the postulate definition. I call this kind of external tool in the geometrical reasoning *mirror diagram*. In general this diagram mirrors the internal imagery and provide the possibility of detecting anomalies. The external representation of geometrical structures often activates direct perceptual operations (for example, identify the parallels and search for the limits) to elicit consistency or inconsistency routines. Sometimes the mirror diagram biases are inconsistent with the task and so they can make the task more difficult by misguiding actions away from the goal. If consistent, they can make the task easier by guiding actions toward the goal. In certain cases the mirror diagrams biases are irrelevant, they should have no effects on the decision of abductive actions, and play lower cognitive roles.

We said that in some cases the mirror diagram plays a negative role and inhibits further creative abductive theoretical developments. As is well-known, Proclus tried to solve the anomaly by proving the fifth postulate from the other four. If we were able to prove the postulate in this way, it would become a theorem in a geometry which does not require that postulate (the future "absolute geometry") and which would contain all of Euclid's geometry. We need only remember that the argument seemed correct because it was proved using a diagram. Nevertheless, in this example the mirror diagram biases are consistent with the task of *justifying* Euclidean geometry and make this task easier by guiding actions toward the goal.[13] But it inhibits the discovery of non-Euclidean geometries.

Furthermore, we will see (cf. below) that external representations like the ones I call *unveiling diagrams* can enhance the consistency of a cognitive process but also provide more radically creative suggestions for new useful information and discoveries. In the case of mirror diagrams the coordination between perception and cognition is central, from both static and dynamic (constructions) points of view; in the case of unveiling diagrams allocating and switching attention between internal and external representation govern the reasoning strategy, by integrating internal and external representation in a more dynamical and complicated way.

Let us illustrate how Lobachevsky continues to develop the absolute geometry. The immediate further step is to define the concept of plane, which is defined as the geometrical locus of the intersections of equal spheres described around two fixed points as centers, and, immediately after, the con-

[13] I illustrated this cognitive and epistemological case in Magnani, 2001b, chapter 7.

cept of straight line (for example BB' in the Figure 3) as the geometrical locus of the intersections of equal circles, all situated in a single plane and described around two fixed points of this plane as centers. The straight line is so defined by means of "finite" parts (segments) of it: we can prolong it by imaging a repeatable movement of rotation around the fixed points (cf. Figure 3, Lobachevsky, 1835-1838, § 25).

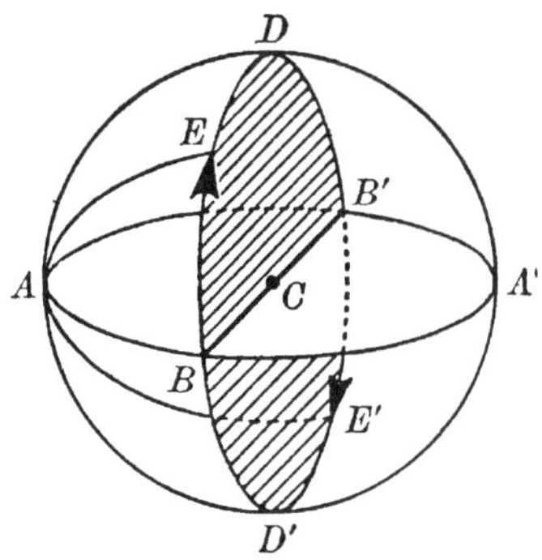


*Figure 3.*

Rectilinear angles (which express arcs of circle) and dihedral angles (which express spherical lunes) are then considered; and the solid angles too, as generic parts of spherical surfaces - and in particular the interesting spherical triangles. $\pi$ means for Lobachevsky the length of a semicircumference, but also the solid angle that corresponds to a semisphere (straight angle). The surface of the spherical triangles is always less than $\pi$, and, if $\pi$, coincides with the semisphere. The theorems about the perpendicular straight lines and planes also belong to absolute geometry.

We have to note some general cognitive and epistemological aspects which characterize the development of this Lobachevskyan absolute geometry.

1. Spherical geometry is always treated together with the plane geometry: the definitions about the sphere are derived from the ones concerning the plane when we substitute the straight lines (geodetics in the plane) with the maximal circles (geodetics in the spherical surface). Lobachevsky says that the maximal circle on the sphere with respect to the other circles presents "properties" that are "similar" to the ones belonging to the straight line with respect to all the segments in the plane (Lobachevsky, 1835-1838, § 66). This is an enhancement, by means of a kind of *analogical* reasoning, reinforced by the external mirror diagrams, of the internal representation of the concept of straight line. The straight line

can be in some sense thought (because it is "seen" and "imagined" in the various configurations provided by the external diagrams) as "belonging" to various types of surfaces, and not only to the plane. Consequently, mirror diagrams not only manage consistency requirements, they can also act in a creative way, providing new perspectives on old entities and structures. The directly perceivable information strongly guides the discoverer's selections of moves by performing the discovery strategy expansion-of-the-scope (of the concept of straight line). This possibility was not indeed available at the simple level of the internal representation. The Figure 4 (*ibid.*, § 79) is another example of the exploitation of the analogy plane/spherical surface by means of a diagram that exploits the perspective of the two-dimensional flat plane.

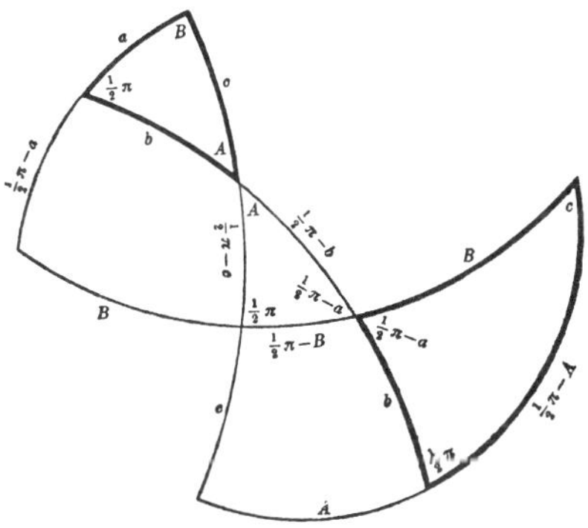

*Figure 4.*

2. In all the previous cases the external representations are constructions that have to respect the empirical attitude described above: because of the fact that the geometrical bodies are characterized by their "finiteness" the external representation is just a mirror of coherent internal images. The "infinite" can be perceived in the "finite" constructions because the infinite is considered only as something potential that can be just mentally and artificially thought: "defined artificially by our understanding". As the modern axiomatic method is absent, the geometer has to conceptualize infinite situations exploiting the finite resources offered by diagrams. In front of the question: "How is it that the finite human resources of internal representations of human mind can conceptualize and formalize abstract notion of infinity?" - notions such as the specific ones embedded in the non-Euclidean assumptions - the geometer is aware we perceive a finite world, act upon it, and think about it. Moreover, the geometer operates in "a combination of perceptual *input*, physical *output*, and internal mental processes. All three are finite. But by thinking

> about the possibility of performing a process again and again, we can easily reach out towards the potential infinite" (Tall, forthcoming) Lobachevsky states: "Which part of the lines we would have to disregard is arbitrary", and adds, "our senses are deficient" and it is only by means of the "artifice" consisting of the continuum "enhancement of the instruments" that we can overcome these limitations (Lobachevsky, 1835-1838, § 38). Given this epistemological situation, it is easy to conclude saying that *instruments* are not just and only telescopes and laboratory tools, but also diagrams.

Let us continue to illustrate the geometer's inventions. In the proposition 27 (a theorem already proved by Euler and Legendre) of the *Geometrical Researches of the Theory of Parallels* (1891[1840]), Lobachevsky states that if $A$, $B$, and $C$ are the angles of a spherical triangle, the ratio of the area of the triangle to the area of the sphere to which it belongs will be equal to the ratio of

$$\tfrac{1}{2}(A + B + C - \pi)$$

to four right angles; that the sum of the three right angles of a rectilinear triangle can never surpass two right angles (Prop. 19), and that, if the sum is equal to two right angles in any triangle, it will be so in all (Prop. 20).

The new definition of parallelism (Prop. 16) introduces the non-Euclidean atmosphere:

> All straight lines which in a plane go out from a point can, with reference to a given straight line in the same plane, be divided in two classes - into *cutting* and *not-cutting*. The *boundary lines* of the one and the other class of those lines will be called *parallel to the given lines* (cit., p. 13).

The external representation is easily constructed like in Figure 5 (p. 13) where the angle $HAD$ between the parallel $HA$ and the perpendicular $AD$ is called the angle of parallelism, designated by $\Pi(p)$ for $AD = p$. If $\Pi(p)$ is $<\tfrac{1}{2}\pi$, then upon the other side of $AD$, making the same angle $DAK = \Pi(p)$ will lie also a line $AK$, parallel to the prolongation $DB$ of the line $DC$, so that under this assumption we must also make a distinction of sides in parallelism. Because of the fact that the diagrams can contemplate only finite parts of straight lines it is easy to represent this new postulate in a mirror image: we cannot know what happens at the infinite nor in the internal representation (because of the limitations of visual buffer) not in the external representation: "in the uncertainty whether the perpendicular $AE$ is the only line which does not meet $DC$, we will assume it may be possible that there are still other lines, for example $AG$, which do not cut $DC$, how far so ever they may be prolonged" (*ibid.*). So the mirror image can support the new non-Euclidean perspective. The idea of constructing an external diagram of a non-Euclidean situation is considered normal and reasonable.

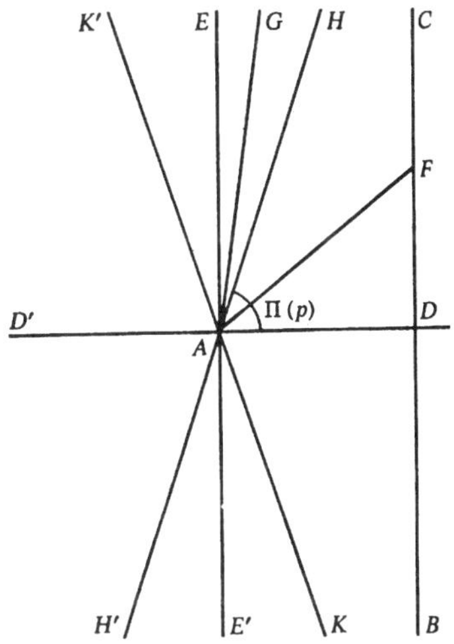


*Figure 5.*

## 4.3 Unveiling diagrams: gateways to imaginary entities

The first step toward the exploitation of what I call *unveiling diagrams* is the use of the notion of non-Euclidean parallelism at the stereometric level, by establishing relationships between straight lines and planes and between planes: Proposition 27 (already proved by Lexell and Euler): "*A three-sided solid angle equals the half sum of surface angles less a right-angle*" (p. 24). Proposition 28, directly derived from Prop. 27: "*If three planes cut each other in parallel lines, then the sum of the three surface angles equals two rights*" (p. 28), cf. Figure 6.

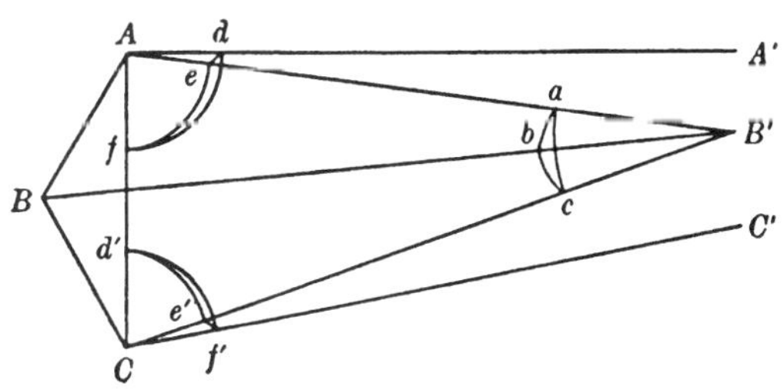


*Figure 6.*

These achievements are absolutely important: it is established that for a certain geometrical configuration of the new geometry (the three planes cut each other in parallel lines that are parallel in Lobachevskyan sense) some properties of the ordinary geometry hold.

The important notions of oricycle and orisphere are now defined to furnish a possible symbolic counterpart able to express a foreseen consistency (as a justification) of the non-Euclidean theory with respect to Euclidean geometry. This consistency is looked at from the point a view of a possible "analytic" solution, that is in terms of verbal-symbolic[14] results.

Such is the case of the Proposition 31. "We call boundary line (oricycle) that curve lying in a plane for which all perpendiculars erected at the mid-points of chords are parallel to each other. [...] The perpendicular *DE* erected upon the chord *AC* at its mid-point *D* will be parallel to the line *AB*, which we call the *Axis of the boundary line*" (pp. 30-31), cf. Figure 7. Proposition 34. "Boundary surface[15] we call that surface which arises from the revolution of the boundary line about one of its axes, which, together with all other axes of the boundary-line, will be also an axis of the boundary surface" (p. 33). Moreover, the intersections of the orisphere by its diametral planes are limit circles. The limit circle arcs are called the sides, and the dihedral angles between the planes of these arcs the angles of the "oriphere triangle".

A part of the surface of the orisphere bounded by three limit circle arcs will be called an orisphere triangle. From Prop. 28 follows that the sum of the angles of an orisphere triangle is always equal to two right angles: "everything that is demonstrated in the ordinary geometry of the proportionality of the sides or rectilinear triangles can therefore be demonstrated in the same manner in the pangeometry"[16] (Lobachevsky, 1929, p. 364) of the orisphere triangles if only we will replace the lines parallel to the sides of the rectilinear triangle by orisphere arcs drawn through the points of one of the sides of the orisphere triangle and all making the same angle with this side. To conclude, the orisphere is a "partial" model of Euclidean plane geometry.

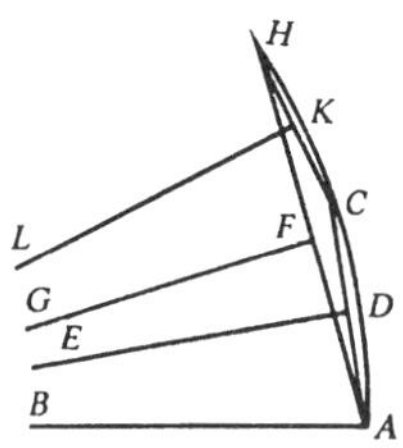


*Figure 7.*

[14] That is not diagrammatic.

[15] Also called limit sphere or orisphere.

[16] Lobachevsky called the new theory "imaginary geometry" but also "pangeometry".

The last constructions of the Lobachevskyan abductive process give rise to two unveiling diagrams (cf. Figures 8 and 10) that accompany the remaining proofs. They are more abstract and exploit "audacious" representations in the perspective of three dimensional geometrical shapes.

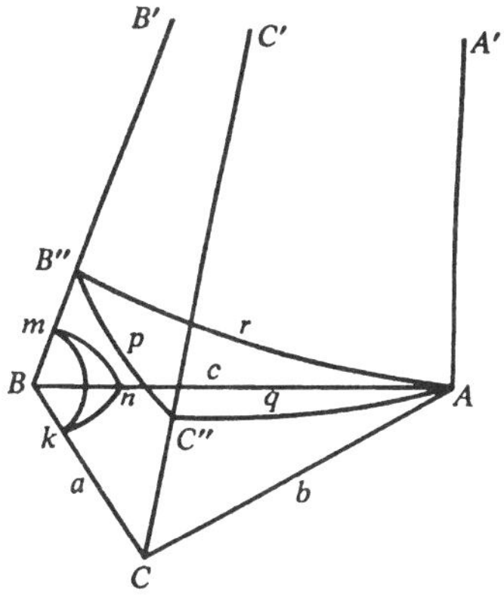


*Figure 8.*

The construction given in Figure 8 aims at diagrammatically "representing" a stereometric non-Euclidean form built on a rectilinear right angled triangle $ABC$ to which the Theorem 28 above (the parallels $AA'$, $BB'$, $CC'$, which lie on the three planes are parallels in non-Euclidean sense) can be applied so that Lobachevsky is able to further apply symbolic identifications; the planes make with each other the angles $\Pi(a)$ at $AA'$, a right angle at $CC'$, and, consequently $\Pi(a')$ at $BB'$[17]: in the same figure we can construct a spherical triangle *mnk*, in which the sides are $mn=\Pi(c)$, $kn=\Pi(\beta)$, $mk=\Pi(a)$ and the opposite angles are $\Pi(a)$, $\Pi(\alpha')$, $\frac{1}{2}\pi$ realizing that with the "existence" of a rectilinear triangle with the sides $a$, $b$, $c$ (like in the case of the previous one) "we must admit" the existence of a *related* spherical triangle (Figure 9), etc. Not only, a *boundary surface* (orisphere) can be constructed, that passes through the point $A$ with $AA'$ as axis, and those intersections with the planes the parallels form a boundary triangle (that is a triangle situated upon the given orisphere), whose sides are $B''C''= p$, $C''A = q$, $B''A = r$ and the angles opposite to them $\Pi(\alpha)$, $\Pi(\alpha')$, $\frac{1}{2}\pi$ and where consequently (this follows from the Theorem 34):

$$p = r \sin \Pi(a),\ q = r \cos \Pi(a).$$

[17] Given that Lobachevsky designates the size of a line by a letter with an accent added, e.g. $x'$, in order to indicate this has a relation to that of another line, which is represented by the same letter without the accent $x$, "which relation is given by the equation $\Pi(x) + \Pi(x') = \frac{1}{2}\pi$" (Prop. 35).

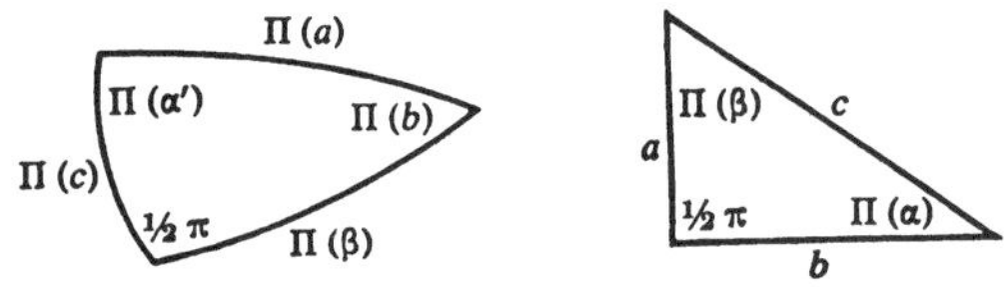


*Figure 9.*

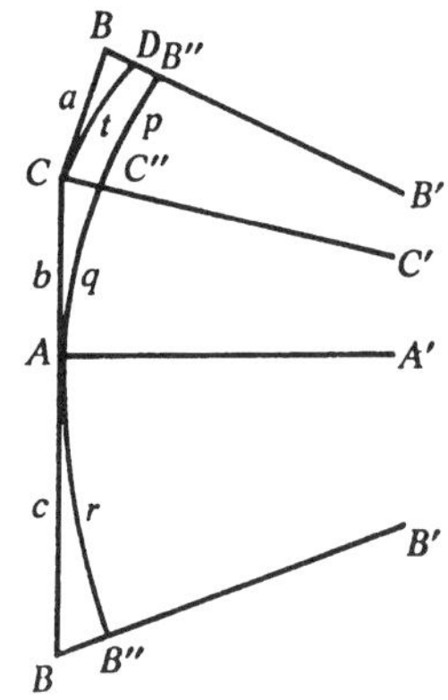


*Figure 10.*

Inside the perspective representations (given by the unveiling diagram of of a *non-Euclidean structure*, cf. Figure 8) an *Euclidean spherical triangle* and the *orisphere* (and its boundary triangle where the Euclidean properties hold) are constructed. The directly perceivable information strongly guides the geometer's selections of moves by eliciting the *Euclidean-inside non-Euclidean* "model matching strategy". This maneuver also constitutes an important step in the affirmation of the modern "scientific" concept of model. We have to note that other perceptions activated by the diagram are of course disregarded as irrelevant to the task, as it usually happens when exploiting external diagrammatic representations in reasoning processes. Because not everything in external representations is always relevant to a task, high level cognitive mechanisms need to use task knowledge (usually supplied by task instructions) to direct attention and perceptual processes to the relevant features of external representations.

This external representation in terms of an unveiling diagram activates a perceptual reorientation in the construction (that is identifies possible further constructions and enhancements); in the meantime the generated internal representation of the external elements activates directly retrievable information (numerical values) that elicit the strategy of building further non-Euclidean structures together with their *analytic* counterpart (cf. below the non-Euclidean trigonometry equations). Moreover, the internal representation of the stereometric figures activates cognitive operations related to the *consistency-searching strategy*. In this process, new "imagery" and strange mathematical entities, like the oricycle and the orisphere, are abduced and *unveiled*, and related to ordinary and unsuspected perceptive entities.

Finally, it is easy to identify in the proof the differences between perceptual and cognitive operations and the differences between *sequential* - the various steps of the constructed unveiling diagram - and *parallel* perceptual operations. Similarly, it is easy to distinguish between the forms that are directly perceptually inspected and the elements that are mentally computed or computed in external symbolic configurations.

To arrive to the second *unveiling* diagram the old diagram is further enhanced by a new construction, by breaking the connection of the three principal planes along the line $BB'$, and by turning them out from each other so that they, together with all the lines lying in them, come to lie in one plane, where consequently the arcs $p$, $q$, $r$ will unite to a single arc of a boundary-line (oricycle), which goes through the point $A$ and has $AA'$ as its axis, in such a manner that (Figure 10) on the one side will lie, the arcs $q$ and $p$, the side $b$ of the triangle, which is perpendicular to $AA'$ at $A$, the axis $CC'$ going from the end of $b$ parallel to $AA'$ and through $C''$ the union point of $p$ and $q$, the side a perpendicular to $CC'$ at the point $C$, and from the end-point of $a$ the axis $BB'$ parallel to $AA'$ which goes through the end-point $B''$ of the arc $p$, etc. Finally, taking $CC'$ as axis, a new boundary line (an arc of oricycle) from the point $C$ to its intersection with the axis $BB'$ is constructed. What happens?

In this case we see that the external representation looses its spatial intuitive interest and its capacity to simulate internal spatial representations. It is not useful to represent it as an internal spatial model in order to enhance the problem solving activity. The diagram of Figure 10 does not have to depict spatial internal coherent forms, it is just devoted to "unveil" the possibility of further calculations by directly activating "perceptual" information that, in conjunction with the non-spatial information and cognitive operations provided by internal representations in memory, determines the subsequent problem solving behavior. This diagram does not have to prompt an internal "spatially coherent" model. Indeed *perception* often plays an autonomous and central role, it is not a peripheral device. In this case the end product of perception coincides with the intermediate data highly analyzed, processed, and transformed, that is prepared for high-level cognitive mechanisms in terms of further *analytic* achievements (the equations).[18]

[18] In other problem solving cases, the end product of perception - directly picked-up - is the end product of the whole problem solving process.

We have to note that of course it cannot be said that the external representation would work independently without the support of anything internal or mental. The mirror and unveiling diagrams have to be processed by perceptual mechanism that are of course internal. And in this sense the end product of the perceptual mechanisms is also internal. But it is not an internal model of the external representation of the task: the internal representation is the knowledge and structure of the task in memory; and the external representation is the knowledge and structure of the task in the environment. The end product of perception is merely the situational information in working memory that usually only reflects a fraction (crucial) of the external representation (Zhang, 1997).

At this point it is clear that the perceptual operations generated by the external representations "mediated" by the unveiling diagrams are central as mechanisms of the whole geometrical abductive and manipulative process; they are not less fundamental than the cognitive operations activated by internal representations, in terms of images and/or symbolic-propositional. They constitute an extraordinary example of complex and perfect coordination between perceptual and cognitive operations.

Let us conclude the survey on Lobachevsky's route to an acceptable assessment of its non-Euclidean theory. By means of further *symbolic/propositional* designations taken from both internal representations followed from previous results and "externalized" calculations, the reasoning path is constrained to find a general "analytic" counterpart for (some aspects of) the non-Euclidean geometry (we skip the exposition of this complicated passage - cf. Lobachevsky, 1891). Therefore we obtain the equations

$$\sin \Pi (c) = \sin \Pi (a) \sin \Pi (c),$$
$$\sin \Pi (\beta) = \cos \Pi (\alpha) \sin \Pi (\alpha)$$

Hence we obtain, by mutation of the letters,

$$\sin \Pi (\alpha) = \cos \Pi (\beta) \sin \Pi (b),$$
$$\cos \Pi (b) = \cos \Pi (c) \cos \Pi (\alpha)$$
$$\cos \Pi (a) = \cos \Pi (c) \cos \Pi (\beta),$$

that express the mutual dependence of the sides and the angles of a non-Euclidean triangle. In these equations of plane non-Euclidean geometry we can pass over the equations for spherical triangles. If we designate in the right-angled spherical triangle (Figure 9) the sides $\Pi(c)$, $\Pi(\beta)$, $\Pi(a)$, with the opposite angles $\Pi(b)$, $\Pi(\alpha')$, by the letters $a$, $b$, $c$, $A$, $B$, then the obtained equations take of the form of those which we know as the equations of spherical trigonometry for the right-angled triangle

$$\sin(a) = \sin(c)\sin(A)$$
$$\sin(b) = \sin(c)\sin(B)$$
$$\cos(A) = \cos(A)\sin(B)$$
$$\cos(B) = \cos(B)\sin(A)$$
$$\cos(c) = \cos(a)\cos(b)$$

The equations are considered to "attain for themselves a sufficient foundation for considering the assumption of imaginary geometry as possible" (p. 44). The new geometry is considered exempt from possible inconsistencies together with the acknowledgment of the reassuring fact that it presents a very complex system full of surprisingly harmonious conclusions. A new contradiction which could have emerged and which would have forced to reject the principles of the new geometry would have been already contained in the equations above. Of course this is not true from the point of view of modern deductive axiomatic systems and a satisfactory model of the non-Euclidean geometry has not yet been built (as Beltrami and Klein will do with the so-called "Euclidean models of non-Euclidean geometry").[19] As for now the argument rests on a formal agreement between two sets of equations, one of which is derived from the new non-Euclidean geometry. Moreover, the other set of equations does not pertain to the Euclidean geometry; rather they are the equations of spherical trigonometry which does not depend on the fifth postulate (as maintained by Lobachevsky himself). Nevertheless, we can conclude that Lobachevsky is not far from this idea of model.

## 5. CONCLUSION

The mirror and unveiling diagrams we have described provide a better understanding of the discovery of the elementary non-Euclidean geometry. They improve and complete the epistemological and cognitive analysis of this important scientific revolution.

Moreover: i) the role of other diagrams (called *optical*) in the calculus and non-standard analysis[20] and in a calculus teaching environment seems relevant. I am preparing a research devoted to detect the details of their didactic effects on the calculus students at the University of Pavia (mathemat-

[19] On the limitations of the Lobachevskyan perspective cf. Torretti, 1978, Rosenfeld, 1988, Boi, 1995.

[20] Cf. footnote 11, above.

ics and engineering curricula); but I am also convinced they can be exploited and studied in everyday non-mathematical applications (finding routes, road signs, buildings maps, for example), in connection with various zooming effects of spatial reasoning; ii) I think the cognitive activities of optical, mirror, and unveiling diagrams can be studied in other areas of model-based reasoning, such as the ones involving creative, analogical, and spatial inferences, both in science and everyday situations so that this can extend the epistemological and the psychological theory.

## REFERENCES

Boi, L., 1991, *Le probléme mathématique de l'espace. Une quête de l'intelligible*, Springer, Berlin.

Boutilier, C. and Becher, V., 1995, Abduction as belief revision, *Artificial intelligence*, 77:43-94.

Call, J., 2000, Representing space and objects in monkeys and apes, *Cognitive Science* 24(3):397-422.

Giere, R., 2002, Models as parts of distributed cognitive systems, in this volume, pp. 227-241.

Glasgow, J.I. and Papadias, D., 1992, Computational imagery, *Cognitive Science* 16:355-394.

Hoffmann, M., 1999, Problems with Peirce's concept of abduction, *Foundations of Science* 4(3): 271-305.

Hoffmann, M., forthcoming, How to get it. Diagrammatic reasoning as a tool of knowledge development and its pragmatic dimension, in: Special Issue *Abductive Reasoning in Science, Foundations of Science*, L. Magnani and N.J. Nersessian, eds.

Hutchins, E., 1995, *Cognition in the Wild*, MIT Press, Cambridge, MA.

Kant, I., 1929, *Critique of Pure Reason* (1781-1787), translation by N. Kemp Smith, MacMillan, New York, reprint 1998.

Lobachevsky, N.J., 1829-1830, 1835-1838, *Zwei geometrische Abhandlungen, aus dem Russischen bersetzt, mit Anmerkungen und mit einer Biographie des Verfassers von Friedrich Engel*, B.G. Teubner, Leipzig. Originally published as O nachalakh geometrii, *Kasanki Vestnik*, Feb.-March, 1829: 178-187; April, 1829: 228-241; Nov.-Dec., 1829: 227-243; March-April, 1830: 251-283; July-Aug., 1830: 571-636, and Novye nachala geometrii, *Uchonia sapiski Kasanskava Universiteta* 3, 1835: 3-48; 2, 1836: 3-98; 3, 1836: 3-50; 1, 1837: 3-97; 1, 1838: 3-124; 3, 1838: 3-65.

Lobachevsky, N.J., 1891, *Geometrical Researches on the Theory of Paralleles* [1840], translated by G.B. Halsted, University of Texas, Austin.

Lobachevsky, N.J., 1897, The introduction to Lobachevsky's *New Elements of Geometry* [1835], translation from the Russian, with a preface by G.B. Halsted, *Transactions of Texas Academy* 2:1-17. Originally published in: N.J. Lobachevsky, Novye nachala geometrii, *Uchonia sapiski Kasanskava Universiteta 3*, 1835: 3-48.

Lobachevsky, N.J., 1929, *Pangeometry or a Summary of Geometry Founded upon a General and Rigorous Theory of Parallels*, in: *A Source Book in Mathematics*, D.E. Smith, ed., McGraw Hill, New York, pp. 360-374.

Magnani, L., 1992, Abductive reasoning: philosophical and educational perspectives in medicine, in: *Advanced Models of Cognition for Medical Training and Practice*, D.A. Evans and V.L. Patel, eds., Springer, Berlin, pp. 21-41.

Magnani, L., 1999a, Creative abduction and hypothesis withdrawal in science, in: *Methodological Aspects of Discovery and Creativity*, J. Meheus and T. Nickles, eds., Kluwer, Dordrecht, forthcoming.

Magnani, L., 1999b, Model-based creative abduction, in: L. Magnani, N.J. Nersessian, and P. Thagard, eds., pp. 219-238.

Magnani, L., 2000, Action-based abduction in science, in: Workshop Scientific Reasoning in Artificial Intelligence and Philosophy of Science, *ECAI 2000 Workshop Notes*, A. Aliseda and D. Pearce, eds., Berlin, pp. 46-51.

Magnani, L., 2001a, *Abduction, Reason, and Science. Processes of Discovery and Explanation*, Kluwer Academic/Plenum Publishers, New York.

Magnani, L., 2001b, *Philosophy and Geometry. Theoretical and Historical Issues*, Kluwer Academic, Dordrecht.

Magnani, L., N.J. Nersessian, and Thagard, P., eds., 1999, *Model-Based Reasoning in Scientific Discovery*, Kluwer Academic/Plenum Publishers, New York.

Magnani, L. and Dossena, R., 2002, Perceiving the infinite and the infinitesimal world: unveiling and optical diagrams and the construction of mathematical concepts, submitted to CogSci2002.

Nersessian, N.J., 1995a, Opening the black box: cognitive science and history of science. Technical Report GIT-COGSCI 94/23, July. Cognitive Science Report Series, Georgia Institute of Technology, Atlanta, GA. Partially published in: *Osiris* 10:194-211.

Nersessian, N.J., 1995b, Should physicists preach what they practice? Constructive modeling in doing and learning physics, *Science and Education* 4:203-226.

Nersessian, N.J., 1998, Kuhn and the cognitive revolution, *Configurations* 6:87-120.

Nersessian, N.J., 1999, *Model-based reasoning in conceptual change*, in: L. Magnani, N.J. Nersessian, and P. Thagard, eds., pp. 5-22.

Newell, A., 1990, *Unified Theories of Cognition*, Harvard University Press, Cambridge, MA.

Norman, D.A., 1993, *Things that Make Us Smart. Defending Human Attributes in the Age of the Machine*, Addison-Wesley, Reading, MA.

Peirce, C.S., 1931-1958 (*CP*), *Collected Papers*, 8 vols., C. Hartshorne and P Weiss (vols. I-VI), eds., and A.W. Burks (vols. VII-VIII), ed., Harvard University Press, Cambridge, MA.

Piaget, J., 1974, *Adaptation and Intelligence*, University of Chicago Press, Chicago, IL.

Ramoni, M., Stefanelli, M., Magnani, L., and Barosi, G., 1992, An epistemological framework for medical knowledge-based systems, *IEEE Transactions on Systems, Man, and Cybernetics* 22(6):1361-1375.

Rosenfeld, B.A., 1988, *A History of Non-Euclidean Geometry. Evolution of the Concept of a Geometric Space* (1976), Springer, Berlin.

Tall, D., forthcoming, Conceptual and formal infinities, *Educational Studies in Mathematics.*

Thagard, P., 1992, *Conceptual Revolutions*, Princeton University Press, Princeton, NJ.

Torretti, R., 1978, *Philosophy of Geometry from Riemann to Poincaré*, Reidel, Dordrecht.

Trafton, J.G., Trickett, S.B., and Mintz, F.E., forthcoming,, Connecting internal and external representations: spatial transformations of scientific visualizations, in: Special Iissue *Model_Based Reasoning: Visual, Analogical, Simulative*, *Foundations of Science*, L. Magnani and N.J. Nersessian, eds.

Zhang, J., 1997, The nature of external representations in problem solving, *Cognitive Science* 21(2):179-217.

Zhang, J. and Norman, D.A., 1994, Representations in distributed cognitive tasks, *Cognitive Science* 18:87-122.

# Deterministic Models and the "Unimportance of the Inevitable"

Claudio Pizzi
*Department of Philosophy and Social Sciences, University of Siena, Siena, Italy, pizzic@unisi.it.*

*Abstract*: In some appreciated essays of the 1980s Jon Elster threw light on the conceptual difficulty which he described as "*the basic paradox of counterfactuals*": the stronger (more deterministic) is the connection between antecedent and consequent of a counterfactual, the weaker is the legitimacy of the antecedent. The aim of the present paper is to show that Elster's thesis rests on several confusions, the most important of which is the one between a theory and a model. When a model is deterministic, only several theoretical extra assumptions allow to qualify a counterfactual supposition as an illegitimate one. In particular, it is argued that only special auxiliary assumptions may allow or forbid drawing non trivial conclusions from suppositions which are "counter-possible"- i.e. such as to deny something which the theory asserts to be necessary - or "counterlegal", i. e. implying the falsity of a law belonging to the model itself.

## 1. INTRODUCTION

Proposed by the author as an essay in the methodology of social sciences, Jon Elster's *Logic and Society* (1978; henceforth LS) soon turned out to be a major source of stimulating reflections for modal logicians and epistemologists. Other Elster's works such as *Explaining Technical Change* (1983; henceforth ETC) contain efficacious integrations of the views exposed in the 1978 book. Chapter VII of *Logic and Society,* "Counterfactuals and the new economical history", contains a critical analysis of various attempts to apply the counterfactual notion of a cause to the history of economics and of tech-

*Model-Based Reasoning: Science, Technology, Values*, edited by
L. Magnani and N.J. Nersessian, Kluwer Academic/Plenum Publishers, New York, 2002

nological innovations. R.W. Fogel, E. Genovese, N. Rosenberg[1] are some of the social methodologists who receive the greatest attention in Elster's investigations. Elster is critical towards the *conditio sine qua non* theories of causation, namely towards the idea that the notion of causation is *definable* in terms of counterfactual statements. He finds that such theories are unable to solve the well known problems of epiphenomena and overdetermination[2]. This handicap, however, in his view does not reduce the importance of counterfactuals in scientific reasoning: "*although counterfactual statements do not exhaust the meaning of causal statements, they can offer true non-definitional characterizations of causal relations in many central cases*" (ETC ,p. 36).

It may be useful to offer a short synthesis of Elster's position on the subject we are concerned with:

1) The basic premise of Elster's methodology is in line with the postpositivistic trend of the 1980s: social scientists always reason in the context of a background theory which fixes the "rules of the game" and thus cannot be abandoned once it is subscribed. "*Max Weber says somewhere that historical materialism is not like a taxi one may enter or leave at will; and the same holds, I believe, for theories in this context*" (1980, pp. 144-7).

2) A possible world is a *vector*, namely a finite configuration of values of the variables admitted by a theory. It is not a full-blooded entity - for instance a set of counterparts or a maximal consistent set - as is currently assumed in modal semantics. "*A possible world is a mere skeleton of a state, as described by a finite number of variables*" (LS, p. 187).

3) A process P is described by a theory T at every moment of time thanks to a specification of vectors (i.e. of possible worlds). We say that a theory T *explains* P when:

    a) The process P is at each moment of discrete time described exhaustively through the variables $x_1 \ldots x_n$ ($\mathbf{x}$ in vectorial notation).
    b) Theory T states that

[1] See bibliographical references.

[2] The problem of epiphenomena - or otherwise stated - the problem of spurious correlation-arises when *C* is sufficient for *A* and necessary for *B*: in this case the counterfactual having not-*A* as antecedent and not-*B* as a consequent is true, but the causal judgment "*A* caused *B*" is false. As a standard example, think of *C* as an episode of atmospheric depression, *A* as the following lowering of the barometer and *B* as the following storm. As regards overdetermination, let us recall the standard example of two killers jointly firing two mortal shots at same victim: they are both sufficient conditions for the death of the victim, but not necessary conditions for it.

i) for every variable $x_i$ there is a set $X_i$ of values of $x$ which are singularly admissible.

ii) there is a set $X$ of jointly admissible values of the variables, where $X$ is a subset of the Cartesian product $X_1 \ldots X_{n.}$ (*static compatibility*).

iii) for each admissible vector **x**($t$) (namely **x** at the moment $t$) there is at least one vector **y** such that T excludes that the process can be represented by **y** at $t + 1$ (*dynamic compatibility*).

iv) the sequence of vectors describing P is compatible with the requirements bi)-biii).

If the aim of knowledge is assumed to be not exclusion (as postulated in iii)) but determination, the theory is deterministic, in the sense that it excludes all the vectors except one as possible values of **x** ($t + 1$).

4) A dynamic counterfactual has the following form: "If the configuration of values at $t$ had been $A_1 \ldots A_{n-k}$, $B_1 \ldots B_k$ then at $t + t'$ the subset $x_{i1}, x_{j2} \ldots x_{ip}$ of the variables would have assumed the values $C_1 \ldots C_p$.

5) According to Elster, counterfactuals lack truth conditions but have assertibality conditions. "*The grounds or conditions for assertability are our currently accepted scientific theories, i.e. currently accepted lawlike universal statements. A counterfactual statement is warranted or assertable if the consequent can be deduced from the antecedent together with some suitably chosen theoretical statements*".[3] In assertable counterfactuals then the consequent follows from the antecedent conjoined with premises representing minimal changes with respect to the actual world, while the derivation is granted by using the laws of the theory itself as inferential tools.

6) The similarity between worlds (vectors) is to be intended as follows: if **x**($t$) is the actual vector, **y**($t$) a hypothetical vector, the distance between **y**($t$) and **x**($t$) is measured by the changes which are needed to find in the past of **x**($t$) the nearest insertion point in which we find a trajectory which leads to **y**($t$). For instance, to quote the well-known discussion produced by Fogel on the question "Which would have been the American National Product in 1890 if the railroad had not been invented?" one has to go back more or less to 1830 (year of the construction of the first american railroad) in order to find the nearest vector from which a trajectory branches leading to a 1890 without railroads. The distance be-

[3] ETC, p. 37

tween 1830 and 1890 is then a measure of the similarity between the actual vector and the counterfactual one.

7) Not every counterfactual question is a *legitimate* one. Legitimacy may be static or dynamic. Dynamic legitimacy depends on the existence of a trajectory of states which is permitted by the background theory. Let us go back to Fogel's mentioned problem and let us suppose that the question is evaluated in the framework of a deterministic theory of technical change as, for instance, the one of N. Rosenberg. Rosenberg maintains that all the transport technologies which would have been invented as substitutive of the actual ones would have been more or less equal to the actual ones. Since no alternative trajectory can be devised, the question is then an illegitimate one: *"If it were possible to answer the question it should not have been put in the first place*" (LS, p.184). In many cases the illegitimacy arises from the presence of preempting causes. It may happen that the cause *C* of an event F preempted another cause *C'* of *F*: so to suppose the absence of both *F* and *C* leads again to *F* thanks to the preempted cause *C'*.

8) In the light of 7), theories have then a double role: as filters (to select the legitimate suppositions) and as deductive machineries. This double role however creates a difficulty since the two roles may be in conflict (*scissors' problem).* The conflict between the two functions yields *"the basic paradox of counterfactual"*. To use Elster's words: *"the stronger (i.e. the more deterministic) the theory T, the better grounded is the conclusion from the antecedent to the consequent, but the more vulnerable is also the legitimacy of the antecedent"*. This paradox may be described also as the paradox of the *"unimportance of the inevitable"*. *"In order to assess the causal importance of some historical event or social fact, the mental removal of that element must be statically and dynamically legitimate; conversely, any element that cannot be subjected to this thought experiment is inherently unfit to causal assessment"* (LS, p. 185). But a deterministic theory stating that processes are determined amounts to saying that they could not be different from what they actually are and this makes every counterfactual question an illegitimate one. At the limit, in a perfectly Laplacean theory of the universe, no branching point would exist in which to insert counterfactual suppositions.

9) To conclude, counterfactual reasoning is possible only in theories which realize a delicate balance between deductive strength and the branching structure of described processes. This conclusion agrees with Elster's general remark that in social sciences *"instead of using deterministic and*

*almost invariably inexact models, we might prefer to use a non-deterministic and exact one"* (LS, p. 183).

# 2. CRITICAL ASPECTS OF ELSTER'S ANALYSIS

## 2.1 Two sources of confusion: conditionals and events

The notion of legitimacy is crucial for Elster's analysis. Even if the words "legitimate/illegitimate" do not belong to the vocabulary of metalogic, in the realm of logic a supposition $S$ added to a theory $T$ might be said to be illegitimate in two different senses: either $T + S$ yields an inconsistency or it yields a trivialization of some kind.

The idea of trivialization appears to be in agreement with most of the intuitions Elster develops about illegitimacy. Let $T$ be a theory and $F_1...F_n$ a certain number of facts belonging to a process which is explained by T in Elster's sense. Let $F_i$ be one of such facts and $\neg F_i$ a counterfactual supposition. Then we propose to say that the supposition $\neg F_i$ is illegitimate when it forces either to reject $T$ or to reject the totality of facts $F_1...F_n$ explained by $T$. For example, let $T$ be a theory which explains the development of the existing animal species by saying that each one is derived from some other by virtue of some deterministic transition. Then the supposition that some species $s$ in the chain did not exist is illegitimate in the proposed sense since it appears to imply, by Modus Tollens, that all other species which in the chain have been preceding or following $s$ have been non existing. Elster's favorite examples of course are drawn from social sciences. In Mill's view, for instance, all counterfactuals are illegitimate since he assumes that all social facts are holistically interconnected, so to remove one means to remove all others.

In a symposium which in 1980 has been devoted to *Logic and Society*[4] many sincere praises have been accompanied by various criticisms. As an example, Steven Lukes pointed out an ambiguity in Elster's treatment of counterfactuals: Elster "*claims to follow the metalinguistic approach, uses the idiom of the ontological-possible-worlds approach but would, so far as I can see, be better advised in terms of his own intuitions and purposes to adopt, unequivocally, the condensed argument approach*" (p. 148).[5]

[4] See the proceedings in *Inquiry*, 1980.

[5] The condensed argument approach, due essentially to J. Mackie, consists in seeing counterfactuals as telescoped arguments, which as such lack a truth value, while they should be properly qualified as assertable or not assertable.

Too much insistence on this critical point may however introduce some unwanted technical limitations. If it is philosophically mistaken to confuse assertability and truth, it is worth while noticing that the "consequentialist" view endorsed by Elster allows to assign a truth-value to conditionals. In counterfactual reasoning we have to do with laws $L_1 \ldots L_n$ and a number of facts $F_1 \ldots F_m$ which are selected from the Background Knowledge with the minimal loss of information in order to be compatible with the supposition. Let $\mathbf{w}(A)$ be the proposition which stands for the conjunction of the selected presupposed statements. $\mathbf{w}(A)$ might be called the *context* of $A$. Clearly an axiom for such $\mathbf{w}(A)$ is the simple

Ax1) $\Diamond A \supset \Diamond(\mathbf{w}(A) \wedge A)$.

along with a rule granting replacement of material equivalents:

Eqw) $\vdash A \equiv B$ only if $\vdash \mathbf{w}(A) \equiv \mathbf{w}(B)$

Since we clearly have among our theorems

$$\Box((\mathbf{w}(A) \wedge A) \supset A)$$

when $A$ implies conditionally $B$ we may introduce a definition of the conditional by putting

(Def>) $A > B =_{\mathrm{Df}} \Box((\mathbf{w}(A) \wedge A) \supset B) \wedge (\Diamond B \supset \Diamond(\mathbf{w}(A) \wedge A))$.

A counterfactual in explicit form has then the form $A \wedge B \wedge (\neg A > \neg B)$. A useful auxiliary operator may be also defined as follows:

(Def*)$A =_{\mathrm{Df}} \mathbf{w}(A) \wedge A$

The latter operator, sometimes called "circumstantial operator", has then among others the following properties:

$$*A \supset A$$
$$\Diamond A \supset \Diamond *A$$

The conditional symbolized by ">" turns out to be non-monotonic, non-transitive and non-contrapositive just as Stalnaker-Lewis conditional operator. But the logic of > as above defined boils down to the logic of the multi-modal system containing $\Box$ and $\mathbf{w}(\text{-})$, and it may be proved that the resulting system is a bimodal extension of system **KT** of modal logic: so there is no technical obstacle to giving truth values 1,0 to conditionals, as it occurs in standard two-valued modal logic. We will call $\mathbf{KT}^{\mathbf{W}}$ the consistent system which is **KT** extended with Ax1 and Eqw.

A second important point concerns the relation between counterfactual statements and causal statements, which Elster outlines in a way which allows to draw mistaken causal conclusions from true counterfactuals or wrong counterfactuals from correct causal statements. If we look at the way in which Elster defines a dynamic counterfactual, it is easy to see that some correlations among variables which we are able to devise may be transferred into incorrect causal relations. For instance, let $B°=b$ the statement asserting that the actual value of the barometer is $b$, $S°=t$ the statement that the actual value of the storm is $t$ and $D°=d$ a statement about the actual value of a preceding atmospheric depression. The covariation of the values of $B$ and the values of $S$ makes true the counterfactual that if the barometer value had been lower than $B°$, then the intensity of the storm would have been higher than $S°$. Clearly, if we give a causal interpretation to such a counterfactual the resulting correlation turns out to be spurious, and gives Elster a good reason to negate support to the counterfactual theories of causation (see note 2).

But here we find something perplexing. Elster himself says that the causal relation holds among events (see ETC: "The causal relation holds between *events*, by virtue of a regular conjunction between events of that kind" (p. 25)), but a value assignment to variables need not be considered an event, unless we define an event in such a way to allow for such an identification. As a matter of fact, in the case of the example the counterfactual question which we have to face is not

1) If it were not the case that $B° = b$, would it be the case that $T° = t$?

But

2) If the event consisting of lowering the barometer had not occurred, would the event consisting of the storm have occurred?

2) is different from 1) in the very sense that the answer depends on the notion of event we theoretically are inclined to endorse. As we are going to show in what follows, there are notions of events which do not allow justification of spurious causal correlations.

## 2.2 Epiphenomena and overdetermination

Before examining the consequences of the above premises we express the general remark that Elster's hostility to counterfactual theories of causality is a weak point in his analysis. Why should we really concentrate so much attention on counterfactuals if we really believe that causal relations are something independent from them?

The problem of epiphenomena which Elster sees as an obstacle has already been mentioned. In this connection, however, we have to remark that if we assume that asserting the occurrence of an event implies asserting the existence of the event-subject then when we say

i) The barometer lowered

this amounts to saying

ii) The barometer lowered and the barometer existed

So if I suppose counterfactually that i) is false then this amounts to denying ii), i.e. to denying a conjunction. In such a case, by De Morgan laws, we are actually supposing a disjunction: "the barometer did not lower" or "the barometer did not exist". The conclusion "the storm would have not taken place" in this case should have to follow by both members of the disjunction. This is however wrong, since the non existence of the barometer does not imply anything about the occurrence of the storm.

As regards the second critical point raised by Elster (overdetermination and preemption) one might defend a "pluralistic" view of causality in the following terms. There is an unlimited number of causal relations having different degrees of complexity: among them we recognize that most simple one is the notion of causal relevance or *conditio sine qua non.* Let *A*, *B*, *C* be variables for token-event, O an "occurrence" operator, t(*A*) the instant in which *A* occurs. Then the minimal notion of causal relation is as follows:

(Def **C**) $A \mathbf{C} B =_{\mathrm{Df}} (\neg \mathrm{O}A > \neg \mathrm{O}B) \wedge \mathrm{t}(A) < \mathrm{t}(B)$.

If the relation takes place between actually occurring events we may define a slightly more complex notion:

(Def **C**°) $A \mathbf{C}^{\circ} B =_{\mathrm{Df}} A \mathbf{C} B \wedge \mathrm{O}A \wedge \mathrm{O}B$

The notion of a *determining* cause is very strong and may be for instance so defined

(Def $\mathbf{C}_{\mathbf{d}}$) $A \mathbf{C}_{\mathbf{d}} B =_{\mathrm{Df}} A \mathbf{C}^{\circ} B \wedge (\mathrm{O}A > \mathrm{O}B)$

The additional information here might be paraphrased in our language by saying that event $B$ is determined by event $A$, or alternatively that event $A$ is *ceteris paribus* sufficient for $B$.

But we have notions at our disposal in scientific language which are 2-place, 3-place ... causal notions: for instance

(Def ( ) $\mathbf{C}^1$) $(A,B)\ \mathbf{C}^1\ C =_{Df} A\ \mathbf{C}^\circ\ C \wedge B\ \mathbf{C}^\circ\ C$ (concomitant cause)

(Def < > $\mathbf{C}^1$) $\langle A,B\rangle \mathbf{C}^1\ C =_{Df} (\neg A > (B\ \mathbf{C}\ C)) \wedge OA \wedge OB \wedge OC \wedge t(A) < t(C) \wedge t(B) < t(C)$ (concurring cause).

To any such notion we may add further specifications yielding more and more complex causal notions.

In this framework overdetermination and preemption mirror simply some special case of concurrency in the defined sense. Let us take the paradigm case of two killers firing two mortal shots to the same victim. The paradox here is that none is a *conditio sine qua non* for the death of the victim (if one of them had not fired the shot the victim would have died anyway), so none is causally relevant for it. We have then the true conjunction

$$\neg(A\ \mathbf{C}^\circ C) \wedge \neg(B\ \mathbf{C}^\circ\ C),$$

even if $A$ and $B$ are both sufficient conditions of $C$: $A > C$ and $B > C$.
The interesting fact is that A and B are both concurring for the death:

$$\langle A,B\rangle\ \mathbf{C}^1\ C \wedge \langle B,A\rangle\ \mathbf{C}^1\ C$$

or otherwise stated

$$(\neg A > (\neg B > \neg C)) \wedge (\neg B > (\neg A > \neg C))$$

What is true is in fact that, in absence of the first shot, if the second had not been fired, the victim would have not died.

So the overdetermining causes are really expressible by counterfactual conditions, even if such conditionals are conditionals with iterated antecedents.

## 2.3 Models and theories for models

We come now to the main equivocation which in our opinion is involved in Elster's analysis, namely to the confusion between models and theories. In the sense which is relevant here, a model is simply a finite set of functional equations expressing basic laws which are usable to calculate vectors in dependence of other vectors. Deterministic models may be seen as sets of functional laws whose form is simply $y = f(x_1 \ldots x_n)$, while nondeterministic models admit different forms of representation: for instance, the function $f$ may be conceived as a set-valued function, or alternatively the form of equation may expressed as $y=f(x_1 \ldots x_n, u_1 \ldots u_n)$, being $u_1 \ldots u_n$ factors of disturbance.[6]

In order to understand the difference between models and theories let us consider so called Simon-Blalock models.[7] Such models are systems of $n$ equations involving $n$ variables which may be manipulated in order to identify causal relations among variables and to eliminate spurious correlations. In the simplest cases, in order to detect a spurious correlation among $x$ and $y$, the key of the procedure is essentially the insertion of a third variable $z$ which is supposed to act as a common cause of both $x$ and $y$. But the refined manipulation techniques devised by Simon are insufficient to always grant a univocal result. The third variable $z$ in many cases may give rise to two causal chains : $x \rightarrow z \rightarrow y$ and $x \rightarrow z \rightarrow y$: among them the choice is impossible by simply using the tools offered by the model. It turns out that *no amount of mere manipulation or controlling of the variables can substitute for* apriori *assumptions*. It is essential then to rest on a certain number of *a priori* assumptions. To quote the main ones listed by Simon: 1) to know which variables do *not* appear in mechanisms; 2) to assume that the behaviour of any system involving a great amount of energy (for instance atmosphere) is practically autonomous with respect to the behavior of variables which involve much less energy; 3) to assume that causes cannot be later than the effects in macroscopical phenomena; 4) to assume that the majority of variables in the world are not connected with the majority of other variables. The last *a priori* assumption, which Simon and Rescher[8] call "the Empty Set Postulate", is equivalent to the refusal of holism and, as we already saw in the case of Mill, is tantamount to giving legitimacy to at least some class of counterfactual suppositions.

Elster's reference to Mill's conception and to Laplacean determinism makes it clear that what he calls a theory is not a set of equations but may be, on the contrary, a philosophical or cosmological *Weltanschauung*. In the sense of the word which we are willing to use here, a theory *T for a model M* is a set of a priori assumptions which allows us to draw from *M* conclusions expressible in a language richer than the language of *M* itself (such are, for instance, causal language, the language of possible worlds, or the language of events).

[6] For the first proposal see Elster, LS, p. 183. For the second see A. Blake and J. Pearl, 1995.
[7] See Blalock, 1961, Simon, 1957, Simon and Rescher, 1966.
[8] See Simon and Rescher, 1966.

An example of a gap between theories and models is to be recorded if we want to follow Elster in seeing events as causal *relata*. A theory allowing us to use the language of events should first of all be a theory concerning the nature of events. As is well known, there are several conceptions of events.

One of the most popular theory of events is due to J. Kim and consists in seeing events as 3-ple [*A*, *a*, *t*] where *A* is a property, *a* is an individual, *t* is an instant.[9] Furthermore, a theory of events should say - what Kim does not do - what it means to be an *occurrence* of an event. Notice that a counterfactual supposition stating that a given Kim-event does *not* occur is ambiguous, since it amounts to a disjunction of suppositions referring to each single member of the 3-ple.

Some *a priori* assumptions which we discussed at the beginning of section 2.2 concern existence presuppositions in talking of events The point clearly depends on the theory of events which we subscribe. In fact, if the background theory of events states that occurrence of events implies the existence of the object *a* which is member of the Kim 3-ple, a counterfactual supposition asserting that [*A*, *a*, *t*] did not occur splits in several alternatives, one of which is that *a* did not exist. As we already stated, from such a supposition one concerning the barometer one cannot derive the conclusion that the following storm did not exist.

We list in what follows three kinds of assumptions which may be relevant as theoretical assumptions to be used in drawing causal conclusions from models.

1) Assumptions concerning the logical closure of knowledge. Does the Background Knowledge include or not include all the logical consequences of the propositions which are retained in it after introducing the counterfactual supposition?

2) Assumptions concerning the revisions of the model. The theory associated to *M* should say if beyond using *M* in our reasoning the operator may use revised variants of *M* required by the counterfactual reasoning itself.

3) Assumptions concerning the structural or non structural character of elements of the Background Knowledge. What I here call *structural* are the data which are assumed to be immune from revision. To use Elster terminology, they are data which are "inevitable" or, in modal terminology, are "necessary".

[9] See Kim, 1973.

With reference to point 3, however, we find a difficult point in Elster's analysis. Speaking of data which are by virtue of the theory "inevitable" (as, for instance, inventions in Rosenberg's theory) Elster draws the conclusion that to suppose their falsity is illegitimate. Let us recall that illegitimacy means either inconsistency or trivialization. Let us call conventionally *facts* all the statements of event-occurrence in the mentioned sense, and let us use symbols $F_1$,$F_2$... for facts. No model $M$ logically implies any empirical fact, so there is no inconsistency in supposing that a single fact did not occur: then we may say that a contrary-to-fact supposition is illegitimate when it trivializes the data, or in other words is such as to force us to remove *all* the background empirical knowledge.

Let us suppose that we have a set of facts in chronological order $F_1$, $F_2$ ... such that a deterministic model $M$ allows to establish for every $n$ both $F_n > F_{n+1}$ and $\neg F_n > \neg F_{n+1}$. Such set of facts is what Elster calls a *process*. Then by the definition which has been given in section 2.2, we may say that every fact of the process is a *determining cause* of every fact which follows it in the given order, and thanks to the transitivity in the calculations of vectors we may also conclude that every fact $F_n$ is a cause of every other fact which follows it in the given order. But, given $\neg F_n$, what about facts which chronologically precede $F_n$? In order remove them we should employ a so-called *backtracking counterfactual*, i.e. something of the form $\neg F_{n+1} > \neg F_n$. Unfortunately we cannot logically pass from $\neg F_n > \neg F_{n+1}$ to its converse $\neg F_{n+1} > \neg F_n$ or from $F_n > F_{n+1}$ to the contrapositive $\neg F_{n+1} > \neg F_n$, since the >-operator, as we already remarked, is non-contrapositive[10]. So removing $F_n$ does not in general imply removing *every* $F_i$. We have to conclude that only some extra assumptions may allow for the contraposition of the corner operator and we will take them in consideration in the next section.

It is anyway wrong to think, as Elster suggests, that there is a straightforward link between the status of necessity (inevitability) which a theory assigns to facts and the subsistence of a necessary connection between them. In this context we have simply to recall the confusion between *necessitas consequentiae* and *necessitas consequentis*. A necessary connection between - say - the fall of a meteorite (a highly improbable fact) and a local ecological catastrophe does not make each of the two events something which is by itself necessary. Conversely, it is true that, for some implication relation ***CI*** , knowing that $A$ is necessary ($\Box A$) and that $B$ is necessary ($\Box B$) implies knowing that $A$ ***CI*** $B$ and $B$ ***CI*** $A$.[11] But if ***CI*** is not-contrapositive, $A$ ***CI*** $B$ is

[10] Let us recall that $(\mathbf{w}(F_1) \wedge F_1) -3\ F_2$ is equivalent to $((\mathbf{w}(F_1) \wedge \neg F_2) -3\ \neg F_1)$, not to $((\mathbf{w}(F_2) \wedge \neg F_2) -3\ \neg F_1)$.

[11] If ***CI*** is strict implication , we have the well known theorem $(\Box A \wedge \Box B) \supset (A -3\ B)$. But this relation is not granted if ***CI*** is some other implication connective.

not sufficient to yield $\neg A$ ***CI*** $\neg B$, and $B$ ***CI*** $A$ is not sufficient to yield $\neg B$ ***CI*** $\neg A$. Thus if $A$ and $B$ are necessary, only the properties of ***CI*** justify that the counterfactual removal of one of them implies the removal of the other.

# 3. LEGITIMACY AND ILLEGITIMACY AS THEORY DEPENDENT

## 3.1 Counterpossible conditionals

In the present section we consider some examples of theoretical assumptions which are sufficient to yield more or less general trivialization results, or in other words results implying illegitimacy of counterfactual hypotheses.

In section 2.1 we have introduced the notion of a context of $A$, symbolized by $\mathbf{w}(A)$ and axiomatized by axiom Ax1 and by two rules. A further *prima facie* innocent assumption about context consists in requiring that the context of two different supposition, equals the context of the conjunction of both. Formally:

($\wedge\mathbf{w}$) $\mathbf{w}(A) \wedge \mathbf{w}(B) \equiv \mathbf{w}(A \wedge B)$

Notice that this clause in a sense imposes a boundary on the compatibility of suppositions. If $A$ and $B$ are incompatible propositions, $A \wedge B$ is impossible and $\mathbf{w}(A \wedge B)$ is also such: ($\wedge\mathbf{w}$) implies then that $\mathbf{w}(A)$ and $\mathbf{w}(B)$ are also incompatible. The effect of ($\wedge\mathbf{w}$) is then that it puts a limit to the derivation from incompatible hypotheses, since this would amount to hypothesizing a contradiction.

Let us now define a new operator which is stronger than strict implication:

(Def => ) $A \Rightarrow B =_{\mathrm{Df}} (A -3\ B) \wedge (\Diamond B \supset \Diamond A)$.

The relation between $\Box$ and => is given by the equivalence

$$\Box A \equiv (B \vee \neg B) \Rightarrow A$$

which may be proved as a theorem of **KT**.

A simple result concerning => is

**T0.** *Let us call a relation* ***CI*** among *A and B* weakly contrapositive *if* $\Diamond\neg B$ *implies* $(A\ \mathbf{CI}\ B) \supset (\neg B\ \mathbf{CI}\ \neg A)$, *and* weakly monotonic *if* $\Diamond(A \wedge R)$ *implies* $((A\ \mathbf{CI}\ B) \supset ((A \wedge R)\ \mathbf{CI}\ B))$. Then => *is transitive, weakly contrapositive and weakly monotonic.* (For a sketch of the proof see Appendix , A2 )

Then ($\wedge$**w**) allows us to prove the following (in a sense surprising) theorem:

**T1.** *If* $\mathbf{w}(A) \wedge \mathbf{w}(B) \equiv \mathbf{w}(A \wedge B)$ is an axiom for **w**(-), $A > B$ *is equivalent to* $A \Rightarrow B$. (see Appendix, A1).

An effect of ($\wedge$**w**) is then that the standard conditional collapses onto a relation which is stronger that strict implication and, as stated in **T0**, is weakly contrapositive

Now let us define "inevitability" in terms of the corner conditional in this way by an obvious definition of $\Diamond°$:

(Def $\Box°$) $\Box° A =_{Df} \top > A$

where $\top$ is an arbitrary tautology.[12] Then obviously we have as **KT** - theorems:

$$\Box A \supset \Box° A$$
$$\neg\Diamond° A \supset \neg\Diamond A$$

Let us suppose then that *A* and *B* are true, $A > B$ is true and that *B* is contingently true: this means $\Diamond°\neg B$ and $\Diamond\neg B$ . Then by **T0** and **T1** we have $\neg B > \neg A$. If we suppose counterfactually not-*B* then, the preceding event *A* is also to be removed and by the transitivity of => all the events preceding *A* in the process are also to be removed. So if the deterministic models explain a chain of contingent facts *A*,*B*,*C*... each one of which is a determining cause for the following, *all* the elements of the chain are removed by any counterfactual supposition about them.

Another triviality theorem following from **T1** as follows.

**T3.** *Suppose* $\mathbf{w}(A) \wedge \mathbf{w}(B) \equiv \mathbf{w}(A \wedge B)$ *is an axiom for* **w**(-). *Then if B is a contingent determining cause of C which is concomitant cause of C along with A is also a cause of A* (see Appendix, A3 ).

[12] Other definitions are of inevitability are of course possible. However, let us remark that $\Box° A$ is equivalent to $(\mathbf{w}(\top) \wedge \top) \Rightarrow A$, so to $\mathbf{w}(\top) \Rightarrow A$ . $\mathbf{w}(\top)$ is then the context of a tautology, which might be identified with the totality of the Background Knowledge .

The meaning of **T3** is that no jointly concomitant causes of any arbitrary event are independent under the mentioned assumption, so **T3** trivializes the notion of a concomitant cause.

**T2** and **T3** depend on the conditions that the fact $B$ is contingent; what happens if we drop this restriction? Limiting our attention to facts as defined in this paper, this is tantamount to endorsing a second assumption to be added to ($\wedge$**w**) , namely

($\Box°F$) if $F_i$ is a fact, $F_i \supset \Box°F_i$

In presence of ($\wedge$**w**) and ($\Box°F$) entertaining the supposition that some F is false means hypothesizing something *per impossibile*, or making what we might call a *counterpossible* supposition.[13] If $A$ and $B$ are facts belonging to the Background Knowledge A it is true, so thanks to ($\Box°F$) by Modus Ponens we have $\Box°A$ and $\Box°B$. Let us remark however that since > collapses on => we have that $\mathsf{T} > A$ collapses on $\mathsf{T} \Rightarrow A$, so $\Box°A$ collapses on $\Box A$ and $\Box°B$ on $\Box B$. But $\Box A$ conjoined with $\Box B$ implies $\Box A \equiv \Box B$. Now the connective defined as $A \rightarrow B =_{Df} A \Rightarrow B \wedge (\Box B \equiv \Box A)$ has the property that $A \rightarrow B$ equals $\neg B \rightarrow \neg A$, which means that the operator $\rightarrow$ is fully contrapositive [14].

The assumption ($\Box°F$) added to ($\wedge$**w**) then yields a full trivialization. In other words it is straightforward to prove:

**T4**. *Suppose $F_i$ and $F_j$ are facts belonging to a deterministic causal process. Suppose that among our logical truths we have all the theses of* ***KT***$^w$ *and also*

$$w(F_i) \wedge w(F_j) \equiv w(F_i \wedge F_j)$$
$$F_i \supset \Box°F_i$$

*Then supposing counterfactually $\neg F_i$ implies $\neg F_j$, for every $F_j$.*

[13] For this terminology see Lewis, 1973. Lewis however uses the term "counterpossible" to denote a supposition which is false in every possible world, such as the negation of a logical or mathematical truth. In Lewis' theory all counterpossible conditionals turn out to be true since Lewis' conditional operator, material implication and strict implication all share the property that a contradiction implies everything. A remarkable virtue of => and > with respect to Lewis' conditional operator is that they are "non-scotian" connectives : it may be checked in fact that for such connectives a contradiction implies only another contradiction. The need for a nostandard implication connectives in treating counterpossible conditionals is stressed also by Cohen, 1988.

[14] For the proof see Pizzi, 1981.

The joint addition of ( $\wedge$**w**) and ($\Box°F$) to $\mathbf{KT^w}$ leads to other trivialization results. What is trivialized in the next theorem is not concomitance but concurrency.

**T5**. *Suppose > has the same properties as =>. Then if A and B are inevitable (i.e. we have* $\Box°A$ *and* $\Box°B$*) none of them may be an overdetermining cause with respect to any event C* (see Appendix, A5).

It would be interesting to study what follows from $\Box°F$ alone or from other definition of inevitability, but this problem will not be discussed in the present paper.

## 3.2 Counterlegal conditionals

In the preceding section we examined the kind of suppositions which are illegitimate in the sense of leading to some kind of trivialization.

But a conceptually very interesting problem arises when we suppose something implying that some of the laws of the model itself are false. The counterfactual suppositions in this case may properly be called *counterlegal* suppositions. If a model $M$ is seen as the conjunction of its laws $L_1...L_n$, the supposition $\neg L_i$ (where $1 < i < n$)implies $\neg(L_1 \wedge...\wedge L_n)$, so $\neg L_i$ implies $\neg M$. Since from Elster's viewpoint $M$ is presupposed as a major premise for every argument, apparently any counterlegal supposition leads not to a trivialization but directly to a contradiction. Notice that this result is independent by any auxiliary assumptions. But if the theory $T$ for the model $M$ contains the assumption that laws are necessary, i.e. $L_i \supset \Box L_i$, the consequence is even more shocking, since it is enough to suppose $\neg\Box L_i$ to obtain $\neg L_i$ , hence $\neg M$. Notice that $\neg\Box L_i$ is not a particularly far-fetched supposition, since it means simply to suppose not that the law $L_i$ is false but that there is a possible world in which $L_i$ is false. (Of such kind, for instance, is the hypothesis that tachions *might* exist ).

The difficulty may actually be reformulated as an antinomy. Since $M$ works as deductive machinery, we cannot use $M$ to conclude that $M$ itself is false. As a matter of fact, counterlegal conditionals have been a source of puzzlement especially for consequentialist theories of counterfactuals. In the framework of Goodman's analysis of counterfactuals, for instance, all counterlegals turn out to be false.[15]

Hans Reichenbach in his *Nomological Statements and Admissible Operations* proposed to assign counterlegals a third truth-value, putting them on

[15] This happens since a counterfactual is true for this analysis only if some $S$ of propositions exists which is compatible with the antecedent $A$ and leads by a law $L$ to the consequent $B$. So when $A$ is itself the negation of $L$ this proviso cannot be satisfied.

the level of meaningless statements.[16] At the end of his rightly celebrated essay he tells the story of the missionary who in a school for Indios in South America puts the following question: "If a white man kills six bears in one day, how many bears does he kill in five days?" The only answer he received was: "a white man cannot kill six bears in one day".[17]

This little story deserves attention. The Indios refused to draw a conclusion from an hypothesis which they thought to be an illegitimate one, and which they believed to be such because it denies a law which puts a limit to the ability of a particular kind of men to do something. However, the answer to the question exists, and it is given by simply multiplying the number six and the number five: so after all we cannot be content with Reichenbach's refusal to assign the truth value "true" to some counterlegal conditionals.

One may guess that, since laws are not events, it is irrelevant to use counterlegal hypotheses in causal reasoning. But things are not so simple. For instance, a dispositional statement contains the reference to laws in an implicit way. If we suppose, for instance, that a certain body lacks one of its dispositions, we are implicitly denying some of the laws which govern the body itself.

Let us suppose for instance that we saw a cat which catches a mouse in conditions in which the event was clearly determined by its strong will to reach the goal. In these conditions if I suppose that the cat did not catch the mouse I am supposing that it had no disposition to do it: in other words I am implicitly denying the "nature" of the cat, which means denying the laws which rule the ordinary behavior of normal cats.

The answer to the puzzle is given again by distinguishing between a model and the theory for the model. If we look back to some of the assumptions listed in section 2.3 we find two which are relevant here. The first to be mentioned is an assumption granting that Background Knowledge is closed under logical rules. This is important since under this assumption our Background Knowledge contains, for every law $L_i$, any law which is logically derived by $L_i$: in particular, every law $L_j$ which is more general, or less specific, than $L_i$.

For instance, Kepler's law

*L: All the planets rotate in an ellipsis one of whose focuses is occupied by the sun*

implies the following more generic variants:

*L': all the planets rotate in a closed orbit around the sun*

[16] 1954, p. 176, n. 18.
[17] See 1954, p. 83.

and
$L''$: *all the planets rotate in orbits having the same shape.*

Now, even if we may agree that $\neg L$ (the negation of law $L$) implies the negation of the model $M$ to which $L$ belongs, it is false that $\neg L$ implies the negation of $L'$ or $L''$, i.e. of its generic variants. So if $L'$ and $L''$ remain part of our Background Knowledge there is no plausible reason to exclude the validity of such counterlegal conditionals as

(++) If Jupiter were to move in a circular orbit also Venus would move in a circular orbit

since (++) is granted by the law $L''$.

Are we using model M in such derivation? Clearly no, since $M$ is falsified by the counterlegal supposition. But here we have to consider the second mentioned theoretical assumption which is relevant here, i.e. the one concerning the revisability of models. A theory for a model $M$ should say in fact whether using $M$ implies using models which are derived from $M$ in the sense that they contain laws which are "less determined" than the laws of $M$. If the answer is positive - and there is no conceptual obstacle to such theoretical assumption - counterlegal suppositions cannot be considered a systematic source of contradictions and in suitable contexts may receive the positive mark of legitimacy.

## APPENDIX

**A1** We first prove in $\mathbf{KT^w}$ extended with ($\wedge\mathbf{w}$) the equivalence $\mathbf{w}(A) \wedge A \equiv A$. Defining the auxiliary operator as $*A =_{\mathrm{Df}} \mathbf{w}(A) \wedge A$ we have to prove then the collapse-equivalence

(Coll) $*A \equiv A$

It is obvious that from ( $\wedge\mathbf{w}$ ) we reach straightforwardly the equivalence

(*d) $*(A \wedge B) \equiv *A \wedge *B$

By (*d) we obtain, thanks to the standard law $(*A \wedge *B) \supset *A$ and replacement of equivalents, the law $*(A \wedge B) \supset *A$.

A remarkable consequence of (*d) is that in the logic of * we may prove $*(A \supset B) \supset (*A \supset *B)$ as follows:

1) $*(A \wedge B) \supset *B$

| | |
|---|---|
| 2) $*(A \wedge (A \supset B)) \supset *B$ | 1), $(A \wedge (A \supset B)) \equiv A \wedge B$ |
| 3) $(*A \wedge *(A \supset B)) \supset *B$ | 2), (*d) |
| 4) $*(A \supset B) \supset (*A \supset *B)$ | 3), standard logic |

As an independent step we perform the following proof:

| | |
|---|---|
| 1) $*(A \wedge \neg *A) \supset *A$ | $*\,(A \wedge B) \supset *A$, $\neg *A/B$ |
| 2) $*(\neg *A \wedge A) \supset \neg *A$ | $*(A \wedge B) \supset A$ |
| 3) $\neg *(A \wedge \neg *A)$ | 1),2), standard calculus |
| 4) $\neg \Diamond *(A \wedge \neg *A)$ | 3), $\neg A/\Box\neg A$ |
| 5) $\neg \Diamond (A \wedge \neg *A)$ | 4), $\Diamond A \equiv \Diamond *A$ |
| 6) $A \supset *A$, | 5), $\neg\Diamond A \supset \neg A$, standard calculus |

The converse of 6), $*A \supset A$, is an obvious consequence of theorem $(\mathbf{w}(A) \wedge A) \supset A$, so we obtain the equivalence $A \equiv *A$.

Now $A > B$ is by definition equivalent to $((\mathbf{w}(A) \wedge A) -3\ B) \wedge (\Diamond B \supset \Diamond A)$, so to $(*A -3\ B) \wedge (\Diamond B \supset \Diamond A)$. Since we have proved the equivalence $*A \equiv A$, then by replacing $*A$ for $A$ we have that $A -3\ B$ is equivalent to $*A -3\ B$ and that $\Diamond B \supset \Diamond *A$ equals $\Diamond B \supset \Diamond A$. So $A > B$ equals $A => B$.

**A2.** Thanks to some routine application of the *tableau* method for **KT**, we may prove that the modal translations of

1) $((A => B) \wedge (B => C)) \supset (A => C)$
2) $\Diamond(A \wedge R) \supset ((A => B) \supset (A \wedge R => B))$
3) $\Diamond\neg B \supset ((A => B) \supset (\neg B => \neg A))$

are **KT**-valid wffs. So they are **KT**-theorems for the completeness of **KT**.

**A3.**Let $A$ and $B$ be events which are concomitant causes of $C$ and let $A$ be such that $A\ \mathbf{C}^d\ C$ and $B\ \mathbf{C}^d\ C$, i.e. that $A$ and $B$ are determining causes of $C$. This implies $\neg OA > \neg OC$ and $\neg OA > \neg OB$. Furthermore, being $A$ a determining cause of $C$, we have $OA > OC$. Since $OA > OC$ implies $OA => OC$ by **T1**, it follows then (i) $OA => OC$, (ii) $\neg OA => \neg OC$ e (iii) $\neg OB => \neg OC$. But being the relation $=>$ weakly contrapositive we have by (i): (iv) $\Diamond\neg OC \supset (\neg OC => \neg OA)$. Thus by transitivity, (iii) e (iv), it follows that $\Diamond\neg OC \supset (\neg OB => \neg OA)$. Hence, if $OC$ is not necessary, we have by **T1** $\neg OB > \neg OA$ so, by definition, $B\ \mathbf{C}\ A$. Hence, if $A$ is a determining cause of $C$ and $C$ is contingent, $A$ is in its turn caused by any other determining cause of $C$.

**A5**. If $A$ and $B$ are such that $\Box°A$ and $\Box°B$, by **T1** we have $\Box A$ and $\Box B$. This implies $\Box\ (A \supset B) \wedge (\Diamond B \supset \Diamond A)$, hence $A \Rightarrow B$ and viceversa. If $A$ and $B$ are concurring causes for $C$ we have by definition of concurrency (°) $\neg A > (\neg B > \neg C)$. Since they are also overdetermining causes it must be true $A > C$ and $B > C$, but also $\neg(A$ C $C)$ and $\neg(B$ C $C)$ (since $A$ and $B$ are not necessary conditions for $C$).

But by virtue of the collapse of > on => we have by (°)

| | |
|---|---|
| 1) $\neg A \Rightarrow (\neg B \Rightarrow \neg C)$ | Hypothesis |
| 2) $\neg A \Rightarrow (\neg B -3\ \neg C))$ | 1) $(A \Rightarrow B) \supset (A -3B)$, |
| 3) $\neg A -3\ (\neg B -3\ \neg C)$ | 2) $(A \Rightarrow B) \supset (A -3B)$, |
| 4) $(\neg A -3\ \neg B) -3\ (\neg A -3\ \neg C)$ | 3), by **KT** |
| 5)$(\Box A \wedge \Box B) -3\ (\neg A -3\ \neg C)$ | $(\Box A \wedge \Box B) -3(\neg A{-}3\neg B)$, 4) **KT** |

Thus if $A$ ad $B$ are inevitable and they are both concurring causes for $C$, we have $\neg A -3\ \neg C$.

Now if $A$ ad $B$ are overdetermining causes we know $\neg(A$ C $C)$, so by **T1** and DeMorgan $\neg(\neg A\ -3\ \neg C) \vee \neg(\Diamond\neg C \supset \Diamond\neg A)$, hence $(\neg A\ -3\ \neg C) \supset \neg(\Diamond\neg C \supset \Diamond\neg A)$. By line 5) and transitivity we should have then that $\Box A \wedge \Box B$ implies $\Diamond\neg C \wedge \Box A$.

But this is impossible, since from the premise $A > C$ (which equals $A \Rightarrow C$ ) and $\Box A$ we have by **KT** $\Box C$, which is incompatible with $\Diamond\neg C$.

We conclude that if two causes are necessary they cannot be overdetermining causes for a common effect.

## REFERENCES


Blalock, H., 1961, *Causal Inferences in Nonexperimental Research*, The University of North Carolina Press, Chapel Hill.

Climo,T.A. and Howells, P.G.A., 1976, Possible worlds in historical explanation, *History and Theory 15:1-20.*

Cohen, D.J., 1988, The problem of counterpossibles, *Notre Dame Journal of Formal Logic*, 29: 91-101.

Elster, J., 1978, *Logic and Society*, John Wiley and Sons, Chichester.

Elster, J., 1983, *Explaining Technical Change*, Cambridge University Press, Cambridge.

Elster, J., 1980, Reply to comments, *Inquiry*, 23: 157-172.

Fogel, R.W., 1964, *Railroads and American Economic Growth*, John Hopkins Press, Baltimore.

Genovese, E., 1965, *The Political Economics of slavery*, Pantheon Books, New York.

Kim, J.,1973, Causation, nomic subsumption, and the concept of event, *Journal of Philosophy*, 70:217-236.

Lewis, D.K., 1973, *Counterfactuals*, Blackwell, Oxford.

Lukes, S., 1980, Elster on counterfactuals, *Inquiry*, 23: 145-155.

Pearl, J. and Blake, A., 1995, Counterfactuals ad policy analysis in structural models, in P. Besnard and S. Hanks, eds., *Uncertainty in Artificial Intelligence 11*, UCLA Cognitive Systems Laboratory, Technical Report (R-232-U), Morgan Kaufmann, San Francisco, CA.
Pizzi, C., 1991, Decision procedures for logics of consequential implication, *Notre Dame Journal of Formal Logic* 32:618-636.
Reichenbach, H., 1954, *Nomological statements and Admissible Operations,* North Holland, Amsterdam.
Rosenberg, N., 1974, Science, invention and economic growth, *Economic Journal* 84:90-180
Simon H.A and Rescher N., 1966, Cause and counterfactuals, *Philosophy of Science* 33:323-340.
Simon, H.A., 1957, *Models of Man*, John Wiley and Sons, New York.

# Mental Models in Conceptual Development

Stella Vosniadou
*Department of Philosophy and History of Science, National and Kapodistrian University of Athens, Athens, Greece, svosniad@compulink.gr*

Abstract: Models are important in science because they can be used as instruments to help in the construction of theories. They are the sources of both predictive and explanatory power and can be used to suggest new hypotheses and aid in scientific discovery. In this paper I argue that the mental models that children construct also have predictive and explanatory power and can be used as mediating mechanisms for the revision of existing theories and the construction of new ones.

## 1. INTRODUCTION

### 1.1 Mental model as sources of predictive and explanatory power

For a number of years now, I have been engaged in research attempting to understand the development of knowledge about the physical world and the learning of science concepts. In this paper, I will focus on one aspect of this research that has to do with the role of mental models in conceptual development and in reasoning about the physical world. More specifically, I will argue that the ability to form mental models is a basic characteristic of the human cognitive system and that the use of models by children is the foundation of the more elaborate and intentional use of models by scientists.

*Model-Based Reasoning: Science, Technology, Values*, edited by
L. Magnani and N.J. Nersessian, Kluwer Academic/Plenum Publishers, New York, 2002

Models are important in science because they can be used as instruments to help in the construction of theories. They are the sources of both predictive and explanatory power and can be used to suggest new hypotheses and aid in scientific discovery. Similarly, I will argue that the mental models that children and lay adults construct are important for conceptual development and conceptual change. They also have predictive and explanatory power and can be used as mediating mechanisms for the revision of existing theories and the construction of new ones.

The paper starts with a discussion of the theoretical framework within which conceptual development is considered. More specifically, it is argued that concepts are embedded within framework and specific theories and that conceptual development involves not only the enrichment but also the revision and radical restructuring of these theories. Mental models are defined as a form of mental representation that preserves the structure of that which it represents. It is argued that mental models can be particularly useful in situations where implicit physical knowledge needs to be exploited for the purpose of answering a question, solving a problem, or in order to understand incoming information. Examples are given of how mental models function in conceptual development and how they are used in the construction and revision of theories about the physical world.

## 1.2 Concepts are embedded in framework and specific theories

Cognitive developmental research has so far confirmed the hypothesis that the knowledge acquisition process starts at birth and that infants proceed rapidly to construct certain fundamental understandings of the physical and social world (e.g., Carey, 1985; Wellman and Gelman, 1992). These fundamental understandings appear to develop in at least two core domains, physics and psychology with biology emerging as a possible third domain (Inagagi and Hatano, 1996).

Many developmental psychologists describe concepts in these core domains as entrenched in naïve or folk theories (although this issue is still debated especially outside developmental psychology). The term "theory" is used here relatively freely to distinguish a coherent body of knowledge that involves causal, explanatory understanding, as compared to a body of knowledge that involves a collection of facts and some procedures for operating on them. It is assumed that children's theories are different from scientific theories because children's theories lack many important characteristics of scientific theories, such as, their systematicity, abstractness, and social/institutional nature. In addition, it appears that children lack metaconceptual awareness of their theories, do not make the explicit distinction between theory and evidence, and do not understand how theories guide the hypothesis testing process (Carey and Smith, 1993; Kuhn, Amsel, and O' Longhlin, 1988).

Some developmental psychologists make the distinction between *framework theories* and *specific theories* following a similar distinction in the history of science between paradigms (research programs, or research traditions) and specific theories (Kuhn, 1962; Lakatos, 1970; Laudan, 1977). In the area of physics, a framework theory would include the basic ontological and epistemological presuppositions that define the concept of the physical object. Such a framework theory of physics would be different, for example, from a framework theory of psychology both in its ontology (physical objects versus psychological beings) and in its causality (mechanical causality versus intentional causality).

Indeed it appears that young children distinguish physical objects from psychological beings from early on. Research with infants has documented the rapid acquisition of a coherent body of physical knowledge concerning primarily certain types of principles, or presuppositions, that further define the ontology of the physical object as well as the notion of mechanical causality. For example, using preferential looking methods, Spelke and her colleagues have shown that 3-month-old infants will perceive three dimensional objects presented against a uniform background as connected bodies that will maintain their cohesion as they move (Spelke, Breinlinger, Makomber, and Jacobson, 1992). Other experiments suggest that 4-month-old infants understand that surfaces move together only if they are in contact and independently only if they are separated (Kellman and Spelke, 1983; Kellman, Spelke and Short, 1986). Infants provide evidence that they start to understand that objects move downward in the absence of support (gravity principle) or continue to move in the absence of obstacles (principle of inertia) between 6 and 8 months (Spelke, 1991; Baillargeon, 1990). In addition, experiments by Leslie on 6-month-old infants' perception of physical causality suggest an early appreciation of some aspects of mechanical causality (Leslie and Keeble, 1987; Leslie, 1988), although, according to Oakes and Cohen (1995), physical causality does not appear until the age of 10 months. Certainly by the end of the first year, infants understand how physical causation works and appreciate the differences between psychological and physical causality (Meltzoff, 1995).

Specific theories are embedded within framework theories and are constrained by them. For example, research with preschool and elementary school children has shown that young children categorize the earth as a physical object (rather than as an astronomical object) and apply to it all the properties of physical objects. The specific theory of the earth is constructed on the basis of children's observations and cultural information they receive in their everyday life under the constrains of the framework theory of physics (Vosniadou, 1994a; Vosniadou and Brewer, 1992). Figure 1 shows the hypothetical conceptual structure underlying children's initial concept of the earth.

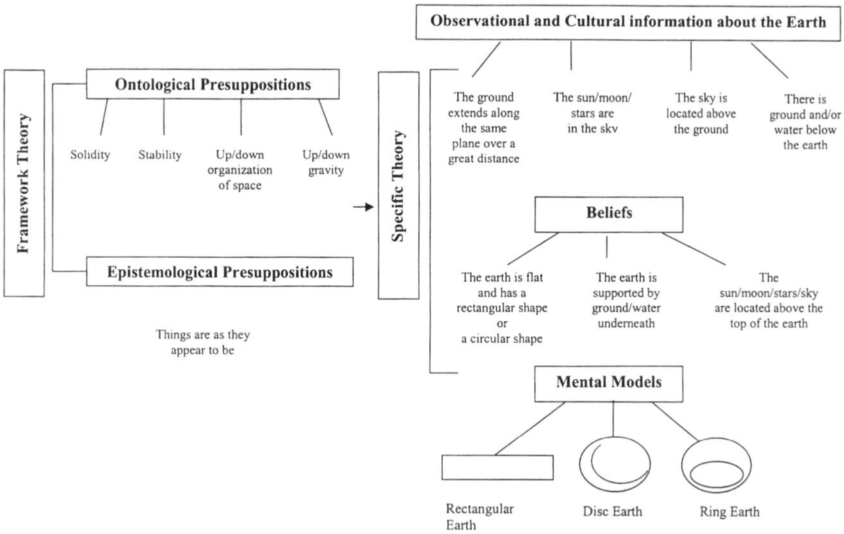


*Figure 1.* Hypothetical conceptual structure underlying children's initial mental models of the Earth.

## 1.3 Mental models

In order to give a full account of conceptual development, another theoretical construct appears to be needed, that of the *mental model*. Mental models are analog representations that preserve the structure of the thing they represent. It is assumed that most mental models are constructed on the spot to deal with the demands of specific situations, although it is possible that some mental models may be stored in long-term memory. When a representation in the form of a mental model is constructed, this representation is designed to be as useful as possible in the situation in which it is involved (in the solution of problem, to answer a question, etc.). However, mental models are also constrained by the framework and specific theories within which they are embedded and thus can be important sources of information about them (see Vosniadou, 1994a).

An important characteristic of mental models is that they can be explored extensively, run in the mind's eye, so to speak, in order to generate predictions and explanations. This point will be developed later in greater detail. Experiments in cognitive psychology have shown the psychological reality of mental models (e.g., Sheppard and Metzler, 1971; Kosslyn, 1994), while experiments in cognitive neuroscience have provided neurological evidence for mental models.

In examining the role of mental models in conceptual development and conceptual change I will focus on two aspects: The first has to do with the *construction of mental models* and the *second with their function*.

## 2. THE CONSTRUCTION OF MENTAL MODELS

Figure 2 shows the mental models of the earth constructed by elementary school children in a study designed to investigate children's understanding of the shape of the earth (Vosniadou and Brewer, 1992).

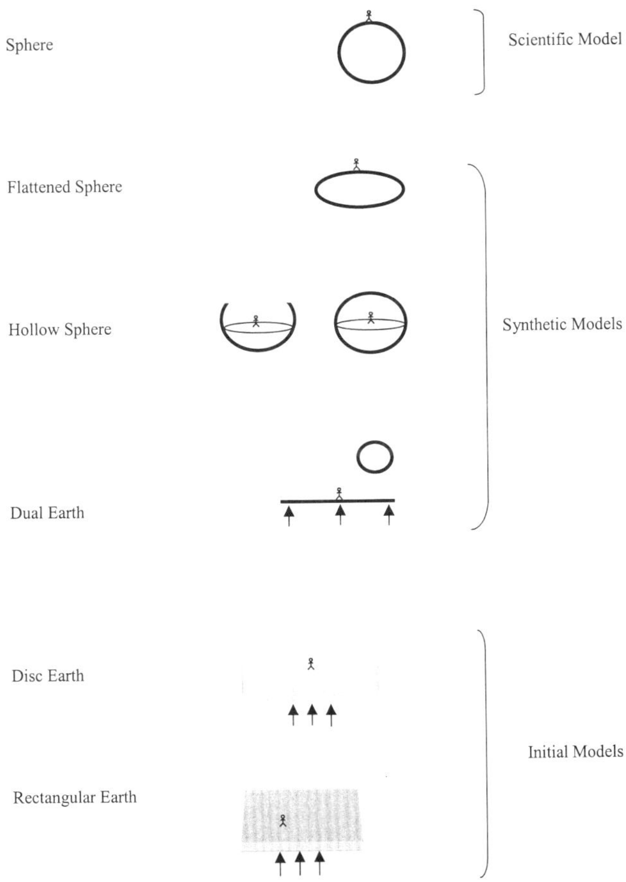


*Figure 2*. Mental models of the Earth.

These mental models were obtained by combining data coming from children's answers to verbal questions with drawings and models made out of play-dough. The children were asked a number of questions about the shape of the earth and where people live on the earth. They were also asked to make drawings and play-dough models of the earth.

One important aspect of the methodology was the use of generative questions. Generative questions cannot be answered on the basis of stored information but require the genuine solution of a new problem. For example, the question "If you were to walk for many days, would you ever reach the end of the earth? Does the earth have an end/edge?" is assumed to be a generative question. This is because most children have not encountered this question before and therefore do not have explicit conceptual knowledge from which they can derive ready-made answer. It is assumed that in order to answer a generative question, our subjects must create a mental representation or a mental model of the earth, and explore it in order to derive from it a relevant answer.

During the data analysis we try to determine whether each individual subject's responses (both verbal and non-verbal) form a coherent pattern consistent with the use of a generic model of the earth. For example, if a given subject draws a flat, rectangularly-shaped earth, creates a similar play-dough model, answers that the earth is flat, and that people could fall off the flat end/edge of the earth, this is taken as adequate evidence to support the claim that this subject is using a model of a flat earth to reason from.

Our studies in the United States (Vosniadou and Brewer, 1992; 1994) but also a number of cross-cultural studies in Greece (Vosniadou et al., 1996), in India (Samarapungavan, Vosniadou and Brewer, 1992), and in Samoa (Vosniadou, 1994b) have shown that a) there is a small number of overall models of the earth that our subjects form, b) about 80-85% of the subjects in our sample use one of these models in a relatively consistent way, and c) there are interesting cross-cultural similarities and differences in the use of these models.

Recently, a study by Ioannidou and Vosmiadou (2001), following a similar methodology investigated children's and adults representations of the layering and composition of the inside of the earth and obtained similar findings. Figure 3 shows the models of the layering and composition of the earth that the subjects of this study formed.

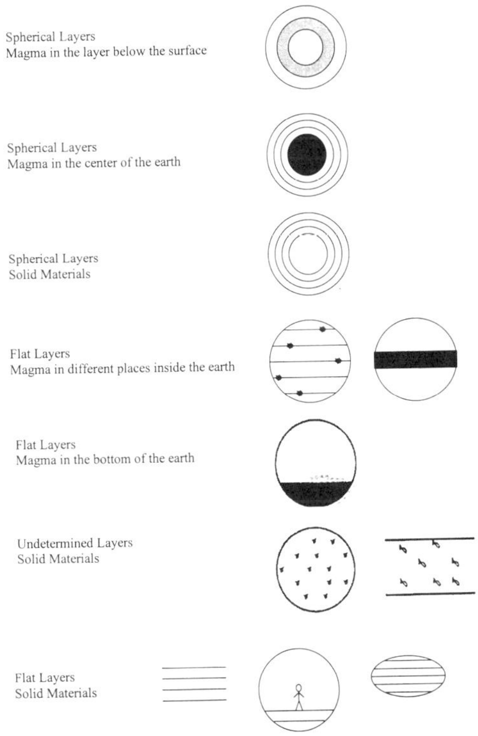


*Figure 3*. Models of the layering and composition of the earth.

# 3. HOW MENTAL MODELS FUNCTION

Mental models can have various functions in the human cognitive system. Here we will discuss three important functions of mental models: a) as aids in the construction of explanations, b) as mediators in the interpretation and acquisition of new information, and c) as tools that allow experimentation and theory revision.

## 3.1 Mental models as aids in the construction of explanations

When people reason about the physical world they often make use of mental models, particularly in situations where the answer cannot be retrieved from previously stored information or can be easily deduced from verbal information. Mental models can help people to draw on implicit physical knowledge that can be then used to answer questions, solve problems, etc. By doing so, the implicit knowledge becomes explicit and conceptual.

It was mentioned earlier that one important aspect of the methodology used in the astronomy studies was the use of generative questions that we believe stimulates the formation of a mental model. Mental models are helpful in these cases precisely because the generative questions cannot be answered on the basis of previously known facts or explanations. In these situations, people tend to draw on their implicit physical knowledge in an effort to find relevant information. The construction of a mental model allows people to represent the relevant physical knowledge in a way that can help them solve the problem at hand.

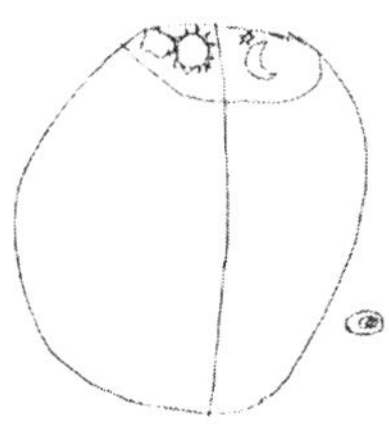

E: So what is the real shape of the earth?
C: Round.
E: Why does it look flat?
C: Because you are inside the earth.
E: If you walked and walked for many days in a straight line, where would you end up?
C: Somewhere in the desert.
E: What if you kept walking?
C: You can go to states and cities.
E: What if you kept on walking?
C: (No response).
E: Would you ever reach the edge of the earth?
C: No. You would have to be in a spaceship if you're going to go to the end of the earth.
E: Is there an edge to the earth?
C: No. Only if you go up.

Later:
E: Can people fall off the end/edge of the earth?
C: No.
E: Why wouldn't they fall off?
C: Because they are inside the earth.
E: What do you mean inside?
C: They don't fall, they have sidewalks, things down like on the bottom.
E: Is the earth round like a ball or round like a thick pancake?
C: Round like a ball.
E: When you say that they live inside the earth, do you mean they live inside the ball?
C: Inside the ball. In the middle of it.

*Figure 4*. Answers to the question regarding the earth end/edge from the point of view of a child with a mental model of the earth as a hollow sphere.

The example shown in Figure 4 demonstrates how one of our subjects, Venica, an 8- year old child, imagines the earth as a hollow pumpkin open at the very top where the various solar objects are drawn, and derives from this model a surprising answer to our question regarding the end/edge of the earth. More specifically, Venica believes that people live inside the earth, at the bottom. From her point of view this hollow earth is open at the very top, where the solar objects are located, but this top is very high up and you need a spaceship to get there. Her answer that "you would have to be in a spaceship if you are going to go to the end of the earth", shows how her mental model of the earth leads her to provide this interesting answer to our question about the end/edge of the earth.

What is important to mention here is that the mental model appears to mediate in the generation of the explanation. The mental model becomes the vehicle through which implicit physical knowledge enters the conceptual system. Once this implicit physical knowledge has entered the conceptual system, it can become explicitly coded and thus available for further theorizing. It is in this respect that mental models can become important sources for the generation of new conceptual knowledge, and can aid in theory construction.

## 3.2 Mental models as mediators in the interpretation of new information and in learning

Mental models can exert important influence on the interpretation of new information. In the examples discussed below, we can see how children's mental models of the earth influence the way information coming from an outside source is interpreted. These examples are taken from a study investigating the development of children's explanations of the day/night cycle (Vosniadou and Brewer, 1994). In this study we presented children with the drawing shown in Figure 5. We told them that this is a drawing showing a person on the earth and that we wanted them to "make it so it is day for that person" and then to "make it night".

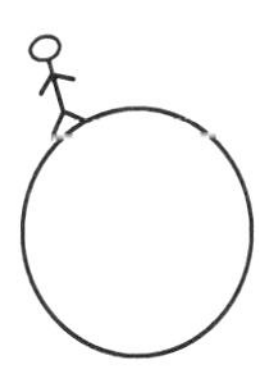

*Figure 5.* Drawing given to the children in order to explain the day/night cycle.

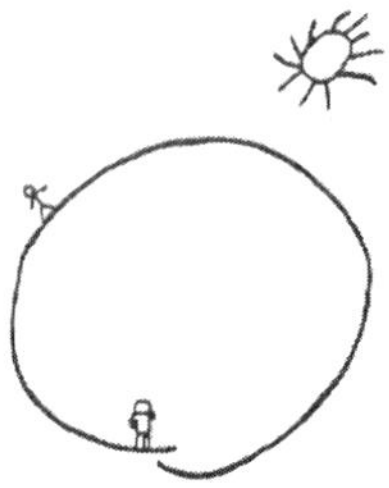

Drawing No. 1

**Tamara (No. 9, Grade 5)**
**The sun is occluded by clouds or darkness**
E: Now can you make it so it is day for that person?
C: He's outside the earth.
E: Where should he be?
C: In here (see Figure 4, drawing 1)
E: …OK now, make it daytime for him.
C: The sun is out here, but it looks like it's in the earth, when it shines…
E: OK. What happens at night?
C: The clouds covered it up.
E: Tell me once more how it happens.
C: Cause at 12 o'clock it's dark.

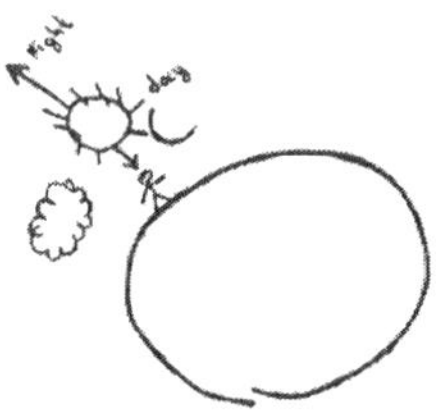

Drawing No. 3

**Allison (No. 52, Grade 1)**
**The sun moves out into space.**
E: Now make it so it is day for that person.
C: (child makes drawing 3 shown in Figure 4) Right here?
E: Whatever you think. Now make it night.
C: It goes in space.
E: Show me. Tell me how it happens.
C: The sun comes back down. It goes into space and when it gets dark the moon comes back out.

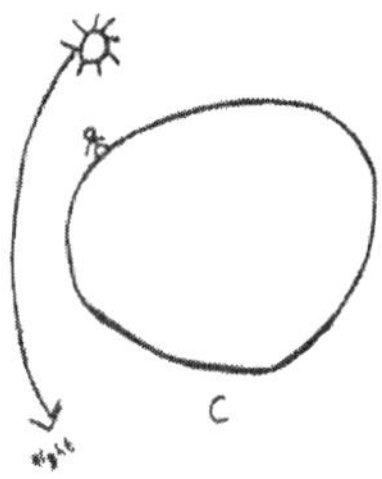

Drawing No. 5

**Timothy (No. 47, Grade 1)**
**The sun goes down to the other side of the earth (and the moon goes up)**
The child makes the drawings shown in Figure 4.
E: Tell me once more how it happens.
C: When the moon comes up and the sun goes down.
E: Where was the moon before?
C: Under the earth.
E: Show me. Tell me how it happens.
C: What time was it when it goes under the earth?
C: Day

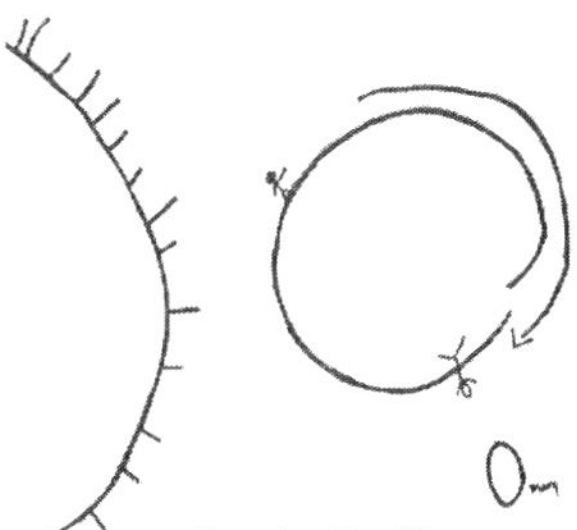

Drawing No. 10

**Robert (No. 5, Grade 5)**
**The earth rotates up/down and the sun and moon are fixed at opposite sides.**
E: Now make it so it is day for that person.
C: (child makes drawing 10 shown in Figure 4).
E: Now can you make it night time?
C: Can I draw him somewhere else? (draws figure at the bottom of the earth)
E: Sure
C: (child draws arrows to show how earth spins)
E: Tell me how it happens.
C: When it was daytime, the earth spinned around to the sun. When it was night time, the earth turned around to where the moon is.

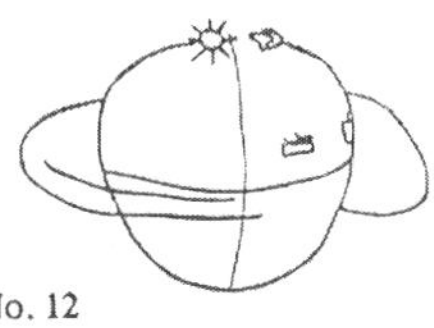

Drawing No. 12

**Venica (No. 33, Grade 3)**
**The earth rotates left/right and the sun and moon are fixed at opposite sides.**
C: (child makes drawing 12, Figure 4).
E: Can you tell me how this happens?
C: When the earth turns around its orbit, this side comes day and the other side comes night.

*Figure 6.* Examples from children's explanations of the day/night cycle (cf. also the previous page).

As can be seen from the examples shown in Figure 6, the children interpreted this drawing in different ways. For example, Tamara (Drawing No. 1) thought that our drawing was wrong because the person was "outside the earth". When we told her to show us where the person should be, she drew the person inside the earth at the bottom and explained the day/night cycle in terms of the sun being covered by clouds.

Allison (Drawing No. 3) accepted our drawing and added the sun to make it day, explaining that the sun "goes in space" and "when it gets dark the moon comes back in". Timothy also accepted our drawing, but he had a very different model of the day/night than Allison. He thought that the sun goes down to the other side of the earth (Drawing No. 5).

Compared to all of the above, Drawing No 10 by Robert is very different. It shows a huge sun, bigger than the earth, and a demonstration of the earth's rotation in an up/down fashion, different from the east/west rotation of the earth envisioned by Venica (Drawing, 12).

It should be mentioned here that children's interpretations of our drawing are consistent with their models of the earth and their beliefs about where people live on the earth. Our studies have shown that children's mental models of the earth act as strong constraints on the way children interpret information regarding the day/night cycle. There was not even a child in our sample that provided interpretations of the day/night cycle based on the earth's axis rotation or revolution around the sun that had not understood the spherical shape of the earth.

It is interesting to note here that most of the children who had a spherical shape of the earth and attributed the day/night cycle to be the earth's axis rotation interpreted this rotation to be an up/down and not an east/west rotation. We think that they do so because the belief in the up/down rotation of the earth is consistent with the phenomenal appearance of the sun to be "up in the sky" as well as with children's pre-scientific explanations of the day/night cycle in terms of the sun going down behind the mountains or at the other side of the earth.

In contrast to the children who have understood that the earth is a rotating sphere, most of the children who form hollow sphere models, like Venica, interpret the rotation to be an east/west rotation, not an up/down one. Again, this makes sense in view of the fact that the hollow-earth children believe in "up/down gravity" and think that the people live inside the earth because they would fell down if they lived outside. For those children it does not make sense to consider the earth's axis rotation to be an up/down one, because then the people would fall down every time the earth turned.

All of the above can be used as arguments for the psychological reality of mental models. It appears that on the one hand mental models are constrained by prior beliefs and presuppositions and on the other, once formed, they themselves act as constraints on the way the new information that enters the conceptual system is interpreted.

## 3.3 Mental models as aids in experimentation and theory revision

In order to show how mental models can aid in experimentation and theory revision I will use two examples. The first is from an unpublished astronomy study I did in Greece with adult illiterates. This was a population of relatively young men and women who for some reason or other had dropped out of school after the third or fourth grade. The women were housewives and the men were construction workers or truck drivers. The man I will talk about was about 30 years old, a truck driver, who was very interested in the interview and said that the questions I was asking him were often the subject of conversations with his friends in the cafes at night after work. He believed the earth to have the shape of an egg and thought that people lived only on the "top part of the earth and not at the bottom" because they would fall.

He made a drawing of the egg-shaped earth on a piece of paper and drew a person standing at the top of this earth. At a later part of the interview I asked him if the earth moved. He said that he knew that the earth rotated and, like most people, interpreted the rotation to be an up/down rotation. When he said that, he made an up/down movement using his hands in order to illustrate the movement of the earth. As he was doing this he came to realize the obvious inference that if the earth rotated up/down then the people who lived at the top of the earth would eventually go down. At this time he stopped and he said that there was something wrong here.

He was absolutely sure that the earth moved but he was also sure that the people could not live at the "bottom" of the earth! Since he could not solve this apparent inconsistency I told him not to worry and we proceeded with the interview. When the interview was almost over, he stopped me and said that he had found a solution to the inconsistency regarding the earth's movement. He said that the earth moved all right but what moved was only the "core" of the earth and not the whole earth. Since only the core of the earth moved up/down, the people on the outside of the earth could go on living undisturbed.

Misconceptions such as the one discussed above, are often creative solutions to the problem of incorporating theoretically inconsistent information into the knowledge base. They demonstrate the constructive nature of human cognition, and are important mediators in the process of conceptual change.

The second example I will use comes from my younger daughter Irene when at the age of four served as a pilot subject for the astronomy studies, At that time, I was in Greece, training a group of students who were going to go to a school to test children. I had brought my daughter and an older child (8 years of age) to be the pilot subjects. We started with Irene who gave consistently flat earth responses to all questions, although she had been exposed to some spherical earth information. When she finished we examined the second child who gave consistently spherical earth responses to the questionnaire. Irene was listening to the other child carefully and at the end of the interview she said to us: "I want to be asked again". We were all surprised by her request but we agreed to have her tested again. So we started asking her the questions from the beginning. One of the first questions in the questionnaire was the question "What is the shape of the earth?" and "Take this play-dough and make a model of the earth that shows its real shape". Irene took the play-dough and with a very happy smile made a spherical earth for us. We continued with the remaining questions one of which involved showing to the subject the picture of a farm-house on what appeared to be a flat farm land. In this question the usual procedure was to tell the subject: "This is a picture of a house on the earth. The earth appears to be flat here. Do you agree?" (Most children would agree). Then we would say: "Can you explain to us why the earth appears to be flat here when you said before that the earth was round?" This question was asked only if the subject had said that the earth is round. Irene looked at us and thought for a while how to answer the question. She then took the spherical earth that she had made with the play-dough and pressed it in her hands flattening it and turning it into a flat pancake, looking again very happy at her solution.

In both of these examples the mental model appears to mediate in the interpretation of information that comes from the outside and plays an important role in how this information is used for theory revision.

## 4. THE AUTONOMY OF MENTAL MODELS

In their recent book *Models as Mediators* Morgan and Morrison (1999), argue that the reason that models can function as important instruments in science is because they occupy an autonomous role, they are *autonomous agents*. "It is because they are neither one thing nor the other, neither just theory nor data, but typically involve some of both – that they can mediate between theory and the world". I would like to suggest that something similar also applies to mental models. In other words, mental models play an important role in conceptual development and conceptual change, because they are not, completely, determined neither by data nor by theory. They retain a degree of independence that is responsible for the fact that they can function in ways that promote theory development.

Is there any empirical evidence in support of this claim? I think that such evidence can be found in the considerable individual and cultural differences in the mental models that exhibit similar constraints. For example, there are a number of different models of the earth that all express the flat ground and up/down gravity constraints, such as the model of the earth as a round pancake/disk, as a flattened sphere or as an egg-shaped sphere with the people living only on the top part, or the model of the earth as a hollow sphere with people living on flat ground inside it. Furthermore, there is evidence that there are systematic cross-cultural differences in the way children interpret the constraint that the earth itself is supported by something. Thus, children from India appear to prefer an earth supported on an ocean of water, an idea not very popular with American children, who think the earth is supported by ground all the way down, or with Greek children, some of whom think that the earth rests on the strong shoulders of Atlas! (Vosniadou, 1994a).

The fact that the same constraints can be embodied in different and distinct mental models is significant if we consider (as was previously argued) that the mental model is the basis on which new information enters the conceptual system. Thus, a child with the model of a disk earth, or a flattened sphere, can explain the day/night cycle in terms of the sun going down to the other side of the earth, but this explanation does not make sense from the point of view of Venica who drew the model of the hollow sphere shown in figure 6, drawing 12. Potentially, such differences in mental models of the earth, regardless of the fact that they started as embodiments of the same theoretical constraints, may give rise to completely different theories.

## CONCLUSION

I have argued that the ability to form mental models is a basic characteristic of the human cognitive system and that the use of models by children is the foundation of the more elaborate and intentional use of models by scientists. The mental models that children and lay adults construct have predictive and explanatory power and can be used as mediating mechanisms for the revision of existing theories and the construction of new ones. This is the reason why they are important in conceptual development and conceptual change.

## REFERENCES

Baillargeon, R., 1990, The development of young infants' intuition about support, paper presented at the $7^{th}$ *International Conference on Infant Studies*, Montreal, Canada.

Carey, S. and Smith C., 1993, On understanding the nature of scientific knowledge, *Educational Psychologist* 28:235-251.

Inagagi, K. and Hatano, G., (1996). Young children's recognition of commonalities between animals and plants, *Child Development* 67:2823-2840.

Ioannidou, I. and Vosniadou, S., 2001, The development of knowledge about the composition and layering of the earth's interior, *Nea Paedia* 31:107-150 (in Greek).

Kellman, P.J. and Spelke, E.S., 1983, Perception of partly occluded objects in infancy, *Cognitive Psychology* 15:483-524.

Kellman, P.J., Spelke E.S., and Short, K., 1986, Infant perception of object unity from translatory motion in depth and vertical translation, *Child Development* 57:72-86.

Kosslyn, S.M., 1994, *Image and Brain: The Resolution of the Imagery Debate*, MIT Press, Cambridge, MA.

Kuhn, D., Amsel, E., and O' Longhlin, M., (1988). *The Development of Scientific Thinking Skills*. London: Academic Press.

Kuhn, T.S., 1962, *The Structure of Scientific Revolutions*, University of Chicago Press, Chicago.

Lakatos, I., 1970, Falsification and the methodology of scientific research programmes, in: I., Lakatos and A. Musgrave, eds., *Criticism and the Growth of Knowledge*, Cambridge University Press, Cambridge, pp. 91-195.

Laudan, L., 1977, *Progress and Its Problems,* University of California Press, Berkeley.

Leslie A.M., and Keeble, S., 1987, Do six-month old infants perceive causality? *Cognition* 25:265-88.

Leslie, A.M., 1988, The necessity of illusion: Perception and thought in infancy, in L. Weiskrantz, eds., *Thought and Language*, Oxford University Press, Oxford, pp. 185-210.

Morgan, M.S. and Morrison, M., eds., 1999, *Models as Mediators: Perspectives on Natural and Social Science,* Cambridge University Press, Cambridge.

Meltzoff, A.N., 1995, Understanding the intentions of others: Re-enactment of intended acts by 18 month-old children, *Developmental Psychology* 319:838-50

Oakes, L.M. and Cohen, L.B., 1995, Infant causal perception, in: C. Rovee-Collier and L.P. Lipsilt, eds., *Advances in Infancy Research*, Ablex, Norwood, NJ, pp. 1-54.

Samarapungavan, A., Vosniadou, S., and Brewer, W.F., 1998, Thinking about the Earth, the Sun and the Moon: Indian Children's Cosmologies, *Cognitive Development* 11:491-521.

Shepard, R.N. and Metzler, J., 1971, Mental rotation of three-dimensional objects, *Science* 191:701-703.

Spelke, E.S., 1991, Physical knowledge in infancy: Reflections on Piaget's theory, in: S. Carey and R. Gelman, eds, *Epigenesis of Mind: Studies in Biology and Cognition*. Erlbaum, Hillsdale, NJ.

Spelke, E.S., Breinlinger, K., Macomber, J., and Jacobson, K., 1992, Origins of knowledge, *Psychological Review* 99:605-632

Vosniadou, S., 1994a, Capturing and modeling the process of conceptual change, *Learning and Instruction* 4:45-69.

Vosniadou, S. (1994b), Universal and culture-specific properties of children's mental models of the earth, in L. Hirschfeld and S. Gelman, eds., *Mapping the Mind*, Cambridge University Press, Cambridge, pp. 412-430.

Vosniadou, S. and Brewer, W.F., 1992, Mental models of the earth: A study of conceptual change in childhood, *Cognitive Psychology* 24:535-85.

Vosniadou, S. and Brewer, W.F., 1994, Mental models of the day/night cycle, *Cognitive Science* 18:123-83.

Vosniadou, S., Archodidou, A., Kalogiannidou, A., and Ioannides, C., 1996, How Greek children understand the shape of the earth, *Psychological Issues* 1:30-51 (in Greek).

Wellman H.M. and Gelman, S.A., 1992, Cognitive development. Foundational theories of core domains, *Annual Review of Psychology* 43:337-75.

# Modeling Core Knowledge and Practices in a Computational Approach to Innovation Process

Stefania Bandini[*] and Sara Manzoni[#]
*[*]Department of Computer Science, Systems and Communication, University of Milano Bicocca, Italy, bandini@disco.unimib.it; [#]Department of Computer Science, Systems and Communication, University of Milano Bicocca, Italy, manzoni@disco.unimib.it*

Abstract. The development of computer based systems supporting the creation and the evolution of core knowledge is challenging, and involves the development of knowledge models dynamically tackling with well-structured notions, formal models, heuristic, tacit and not formalized knowledge, and experience. The main aim of this work is to present the experience of modeling and the implementation of a knowledge based system (P-Race) designed and developed to support the chemical formulation of rubber compounds of tire tread, in order to take part in motor racing. Because of the different competence involved in the decision making process (the compound designer - who owns a large part of the chemical core knowledge of a tire company and the race engineer), multiple knowledge representations have been adopted, and integrated into a unique Case-Based Reasoning (CBR) computational framework. Moreover, a dedicated formalism for the representation of model-based knowledge for chemical formulation (called Abstract Compounds Machine - ACM) has been created. It allows the core knowledge about rubber compounds to be explicitly represented, computed and integrated in the CBR architecture.

## 1. INTRODUCTION

*Innovation* is a key factor for the progress of mankind. Defining its role is not straightforward, and human activities concerning innovation often conflict with conservative tendencies aimed to avoid the change. On the other hand, change is necessary, because of the intrinsically dynamic and complex nature of the world. Moreover, the pressing evolutionary pace imposed by

*Model-Based Reasoning: Science, Technology, Values*, edited by
L. Magnani and N.J. Nersessian, Kluwer Academic/Plenum Publishers, New York, 2002

technology characterizing our era stresses this dynamics, posing change as a necessity, rather than a choice. Within this scenario, innovation becomes the field where several players face each other, and only those able to find the right equilibrium between the tendency to conserve and the necessity of change will continue playing.

If we consider innovation in the production world, it is possible to see innovation as a *process,* running in parallel with production processes requiring procedures, protocols, structures, and investments to be suitably designed, engineered, applied, and kept under strict control to achieve efficiency. Production processes must guarantee the reproducibility and the reuse of already existing successful products, that is, concrete solutions to previously defined sets of needs, requirements, and constraints, according to which their performance is evaluated. Instead, innovation *is* change, and implies the creation of new products, involving the alteration of the strong requirements and constraints imposed by industrial production processes, at the same time satisfying new market needs or *inducing* them.

Inducing needs on the market means "*creating products that customers need but have not yet even imagined*", Prahalad and Hamel (2000), and the ability of infusing products/needs is becoming a critical task in the global market (where boundaries are changing ever more quickly, and targets are elusive and their capture is at best temporary). The competitive advantage roots its essence in this framework, and innovation plays a leading role on this stage, as the core ability creating new products or significantly modifying *by adaptation* products without ignoring production constraints. Designing new products by adaptation is the very challenge of competitive advantage. This means valorizing and exploiting to the maximum degree the core competencies owned by a company, namely focusing on its own core knowledge.

For a company, the successful management of its core competencies (as its most valuable not tangible asset) implies focusing on knowledge management (that is, on the ability to identify, cultivate and exploit knowledge instead of merely restructuring, decluttering and delayering the company organization), (Liebowitz, 1999; Davenport and Prusak, 1998).

From the knowledge management perspective, pointing out a core competence - as Prahalad and Hamel (2000) suggest - means identifying the competence that: provides potential access to a wide variety of markets; gives a significant contribution to the - actual or potential - customer's benefits of the end product; and is difficult to imitate for competitors. The concrete tie between identified core competencies and the end product is the *core product*, namely, those components or subassemblies that actually contribute to the value of the end product. For instance, in computer industry, a new CPU is a core product; likewise, a new sculptured tread is a core product in tire industry.

In this conceptual framework it is easy to consider core knowledge as the organized set of core competencies, whose value is greater than the mere sum of all its elements. Core knowledge directly supports the core business.

Another basic characteristic of core competence is its experiential nature. Experience is derived from the direct application of knowledge in problem solving on a specific domain, and allows to structure explicit knowledge and to accumulate tacit knowledge. The latter is the most valuable one, and can be considered a measurement of the skills of an expert. Experience means having dealt with several *cases* during time, regardless whether successful or not. The role of failure in the accumulation of experience is in fact a key factor for learning. However, when core and experienced competencies are embedded in some organized environment, the phenomenon of *failure removal* is very common, because of conventional and psychological factors. Reconsidering calmly unsuccessful cases, in fact, is not a neutral procedure in organized production human environments, since it is related to career and retributive issues, and not only to social or psychological behaviors in the group. Moreover, to "treasure" knowledge derived from experience in a production environment is a very hard problem in knowledge management. The idea that structured collections of documents and reports on cases (often supported by automated systems, i.e. databases) can provide a solution to the problem by themselves is just the tip of an iceberg (for more details on the role of experience in business, see Pine and Gilmore (1999)).

Core competencies working by solving, by learning from experience, and by sharing innovation problems, are in several cases defined as "communities-of-practices", Seely Brown and Duguid (2000). In most of the cases, innovation depends on these communities, and represents the interface between the organization and its environment. If this interface perceives change as the very challenge in competition, the focus on knowledge is naturally guaranteed. On the contrary, if conflicts emerge, the core knowledge is just a fragmentary collection of notions. Within this framework, shared experience can be viewed as the "melting pot" where the communities-of-practices, producing innovation, can create core knowledge. However, core knowledge coming from shared experience is the most difficult to be captured, structured, represented and managed, and computational approaches to knowledge management find here a major challenge.

Managing core knowledge in a community-of practices often implies the deep analysis of the nature of such process of knowledge creation, that means to support the possibility of discover new modeling practices, such as in scientific environment. In the case of new products design, models are revised or, sometimes, replaced by new methods of creative changes such as in scientific discovery process. Performing creative reasoning within a computational framework in order to support innovation process requires to investigate computational methods that allow discovery to emerge.

Model-based reasoning in a more general conceptual and computational framework which captures also experiential and practice knowledge promotes creative change because it is effective in abstracting, generating, and integrating constraints in ways that produce novel results (Magnani et al., 1999). In the following subsection, a general computational framework allowing both experiential and model-based reasoning will be introduced.

## 1.1 Knowledge management and Case-Based Reasoning

The computational approach to knowledge management concerns the design, the implementation and the application of computer-based systems to support the representation, the sharing and the computation of data and knowledge within an organization. Several branches of computer science are devoted also to the development of computer-based systems supporting knowledge management (networking, data-mining, databases, neural networks, computer systems for cooperative working, artificial intelligence, and so on). In the case of core knowledge management and its role in innovative product design, we focus on the artificial intelligence approach, and in particular, on the Case-Based Reasoning approach to knowledge management (Aha and Munoz Avila, 1999; Aha and Weber, 2000). The latter is a very promising research topic, with increasing importance for the development of computer-based solutions to knowledge management, Davenport and Prusak (1998).

Case Based Reasoning (CBR) (Kolodner, 1993; Slade, 1991) is a problem-solving paradigm based on specific knowledge about previously experienced real problem situations (*cases*). From a conceptual standpoint, reasoning on cases means to *identify* the current problem situation, to *find a past case similar* to the new one, to use that case to *suggest* a solution to the current problem, to *evaluate* the proposed solution either by suggesting the *adoption* of the solution or the *revise by adaptation* of the solution to the current situation, and finally to *update* some *case memory* by incremental learning with this experience.

From a computational standpoint, Case Based Reasoning is a cyclic and integrated problem solving process. At the highest level of generality Aamodt and Plaza (1994), the CBR cycle may be divided into the following four steps (Figure 1):

- RETRIEVE the most similar case(s);

- REUSE the information and knowledge in the case retrieved to solve the problem;
- REVISE the proposed solution;
- RETAIN the parts of this experience likely to be useful for future problem solving by incorporating them into the existing knowledge-base.

CBR allows experiential knowledge to be captured when models cannot or are hard to be formalized. Experience is the center of CBR, and CBR deals with it through its main idea of *case*, that is, an episode where a problem or a problem situation was partially or totally solved. Several applications of CBR have been developed, like, for example, classification, diagnosis, planning, decision support, information retrieval, configuration, and design. Case-Based Reasoning supporting *design* is an increasing application area, since most of design knowledge comes from the experience of previously solved situations (Maher *et al.*, 1995; Borner, 1998). Successful examples of application of CBR to chemical products' design can be found in (Cheetham and Graf, 1997; Craw *et al.*, 1998; Craw *et al.*, 2001).

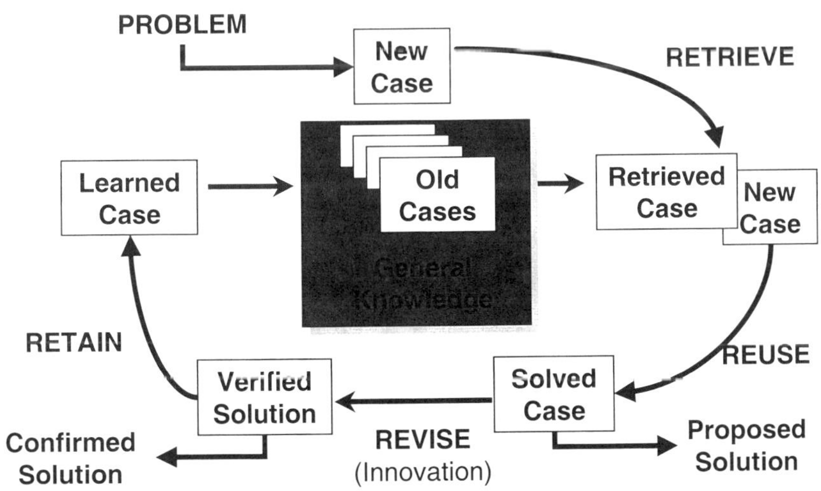


*Figure 1.* The general structure of the Case-Base Reasoning Cycle.

From the point of view of knowledge management for the support of innovation through design and adaptation of core products, CBR offers a computational paradigm to study knowledge models involved in competitive product design and to develop applications dedicated to communities-of-practices. As it will be presented in the following, the REVISE module in a CBR framework is a model reasoning module allowing chemical product innovation to be performed. This model is the interpretation of the chemical-physical knowledge shared in a community-of practice. Moreover, this model allows new potential chemical product to be automatically generated in order to satisfy salient constraints of the target domain.

The main aim of this paper is to present the experience of the modeling and implementation of a knowledge based system (P-Race) designed and developed to support the chemical formulation of rubber compounds of tire tread, in order to take part (and win) in motor racing. Because of the different competencies involved in the decision-making process (the compound designer - who owns a large part of the chemical core knowledge of a tire company - and the race engineer), multiple knowledge representations have been adopted, and integrated into a unique Case-Based Reasoning computational framework. The latter captures the episodic knowledge characterizing most of reasoning activity of the race engineer, and allows incremental learning to support the dynamical process of experience growth and knowledge creation. Moreover, a dedicated formalism for the representation of model-based knowledge for chemical formulation (called Abstract Compounds Machine - ACM) has been created. It permits the core competence about rubber compounds to be explicitly represented, computed and integrated in the CBR architecture.

The most meaningful and innovative contribution of P-Race consists of a CBR architecture where the adaptation step is performed by the ACM chemical formulation model. P-Race has been designed in order to support the computer-based coordination and sharing of knowledge within an organized structure, namely, in a knowledge management perspective of a well defined community-of practices. The P-Race system has been developed for the Motor-sports Department of Pirelli Tires, where it is currently in use.

## 2. PRACTICES AND INNOVATION: A DOMAIN

In motor racing the role of tires is crucial. Among the parts that must be assembled in order to build a tire, tread is one of the most important. It is a chemical compound represented by a *recipe*, which determines its major properties. The basic material composing tread obtained by the recipe is called in jargon “batch”. Tread batch comprises a set of ingredients: artificial or natural elastomers (rubber), active fillers (carbon black, silica), accelerants, oils, and some others. All these ingredients are essential for the acquisition of the desired chemical-physical properties determining the needed performance. The knowledge about the recipes is the chemical formulation of rubber compounds, and is a large part of core knowledge of a tire company (while the knowledge about the structure of the tire is another large portion). Tread batch is a core product and any innovation of it represents an innovation involving the whole tire. Any innovation on tread batch influences the production process, both in the case of tires dedicated to car racing, and all the other products of a tire company (*large-scale products*, as tires for cars, trucks, motorbikes, and so on). Motor-sports tire production is a production niche: generally the production yield is limited and it doesn't represent a sizable slice of in core business of a tire company. Notwithstanding this, the role of the motor-sports department is anyway crucial since, in many cases, innovation of core products starts from innovation experienced in racing - that is, the most demanding case of tire employment, in very particular test conditions.

For instance, in large-scale production, a set of standard lab tests is usually performed in order to obtain the best performance from a tread batch. In motor racing, however, only few tests can be performed, because of the particular raw materials used, and the characteristics of some basic chemical ingredients of the batch. The global performance of tread can be verified therefore only during the trials or directly during the race. The evaluation of the performance is not absolute, but depends on several factors. The most important of them, characterizing each single race, concerns car set-up, road ground, geometrical profile and severity of the track, weather conditions, and racing team.

Quite obviously, the skills of the people involved in the design of motor racing tires (a real community-of-practices composed by race engineers, tire designers and compound designers) consist of their experience on the field and their knowledge about a very complex decision making problem. More in detail, the race engineer and the compound designer make final decisions about the reuse, the innovation, or the creation of the tread batch. Since their choice is bound to a single race, it is usually strongly dependent on performances and results obtained in previous races on "similar" tracks (usually in previous seasons of the championship, such as Sports Racing World Cup, American Le Mans Series, or others). Thus, the choice of a tread for a particular race depends on the results of previously solved cases: that is the general problem solving mechanism used by race engineers and compound designers is strongly based on reasoning about past cases in order to solve a new case. Also, the use of episodic knowledge is one of the main characteristics determining the choice of the tread batch for motor racing.

Moreover, decisions about the choice of a tread often involve innovation, that is the adaptation of a previously used tread to meet new requirements in order to improve its performance. Improvement is a constant: even if in a previous race the choice had led to success, the improvement of some required performance (grip, warm-up, thermal and mechanical stability, resistance to wear) is anyway advisable, because of possible improvements in the products of other tire makers (competitors).

It is easy to transpose this example to the case of large-scale production: the competitiveness on the tire market strongly depends on the innovation of tire core products (such as tread). The innovative design of the rubber compound of a tread is a fundamental activity in defining and capturing new markets through the improvement of tire performance.

The development of a knowledge management project for the support of designing of rubber compounds for tread batches dedicated to must take in account the following main factors:

- how to structure significant information and documents concerning the activities involved (e.g., weather and track conditions, morphological features of the circuit, data about the type of the race, data concerning of the car team, the adopted recipes, time measurements for each test or race, comments about the race, and so on);
- how to acquire, classify and represent knowledge allowing information to be used (namely, knowledge engineering activity);
- how to capture both the experiential and the model based knowledge involved;
- how to propose reuse of previously adopted solutions to analogue situations;
- how to support the innovative adaptation of previously adopted solutions;
- how to support the process of knowledge creation;
- how to support the incremental learning deriving from the accumulation of experience;
- how to spread information, documents and knowledge among the members of the community-of-practices;
- how to address the right information or knowledge to each member in order to personalize the individual work;
- how to maintain the knowledge management process.

The computational approach supporting this kind of practices and capturing all the factors mentioned above, is the Case-Based Reasoning approach. Moreover, in CBR, the most challenging issue is the modeling of the adaptation step, where a previously adopted solution (i.e., a chemical formulation) must be adapted to the new current problem (a race). Model-based reasoning methods have been integrated with CBR techniques for implementing adaptation step and will be illustrated in Section 4.

In the following section the general architecture of the P-Race developed system will be described, focusing on the role of the fuzzy Case Memory Manager (allowing a reasoning mechanism involving qualitative knowledge to be represented and computed) in Subsection 3.1. Finally, some concluding remarks are presented in Section 5.

## 3. THE CBR-BASED GENERAL ARCHITECTURE

Figure 2 shows the general architecture of the P-Race system. It can be seen that the structure mirrors the CBR cycle (Section 1.1). In particular, it can be divided into three main parts (A, B, and C, drawn by dotted lines in the figure).

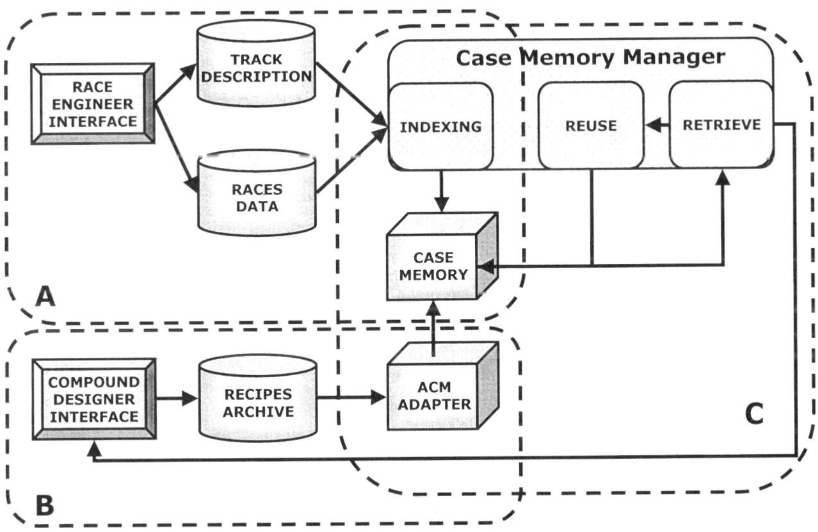


*Figure 2.* General architecture of the system.

Part A contains the main components dedicated to the race engineer:

a *database* containing all meaningful data about past racing activity (dates, championships, cars, teams, drivers, trial and race times, coded recipes of the used tread batch, coded information about the tire structure, and so on);

- a dedicated *Knowledge Acquisition* (KA) module, for the description of the track where race takes place;
- a *user interface* for race engineers.

Part B is made of components supporting the activity of the compound designer:

- a dedicated *integration interface* with the recipes database and other confidential data contained in the information system of the company;
- the *Abstract Compound Model (ACM)* module, that adapts retrieved solutions to the current problem;
- a *user interface* dedicated to compound designers.

Finally, the Part C constitutes the Case-Based Reasoning core, and comprises:

- the *case memory*, where the pragmatic features of races are indexed and structured in cases;
- the *case memory manager* indexing data from track descriptions and races database in form of cases, and evaluating the similarity between the current case and the stored ones.

The race engineer can interact with the system in three different ways, that is, he can input a track description, update the database, or activate the case-based engine. The first two activities have the purpose to provide the system with all the information needed to support the race engineer's choice of the most suitable tire for a given race. The Knowledge Acquisition (KA) module has been designed in order to let users express their knowledge about tracks both in qualitative and quantitative ways, avoiding as much as possible subjective descriptions. The module shown in Figure 3 allows the representation of a track be decomposed in *blocks*. Each block is characterized by a type (stretch, bend, chicane, and so on), and by the level of stress tires have to withstand (determined also by the road ground), expressed with numerical quantities. The result is a representation of the tracks that captures the experience and the knowledge of the race engineer in terms of crucial information about the track. The main ones are, for example, the features of track bends according to their severity and the required tire performance, the characteristics of the track surface, the thermal variation from a straight stretch to a bend and vice versa. This type of representation allows the system to compare tracks according to their morphological profiles in terms of race heuristics (e. g., initial and final speed in a bend, gear used in a given part of the track, weight supported by each wheel, and so on).

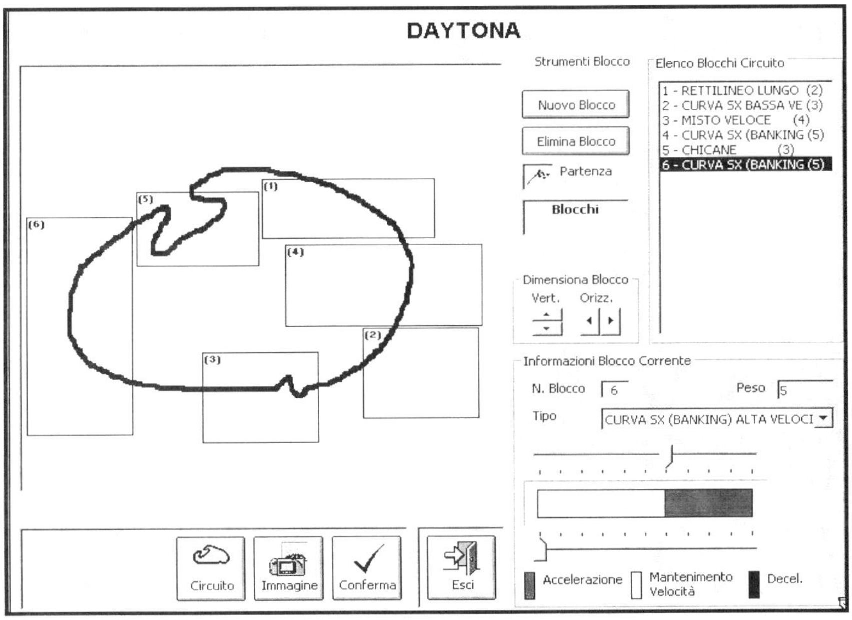


*Figure 3.* The user interface dedicated to knowledge acquisition (KA module).

Another dedicated interface allows the race engineer to update the database containing data about the racing activity (dates, kind of championship, car, team, drivers, trial times, warm-up times, race times, coded recipe of the used tread batch, coded information about the tire structure, tire performances, and so on). This type of activity is usually done directly on the field, that is, at the circuit during the competition.

The system uses these data about tracks and races to support the race engineer in the solution of new problems. The reasoning process starts with the representation of the current problem as a new case to be solved. A case represents a set of chronometrical measurements, concerning a race or a trial, relevant for the performance or the technical solution adopted. As in any CBR system, the three major parts of a case are problem/situation description, solution, and outcome. In P-Race, the description of the current problem contains both qualitative and quantitative information (date, time and location of the event, weather forecast and track conditions) used by the system to retrieve from the case memory the most similar cases. The solution for a case describes the coded recipe of the batch used in that case, while the outcome represents the resulting state in terms of performances obtained when the solution was applied. Starting from the description of the current problem, the system examines the case memory containing past problems already solved, and proposes a list of solutions (the most similar cases) to the race engineer. The main task of the retrieval algorithm is to apply a function giving a measure of similarity among cases. In the P-Race system, the similarity function has been defined as the weighted sum of differences between attributes, some of which are the result of a fuzzy interpretation of the users' inputs. Case retrieval is based on knowledge about tracks, weather conditions, and type of track surface.

The list of solutions proposed by the system could, at this point, include a feasible solution for the problem at hand that could be directly applied. Otherwise, an adaptation process has to modify one of the solutions proposed. The system also supports users in feasibility evaluation, reporting in a structured way the outcomes of proposed cases, including all documents associated to each case (comments of race engineers and drivers after a race or a test on track; quality-values vectors stating results in terms of performances obtained applying the solution; and so on). Thus, the system offers the view of all the current conditions the user needs to make his decision on which performance must be reached. At this point, if some modification to the basic recipe is needed, in order to improve or achieve some desired performance, the adaptation process is invoked. Adaptation could be necessary, for instance, when the proposed solution contains ingredients no longer available for tire production or when the past use of the solution had led top undesired outcomes.

A dedicated module (whose content will be described in the next section) of the P-Race system has been developed in order to support the compound designer in the chemical formulation of a batch. It provides access to the information contained in a database of recipes and other confidential data about raw materials. The formulation of a new compound is guided by the requests of the race engineer that asks for the improvement of some performance of an existing batch. Starting from the solution and the outcome of a case retrieved by the P-Race module dedicated to race engineer, the compound designer examines the recipe and the race conditions in order to fulfil the race engineer's requests.

The decision process of the compound designer can be divided in three separate stages:

- *Batch evaluation*: the expert is usually able to assess the performance of a compound from its parameters. The evaluation of the compound designer is usually different from the one given by the race engineer: the former judges according to his theoretical knowledge about materials, while the latter examines the results of the races.
- *Definition of the objective*: starting from to the results of the analysis of the previous point, and to the information about the race context, the ex-

pert decides which property of the batch has to be changed in order to obtain the desired performance. The properties usually involved are grip, thermal stability, mechanical stability, and warm-up. At the end of this stage, the compound designer has a set of possible options leading to the needed improvement of the performance.

*Choice of the ingredient*: finally, the compound designer describes the batch recipes contained in the archive of the company as lists of ingredient together with their respective quantities. Then, according to the chemical and physical properties of the raw materials, chooses an ingredient, and decides whether (and how) its quantity has to be changed in quantity or if the ingredient must be substituted by another one.

P-Race supports the activities described above with a dedicated adapter module, called *ACM Adapter*. The ACM Adapter activates the integration interface with the recipes archive of the company in order to provide to the ACM component the decoded recipe expressed in terms of quantity of ingredients (see the *Description Rules* in the next section). In other words, it modifies the recipe of the proposed batch in order to improve the performance observed in the outcomes of the past case, or to obtain new performances in relation with the description of the new case.

## 3.1 Case Memory Manager

A fuzzy technique, Zadeh (1996), has been integrated in the Case Memory Manager in order to index data in form of cases acquired from the KA module for the description of tracks and for the description of the current problem. Fuzzy rules play an important role also in the evaluation of the similarity between the current case and the stored ones. This kind of description has been chosen because the representation of knowledge about real problems has also to consider flexibility aspects. In fact, in these situations the problem description might be influenced by the expert's personal taste, and, given the nature of the problem, it could also include incomplete, imprecise and uncertain knowledge, Jaczinsky and Trousse (1994). Other examples on the combination of FL and CBR approaches can be found in (Weber-Lee *et al.*, 1995; Main *et al.*, 1996; Hansen and Riordan, 1998).

As previously mentioned, the information the race engineer considers to choose the tire tread mainly regards the morphological features of tracks, the characteristics of the track surface, the thermal variation from a straight stretch to a bend (and vice versa), the weather and track conditions, and the temperature of the track surface. The knowledge involved is formalized through formal models not always explicit (e.g., the mathematical/geometrical description of the track). Moreover, conventional models are often described by natural language describing the experience and the knowledge owned by all the members of the Motor-sports Department team.

Fuzzy rules have been introduced to derive a degree of membership to classes of values that define the severity of the circuit on the basis of track descriptions provided by the KA module. This set of rules defines the degree to which a circuit belongs to the fuzzy set of "severe" circuits. Race engineers would describe morphological severe circuits as characterized by many sharp curves and with frequent changes in speed. Fuzzy sets have been thus constructed around these concepts. The result is a representation of the tracks that captures the experience and the knowledge of the race engineer in terms of crucial information about the track, and allows the user to compare tracks according to their morphological profiles. For instance, experts consider Daytona and Las Vegas circuits as belonging to the same concept frame ("medium severity"). Their opinion is due to the similarity between road grounds (both "quite smooth") and to the morphological profile of the tracks (both synthetically described as "mostly oval"). So, in the selection of the proper tire for a race taking place in these locations, race engineers would consider this information and could choose a batch with low resistance to wear.

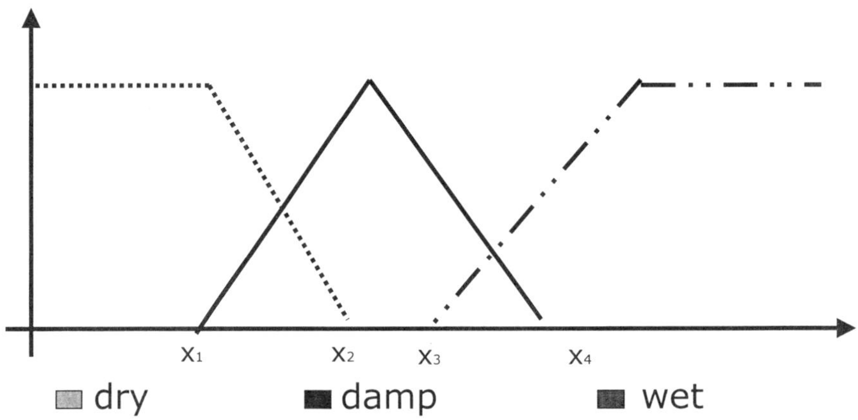


*Figure 4.* Fuzzy representation of the ground humidity of the track.

A second set of fuzzy rules has been integrated in order to handle uncertainty involved in the description of the target problem (for more details on the fuzzy approach to the P-race system, Bandini and Manzoni, 2001). In particular, during the description of the target problem the race engineer has to make a prediction about weather and road conditions for the day of the race. These predictions influence the decision, for instance, about the type of tires (slick, rain or intermediate) that has to be proposed to race teams. For instance, suppose a situation where weather forecasts say that the day of the race will be moderately rainy but the race engineer predicts to have dry road ground. The race engineer indicates as "moderately rainy" a day in which the degree of sun irradiation is comprised between 15 Watt/m$^2$ and 44 Watt/m$^2$. Moreover "dry" for the road ground means that less than 16 for the percentage of humidity. Under these conditions the race engineer would decide to use slick tires. The P-Race system can provide a correct description of the target problem and propose a correct solution to the race engineer by using fuzzy membership functions for temperature, weather and track conditions. An example of the representation with fuzzy rules of the description of the features of a problem is outlined in Figure 4.

The figure shows how the system interprets track conditions through the percentage of humidity of the soil, that is:

```
– if (%humidity<16) →
  track(dry) = 1 and track(dump)= track(wet) = 0
– if (%humidity > 45) →
  track(wet) = 1 and track(dump)= track(dry) = 0
– if (16 < %humidity < 45) →
  degree of membership for track conditions to 'dry',
  'dump'    and 'wet' sets are all greater than zero
```

Usually, a Case-Based reasoning mechanism must find a set of cases similar to the current target problem. As a matter of fact, most CBR systems are based on similarity relations between the target and the cases. On the other hand, these relations are vague by nature. In the CBR cycle, during the analysis of the similarity among cases (Initially Match Process during the Retrieve Step, Kolodner, 1993), crisp classification methods cannot always be used in order to improve the performance and the efficiency of the CBR system. The retrieval of previously solved problems similar to the current one is a two-step process. The Initially Matching process can be described as a function with a domain represented by the Case Base and the target problem specification, and a co-domain represented by the collection of cases with severity index belonging to the same set of values of the target problem (a three-valued severity range has been defined for this purpose: low, medium and high severity). The task of this function is to filter the Case Base and to single out interesting cases. During this step the set of cases is thus reduced, and the remaining ones are compared to the current case in the second step of the similarity computation (by the *Similarity Function*). In order to compute this function, Similarity Metrics has been developed.

A fuzzy approach is introduced also in this step of the CBR cycle, in order to measure the similarity between cases. The main concern in the design of the similarity algorithm has been the implementation of the membership function. Given as input two cases (the target problem and another case), it computes the Similarity Degree between the two cases as a value between 0 and 1. More in detail, the Similarity Function between the case $c=\left(f_1^c, f_2^c, \ldots, f_n^c\right)$ and the target problem $t=\left(f_1^t, f_2^t, \ldots, f_n^t\right)$ is given by the weighted sum

$$\sum_{i=1}^{n} \frac{w_i \times SIM\left(f_i^t, f_i^c\right)}{\sum_{i=1}^{n} w_i}$$

where $w_i$ is the weight associated with the $i$-th feature. The latter can assume two values: *MatchWeight*, if $SIM\left(f_i^t, f_i^c\right)$ is greater than a *Similarity Threshold* and *NoMatchWeight* otherwise. A *NoValueWeight* has also been introduced for features whose is value not specified by the user. MatchWeight, NoMatchWeight, NoValueWeight and Similarity Threshold are constants. $SIM\left(f_i^t, f_i^c\right)$ is the measurement of the difference between the i-th feature of the target problem and the i-th feature of the compared case. To compute this value, the system builds a Gaussian curve with mean value $f_i^t$ and fixed standard deviation σ:

$$SIM\left(f_i^t, x\right) = e^{-\frac{\left(x - f_i^t\right)^2}{\sigma^2}}$$

where $SIM\left(f_i^t, f_i^c\right)$ is the value of the Gaussian curve for $f_i^c$.

## 4. INNOVATING BY ADAPTATION: THE ACM MODEL

Chemical formulation is the design of a chemical compound. Basically, a chemical compound can be described by a "recipe", defining the quantities of some basic ingredients that have to be combined in order to achieve a required property, being "properties in action" the final performance of a compound. Thus, any change in the properties required implies that some modifications have to be made in the chemical formulation of the compound, that is, the corresponding recipe has to be revised. Therefore, the main purpose of product revise is to satisfy some new performance requirements for the product. This is the general characteristic of the innovation of a core product.

The Abstract Compound Machine (ACM) is a model created for the representation and the computation of the chemical formulation of a compound. In the ACM model, a *recipe* of $n$ ingredients is a finite non-ordered set $\{Q_1,\ldots,Q_n\}$, where each element $Q_i$ represents the quantity of the $i$-th ingredient. A given ingredient belongs to one or more families of ingredients. Each family $F_k$ is described by a set $\{A_1^k \ldots A_m^k\}$ of attributes. Each ingredient, that is each element $i$ of a family $F_k$, is thus described by a value $V_{ij}^k$ for each of its own attributes $A_j^k$. If an ingredient $i$ does not belong to a family $F_k$, the corresponding values $V_{ij}^k$ are undefined.

For each attribute $A_{ij}^k$ a constant of tolerance $T_j^k$ is defined. The latter is used in the comparison of two attribute values; two values ($V_{i'j}^k$ and $V_{i''j}^k$ respectively the $j$-th attribute values for ingredients $i'$ and $i''$ belonging to family $F_k$) are considered different only if $\left|V_{i'j}^k - V_{i''j}^k\right|$ is greater than $T_j^k$. These constants are necessary, given the empirical nature of attribute values, and also are used to cope with possible errors deriving from empirical measurements (e.g. lab tests measurements).

Starting from a recipe $R$, a *revised* recipe is a recipe $R'$ where some quantities have been changed. Compound revision follows the application of four sets of rules:

1. *Description Rules*, describing a product already developed as a recipe according to the ACM model, that is, as a vector of quantities of ingredients.
2. *Performance-Properties Rules*, defining which changes are needed in the properties of the recipe, in order to obtain a given change in performance.
3. *Ingredients-Properties Rules*, defining which attributes of the ingredients of a recipe are involved in the modification of the properties of the recipe.
4. *Formulation Rules*, generating a revised recipe $R'$ starting from $R$. Three types of formulation rules have been defined:

   - *Substitution*, replacing the quantity of an ingredient $i$ with an equal quantity of another ingredient $l$ of the same family $F_j^k$ (chosen by the

Ingredients-Properties Rules), in order to change the value of one or more attributes ( $V_{ij}^{k}$ ):

$$\text{if } (Q_i \neq 0)(i \in F_k)(l \in F_k)\left(\left|V_{ij}^{k} - V_{lj}^{k}\right| > T_j^{k}\right) \text{ then}$$
$$\{Q_1, Q_2 \ldots, Q_{i-1}, Q_i, Q_{i+1}, \ldots, Q_l, \ldots, Q_n\} \rightarrow$$
$$\{Q_1, Q_2 \ldots, Q_{i-1}, 0, Q_{i+1}, \ldots, Q_i + Q_l, \ldots, Q_n\}$$

- *Increase in quantity*, adding to the quantity of an ingredient a given constant $U_k$, defined according to the family $F_k$ of the ingredient:

$$\text{if } (Q_i \neq 0)(i \in F_k) \text{ then}$$
$$\{Q_1, Q_2 \ldots, Q_{i-1}, Q_i, Q_{i+1}, \ldots, Q_n\} \rightarrow$$
$$\{Q_1, Q_2 \ldots, Q_{i-1}, Q_i + U_k, Q_{i+1}, \ldots, Q_n\}$$

- *Reduction in quantity*, decreasing the quantity of an ingredient by a constant $U_k$, defined as in the previous point:

$$\text{if } (i \in F_k)(Q_i > U_k) \text{ then}$$
$$\{Q_1, Q_2 \ldots, Q_{i-1}, Q_i, Q_{i+1}, \ldots, Q_n\} \rightarrow$$
$$\{Q_1, Q_2 \ldots, Q_{i-1}, Q_i - U_k, Q_{i+1}, \ldots, Q_n\}$$

Compounds adaptation follows the application of ACM rules. The knowledge base has been partitioned into *knowledge sources* corresponding to ACM rules. As shown in Figure 5, knowledge sources activation starts from the application of Description Rules that split the coded batch representing the solution for the retrieved case into the quantities of its ingredients. This step invokes the integration interface to the Pirelli Tires Archive. Then, the Performance-Properties Rules knowledge source is activated, in order to determine the needed properties of the product starting from the required performance. From the global properties of the product, the Ingredients-Properties Rules knowledge source finds out which ingredients have to be considered in order to obtain a variation of the properties satisfying the required performance. Finally, the Formulation Rules knowledge source formulates the modified recipe applying Substitution, Increase in quantity or Decrease in quantity Rules.

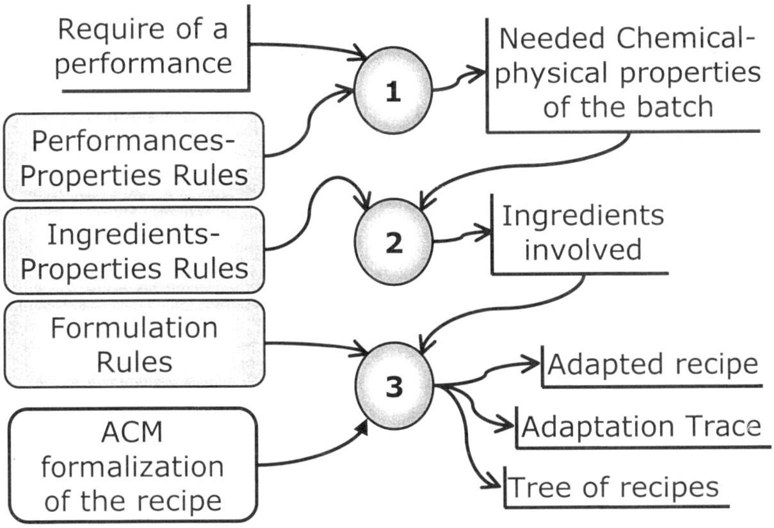


*Figure 5.* The adaptation cycle.

To conclude, some sample rules developed for the chemical formulation of rubber compounds for car racing are listed below in natural language:

– Description Rules:
  `if compound(HSRXX) then recipe(get recipe(HSRXX))`
  (“*it retrieves from the enterprise product archive the chemical formulation for compound HSRXX and produces the representation that will be used by the other sets of rules*”).

– Performance-Properties Rules:
  `if desired_performance(increase_thermal_stability)`
  `then desired_property(high_increase_hysteresis)`
  (“*in order to increase thermal stability, hysteresis must be decreased*”).

– Ingredients-Properties Rules:
  `if desired_property(high_increase_hysteresis)`
  `then interested_ingredient (polymer) and inter-`
  `ested_property(transition_glass)`
  (“*n order to increase hysteresis, transition glass of polymer is involved*”).

– Substitution Rules, one of the types of Formulation Rules:
  `if higher(transition_glass, new_polymer, old_polymer)`
  `then insert(new_polymer) and delete(old_polymer)`
  (“*if it is available a polymer which could increase hysteresis more than the one now present, apply substitution*”).

- Increase in quantity, another type of Formulation Rules:
  ```
  if quantity(ingredient, q) and increase(hysteresis,
  ingredient)
  then increase_quantity(ingredient, q)
  ```
  ("*if the recipe contains an ingredient which could increase hysteresis, increase its quantity*").

# 5. CONCLUDING REMARKS

The P-Race system is a Case-Based Reasoning system developed for the Motor-sports Department of Pirelli Tires where it is currently in use (for more details about application benefits of P-Race see Bandini and Manzoni, 2000). The chemical formulation component (that contains the most valued core competence of the company) has been adopted after an experimental campaign carried out in order to test the quality of system responses based on past solutions. The campaign was structured into two stages. In the first one, the chemical formulations (recipes) suggested by the system have been tested on a set of 250 cases. Cases have been selected from two entire championships (about 30 races from American Le Mans Series, Ferrari Challenge and Sports Racing World Championship), focusing on some of the most competitive teams (Ferrari 333 SP, Riley & Scott, Porsche GT3R, Lola). The recipes adopted by the Motor-sports Department in these real cases have been compared with those proposed by the system. The suggestions provided by the system about the adoption of some ingredients of the recipes have been considered satisfactory by the experts, with no macroscopic mistakes detected. This first stage also allowed the fine-tuning of the adaptation rules.

The second stage of the campaign has been planned in order to evaluate the system's support to race engineers and compound designers in their decision-making processes about tire selection and chemical formulation. The second stage was carried out during the same championships and the results have been judged very good.

From the knowledge management viewpoint, the P-Race project allowed valued core knowledge to be formalized, structured and shared. More in detail:

- the organization of information and data concerning a crucial activity involving core competencies in a conceptual framework, corresponding to the main characteristics of the episodic knowledge involved (Case-Based Reasoning);

- the formalization of a knowledge model (the Abstract Compound Machine) for the chemical formulation of rubber compounds, one of the most valued core competence of the entire company;
- the development of an automated support proposing solutions to the decision making process, i.e. innovative changes in a recipe in order to satisfy some required performance; this fact directly implies the role of innovation by adaptive design;
- the creation a computational framework shared by all the members of the Motor-sports Department.

Future developments of the system will include the integration of P-Race with software systems devoted to the acquisition and the description of track data with telemetric devices, in order to integrate heuristic/qualitative knowledge and quantitative instrumental measurements. The application of the system dedicated to racing will be extended also to rally racing. Moreover, the general ACM model will be adopted also in a new knowledge management project for large-scale production dedicated to Trucks Business Unit.

## REFERENCES

Aamodt, A. and Plaza, E., 1994, Case-based reasoning: Foundational issues, methodological variations, and system approaches, *Artificial Intelligence Communications* 7, 1:39-59.

Aha, D.W. and Munoz Avila, H., eds., 1999, *Exploring Synergies of Knowledge Management and Case-Based Reasoning*, AAAI Workshop, Technical Report WS-99-19.

Aha, D.W., and Weber, R., eds., 2000, *Intelligent Lessons Learned Systems: Papers from the AAAI Workshop*, Technical Report WS-00-03, AAAI Press, Menlo Park, CA.

Bandini, S. and Manzoni, S., 2000, A knowledge-based system for the design of rubber compounds in motor racing, in: *Proceedings of the 14$^{th}$ European Conference on Artificial Intelligence*, W. Horn, ed., IOS Press, Amsterdam, pp. 691-695.

Bandini, S. and Manzoni, S., 2001, Application of fuzzy indexing and retrieval in case based reasoning for design, *16$^{th}$ ACM Symposium on Applied Computing*, ACM Press, New York, pp. 426-466.

Borner, K., 1998, CBR for design, in: *Case-Based Reasoning Technology*, M. Lenz, B. Bartsch-Sporl, H. Burkhard, and S. Wess, eds., LNCS/LNAI 1400, Springer, Berlin, pp. 201-233.

Cheetham, W. and Graf, J., 1997, Case-based reasoning in color matching, in: *Case-Based Reasoning Research and Development*, D.B. Leake, and E. Plaza, eds., Proceedings 2$^{nd}$ International Conference on Case-Based Reasoning, Springer, Berlin, pp. 1-12.

Craw, S., Wiratunga, N., and Rowe, R., 1998, Case-based design for tablet formulation, in: *Proceedings of the 4$^{th}$ European Workshop on Case-Based Reasoning*, B. Smyth and P. Cunningham, eds., Springer, Berlin, pp. 358-369.

Craw, S., Jarmulak, J., and Rowe, R., 2001, Learning and applying case-based adaptation knowledge, in: *Proceedings of the 4th International Conference on Case-Based Reasoning*, D.W. Aha, I. Watson, and Q. Yang, eds., Springer, Berlin, pp. 131-145.

Davenport, T.H. and Prusak, L., 1989, *Working Knowledge*, Harvard Business School Press, Boston.

Hansen, B.K. and Riordan, D., 1998, Fuzzy case-based prediction of ceiling and visibility, in: *Proceedings of the 1st Conference on Artificial Intelligence of the American Meteorological Society*, pp. 118-123.

Jaczynski, M. and Trousse, B., 1994, Fuzzy logic for the retrieval step of a case-based reasoner, in: *Proceedings Second European Conference on Case-Based Reasoning*, LNAI 984, J. Haton, P. Keane, and M. Manago, Springer Verlag, Berlin, pp. 313-322.

Kolodner, J., 1993, *Case-Based Reasoning*, Morgan Kaufmann, San Mateo.

Liebowitz, J., ed., 1999, *Knowledge Management Handbook*, CRC Press, Boca Raton.

Magnani, L., N.J. Nersessian, P. Thagard, eds., 1999, *Model-Based Reasoning in Scientific Discovery*, Kluwer Academic/Plenum Publishers, New York.

Main, J., Dillon, T.S., and Khosla, R., 1996, Use of fuzzy feature vectors and neural networks for case retrieval in case-based systems, in: *Proceedings of the Biennal Conference of the North American Fuzzy Information Processing Society*, IEEE, New York, 438-443.

Maher, M.L., Balachandran, M., and Zhang, D.M., 1995, *Case-Based Reasoning in Design*, Laurence Erlbaum, Hove.

Prahalad, C.K. and Hamel, G., 2000, The core competence of the corporation, in: *Strategic Learning in Knowledge Economy: Individual, Collective and Organizational Learning Process*, R.L. Cross and S.B. Israelit, eds., Butterworth Heinemann, Boston, pp. 3-22.

Pine, B.J. and Gilmore, J.H., 1999, *The Experience Economy*, Harvard Business School Press, Boston.

Seely Brown, J. and Duguid, P., 2000, Organizational learning and communities-of-practice: toward a unified view of working, learning and innovation, in: *Strategic Learning in Knowledge Economy: Individual, Collective and Organizational Learning Process*, R.L. Cross and S.B. Israelit, eds., Butterworth Heinemann, Boston, pp. 143-166.

Slade, S., 1991, Case-based reasoning: a research paradigm, *AI Magazine*, Vol. 12, Spring 1991, 1:42-55.

Weber-Lee, R., Barcia, R.M., and Khator, S.K., 1995, Case-based reasoning for cash flow forecasting using fuzzy retrieval, in: *Case-Based Reasoning Research and Development*, M. Veloso and A. Aamodt, eds., *First International Conference on Case-Based Reasoning*, Sesimbra, Springer Verlag, Berlin, pp. 510-519.

Zadeh, L.A., 1996, Fuzzy Logic = Computing with Words, *IEEE Transactions on Fuzzy Systems*, Vol. 4, 2:103-111.

# Author Index

# Subject Index